Forever Chemicals

Environmental and Occupational Health Series
Series Editors:

LeeAnn Racz
Senior Engineer, ToxStrategies, Inc., Fort Walton Beach, Florida

Adedeji B. Badiru
Professor and Dean, Graduate School of Engineering and Management, AFIT, Ohio

Handbook of Respiratory Protection:
Safeguarding Against Current and Emerging Hazards
LeeAnn Racz, Robert M. Eninger, and Dirk Yamamoto

Perfluoroalkyl Substances in the Environment:
Theory, Practice, and Innovation
David M. Kempisty, Yun Xing, and LeeAnn Racz

Total Health Exposure
An Introduction
Kirk A. Phillips, Dirk P. Yamamoto, and LeeAnn Racz

Forever Chemicals:
Environmental, Economic, and Social Equity Concerns with PFAS in the Environment
David M. Kempisty and LeeAnn Racz

More Books Forthcoming
More about this series can be found at: https://www.crcpress.com/Environmental-and-Occupational-Health-Series/book-series/CRCENVOCCHEASER

Forever Chemicals

Environmental, Economic, and Social Equity Concerns with PFAS in the Environment

Edited by

David M. Kempisty and LeeAnn Racz

CRC Press
Taylor & Francis Group
Boca Raton London New York

CRC Press is an imprint of the
Taylor & Francis Group, an **informa** business

First edition published 2021
by CRC Press
6000 Broken Sound Parkway NW, Suite 300, Boca Raton, FL 33487-2742

and by CRC Press
2 Park Square, Milton Park, Abingdon, Oxon, OX14 4RN

Access the Support Material: https://www.routledge.com/9780367456405

ISBN: 978-0-367-45640-5 (hbk)
ISBN: 978-1-032-01366-4 (pbk)
ISBN: 978-1-003-02452-1 (ebk)

Typeset in Times
by SPi Global, India

Dedicated to those who push the boundaries of ensuring abundant, clean water is available to all.

Contents

Foreword xi
Editors xiii
Contributors xv

SECTION I Social Concerns

Chapter 1 PFAS: Today and Tomorrow 3

LeeAnn Racz and David M. Kempisty

Chapter 2 Fluorine Free Foams: Transitioning Guide 23

Ian Ross, Peter Storch, Ted Schaefer, and Niall Ramsden

Chapter 3 PFASs in Consumer Products: Exposures and Regulatory Approaches...51

Simona Andreea Bălan and Qingyu Meng

Chapter 4 Regulatory Implications of PFAS 87

J. Alan Roberson

Chapter 5 The Analytical Conundrum 99

Taryn McKnight

Chapter 6 Landfills as Sources of PFAS Contamination of Soil and Groundwater 119

Nanthi Bolan, Son A. Hoang, Yubo Yan, Sammani Ramanayaka, P. Koliyabandara, Gayathri Chamanee, Raj Mukhopadhyay, Binoy Sarkar, Ilasintha Wijesekara, Meththika Vithanage, and M.B. Kirkham

SECTION II Toxicology and Epidemiology

Chapter 7 Exposure to PFAS: Biomonitoring Insights 145

Kayoko Kato, Julianne Cook Botelho, and Antonia M. Calafat

Chapter 8 State of the Science for Risk Assessment of PFAS at Contaminated Sites 161

Jeanmarie Zodrow, Jennifer Arblaster, and Jason Conder

SECTION III Remediation

Chapter 9 Advances in Remediation of PFAS-impacted Waters 189

John Anderson, Pingping Meng, Tim Sidnell, and Ian Ross

Chapter 10 Effectiveness of Point-of-Use/Point-of-Entry Systems to Remove PFAS from Drinking Water 211

Craig Patterson, Jonathan B. Burkhardt, Donald Schupp, E. Radha Krishnan, Stephen Dyment, Steven Merritt, Lawrence Zintek, and Danielle Kleinmaier

Chapter 11 Removing PFAS from Water: From Start to Finish 235

Caitlin Berretta, Thomas Mallmann, Kyle Trewitz, and David M. Kempisty

Chapter 12 Rapid Removal of PFAS from Investigation-derived Waste in a Pilot-scale Plasma Reactor 255

Raj Kamal Singh, Nicholas Multari, Chase Nau-Hix, Richard H. Anderson, Stephen D. Richardson, Thomas M. Holsen, and Selma Mededovic Thagard

Chapter 13 Recent Advances in Oxidation and Reduction Processes for Treatment of PFAS in Water 271

Yaal Lester

Chapter 14 Novel Cyclodextrin Polymer Adsorbents for PFAS Removal 291

Yuhan Ling, Gokhan Barin, Shan Li, and Matthew J. Notter

Chapter 15 Modeling Water Treatment Performance and Costs for Removal of PFAS from Drinking Water 315

Jonathan B. Burkhardt, Richard H. Anderson, Rajiv Khera, Levi M. Haupert, Patrick Ransom, David G. Wahman, Page Jordan, Jonathan G. Pressman, Marc A. Mills, and Thomas F. Speth

Index 339

Foreword

As we confront a viral pandemic unlike anything seen in our lifetimes, and people struggle to find ways to avoid being exposed, the entire world is also slowly becoming aware of an even more pervasive health threat to which virtually every human on this planet – and almost every living creature – is *already* exposed. And no vaccine or medication can stop this exposure. It will persist – and can continue to build up in our bodies – for years and years, and will stay in our environment virtually forever. Our babies are exposed even before they are born. These exposures present challenges unlike anything any of us have ever seen before: Scientific, regulatory, legal, social, and political challenges. Yet, these challenges also present us with unprecedented opportunities to explore new ideas and pathways for addressing problems. Welcome to the new and ever-evolving world of the "forever chemicals" known as PFAS – the thousands of completely man-made per- and polyfluoroalkyl substances.

Although PFAS chemicals were first invented and started being released into our environment over 70 years ago, the scientific and regulatory communities (and the general public) did not even begin to learn of their existence – or the scope and severity of the threat they pose to human health and the environment – until relatively recently. It took decades of work by exposed community members for that information to disseminate from internal manufacturers' files to the rest of the world. As the information slowly trickled out, relating primarily to two PFAS chemicals (PFOA and PFOS), scientific and regulatory demands for restrictions and phase-outs of a select few PFAS chemicals were implemented. In turn, manufacturers, in order to avoid regulatory oversight, simply shifted production and use to other PFAS with fewer carbons or slight chemical modifications. Now, the scientific and regulatory communities are expressing serious concerns about the extent to which these other PFAS chemicals might present similar – or possibly even greater – threats to human health or the environment.

In the meantime, hundreds of millions of people all over the planet are learning that their drinking water or properties have been contaminated with PFAS – or that they themselves, their families, and their communities have these chemicals in their blood and bodies. They are demanding information – and action – in response. They want to know what these chemicals are doing to them or will do to them over time as they build up in their blood and bodies and linger in their veins – like a ticking time bomb. They want to know how to get rid of these chemicals from their drinking water, from their soils and properties, from the air they breathe, and from their bodies. They want to know how they were exposed – and how they may continue to be exposed. They want to know what products these chemicals were in – and which companies are making products without them so they might have a chance to begin reducing any additional exposures. And they want to know who is going to be held responsible for the enormous costs associated with these exposures and contamination. And how are we going to insure that the appropriate steps are taken – whether by regulation, legislation, litigation, or otherwise – to make sure that those truly responsible are held accountable – and that something like this *never happens again.*

Over the last 22 years, I have done what I can to try to help find answers to these questions and help make sure as much information as possible regarding what is now known, what has been known, and what we still need to know about PFAS is available to as many people as possible, including those working within the legal, scientific, and regulatory communities. This includes developing and encouraging exploration of innovative ways to address PFAS problems outside of traditional systems and frameworks, and finding ways to bring complex scientific and technical information to the public in ways that are accessible and understandable. Efforts like the C8 Health Project that collected blood samples and medical information from 70,000 PFAS-exposed community residents. And the creation of the C8 Science and Medical Panels with independent scientists and doctors charged with exploring and resolving complex questions regarding the links between PFAS and serious human disease, and associated diagnostic testing. The recent release of a feature-length documentary *The Devil We Know*, and a major Hollywood film *Dark Waters*, has helped elevate awareness and understanding of the PFAS problem, worldwide. And I wrote a book, *Exposure*, to help summarize and memorialize the history and facts on how this situation developed – and hopefully encourage innovative and creative ideas on how to fix these issues and make sure nothing like this ever occurs again.

There are many ways in which all of us can help continue the work necessary to understand and respond to the ongoing public health and environmental crisis posed by PFAS exposure. I am honored to be able to participate in a book like this that brings together authors from such a diverse array of backgrounds, training, and perspectives to highlight, discuss, and tackle these important issues. We may not all agree on what the appropriate path forward should be to solve these problems, but it is incredibly important that we are all sharing our views and having these discussions. I want to personally thank each of you reading this book for helping to continue and expand our collective understanding and awareness of all of these issues. Each and every one of us – armed with this knowledge and understanding – has the power to make sure these "forever chemicals" do not become "forever problems."

Robert A. Bilott
Partner with Taft Stettinius & Hollister LLP
Author of *Exposure: Poisoned Water, Corporate Greed, And One Lawyer's Twenty-Year Battle Against DuPont*

Editors

David M. Kempisty is Senior Business Development Manager for Emerging Contaminants at Evoqua Water Technologies. He is a licensed engineer focusing on PFAS and contaminant removal technologies. As a member of Evoqua's Environmental Solutions Technology and Innovation Team, he is involved with bench- and pilot-scale testing of activated carbons, ionic exchange resins, novel adsorbents, and innovative technologies. Prior to joining Evoqua, Dr. Kempisty served 22 years in the United States Air Force as a bioenvironmental engineer in a variety of capacities around the globe. As director of the Air Force graduate school's Environmental Engineering and Science Program, his research investigated per- and polyfluorinated alkyl substances (PFAS) and included toxicological studies, remedial investigations, and life cycle considerations of treatment alternatives. His research was recognized with awards from the U.S. Environmental Protection Agency's Office of Research and Development in 2017 and 2018. Dave earned his PhD from the University of Colorado-Boulder in Civil and Environmental Engineering in 2014 where he focused on removal of PFAS from groundwater using granular activated carbon. Dr. Kempisty earned his MS from the Air Force Institute of Technology and his BS from Michigan Technological University. Dr. Kempisty was lead editor of the text, *Perfluoroalkyl Substances in the Environment*, published in 2018, and has authored over 25 peer-reviewed papers and presentations on a variety of environmental topics. He maintains active memberships to multiple local and national professional organizations.

LeeAnn Racz is a Senior Engineer at ToxStrategies, Inc. She has also been the President and Founder of ClearView Environmental Engineering and Consulting, LLC, providing environmental and occupational health services to clients around the world. Prior to this initiative, she served for 23 years as a bioenvironmental engineer and civil engineer in the US Air Force. Some highlights of Dr. Racz's military assignments include squadron commander, chief of consultative services at the US Air Force School of Aerospace Medicine, director of the Graduate Environmental Engineering and Science Program at the Air Force Institute of Technology. Dr. Racz is a licensed professional engineer, certified industrial hygienist, and board-certified environmental engineer. She earned a BS in environmental engineering from California Polytechnic State University, San Luis Obispo, an MS in biological and agricultural engineering from the University of Idaho, and a PhD in civil and environmental engineering from the University of Utah. She has authored dozens of refereed journal articles, conference proceedings, magazine articles, and presentations, and edited six handbooks. Dr. Racz is a member of several professional organizations and honor societies and has received numerous prestigious teaching and research awards.

Editor

Contributors

John Anderson
Arcadis
Manchester, United Kingdom

Richard H. Anderson
Air Force Civil Engineer Center
Lackland Air Force Base
Texas, USA

Jennifer Anne Arblaster
Geosyntec Consultants
Milton, VT, USA

Simona Andreea Bălan
California Department of Toxic Substances Control
Sacramento, CA, USA

Gokhan Barin
Cyclopure, Inc.
Skokie, IL, USA

Caitlin Berretta
Evoqua Water Technologies
Arlington, VA, USA

Rob Bilott
Taft, Stettinus and Hollister, LLP
Cincinnati, Ohio

Nanthi Bolan
Global Centre for Environmental Remediation, University of Newcastle
Newcastle, Australia

Julianne Cook Botelho
Centers for Disease Control and Prevention, Division of Laboratory Sciences, National Center for Environmental Health
Atlanta, GA, USA

Jonathan B. Burkhardt
US Environmental Protection Agency
Cincinnati, OH, USA

Antonia M. Calafat
Centers for Disease Control and Prevention, Division of Laboratory Sciences, National Center for Environmental Health
Atlanta, GA, USA

Gayathri Chamanee
Department of Natural Resources, Sabaragamuwa University
Belihuloya, Sri Lanka

Jason Conder
Geosyntec Consultants
Huntington Beach, CA, USA

Michelle Crimi
Clarkson University
Potsdam, NY, USA

Joseph DiMisa
Air Force Institute of Technology
Wright-Patterson Air Force Base
Dayton, OH, USA

Stephen Dyment
US Environmental Protection Agency, Office of Research and Development
Denver, CO, USA

Isaac Emery
Informed Sustainability Consulting, LLC
Seattle, WA, USA

Brittany Fain
University of Nebraska Medical Center
Omaha, NE, USA

Levi M. Haupert
US Environmental Protection Agency
Cincinnati, OH, USA

Son A. Hoang
University of Newcastle
Newcastle, Australia

Thomas M. Holsen
Department of Civil and Environmental Engineering, Clarkson University
Potsdam, NY, USA

Mark S. Johnson
US Army Public Health Center
Aberdeen Proving Ground
Aberdeen, MD, USA

Page Jordan
US Environmental Protection Agency
Cincinnati, OH, USA

Kayoko Kato
Centers for Disease Control and Prevention, Division of Laboratory Sciences, National Center for Environmental Health
Atlanta, GA, USA

David M. Kempisty
Evoqua Water Technologies
Colorado Springs, CO, USA

Rajiv Khera
US Environmental Protection Agency
Washington, DC, USA

M.B. Kirkham
Department of Agronomy, Kansas State University
Manhattan, KS, USA

Danielle Kleinmaier
US Environmental Protection Agency, Region 5
Chicago, IL, USA

P. Koliyabandara
Ecosphere Resilience Research Center University of Sri Jayewardenepura
Nugegoda, Sri Lanka

E. Radha Krishnan
Aptim Federal Services, LLC
Cincinnati, OH, USA

Fiona Laramay
Clarkson University
Potsdam, NY, USA

Yaal Lester
Environmental Technologies, Department of Advanced Materials, Azrieli College of Engineering
Jerusalem, Israel

Shan Li
Cyclopure, Inc.
Skokie, IL, USA

Yuhan Ling
Cyclopure, Inc.
Skokie, IL, USA

Thomas Mallmann
Evoqua Water Technologies
Rockford, IL, USA

Eric Mbonimpa
Air Force Institute of Technology
Wright-Patterson Air Force Base
Dayton, OH, USA

Taryn McKnight
Eurofins Environment Testing America
West Sacramento, CA, USA

Pingping Meng
North Carolina State University
Raleigh, NC, USA

Qingyu Meng
California Department of Toxic Substances Control
Sacramento, CA, USA

Steven Merritt
US Environmental Protection Agency, Region 8
Denver, CO, USA

Marc A. Mills
US Environmental Protection Agency
Cincinnati, OH, USA

Raj Mukhopadhyay
ICAR-Central Soil Salinity Research Institute
Karnal, India

Nicholas Multari
Plasma Research Laboratory, Department of Chemical and Biomolecular Engineering, Clarkson University
Potsdam, NY, USA

Allison M. Narizzano
US Army Public Health Center
Aberdeen Proving Ground
Aberdeen, MD, USA

Chase Nau-Hix
Plasma Research Laboratory, Department of Chemical and Biomolecular Engineering, Clarkson University
Potsdam, NY, USA

Diego Nocetti
Clarkson University
Potsdam, NY, USA

Matthew J. Notter
Cyclopure, Inc.
Skokie, IL, USA

Craig Patterson
US Environmental Protection Agency
Cincinnati, OH, USA

Jonathan G. Pressman
US Environmental Protection Agency
Cincinnati, OH, USA

Michael I. Quinn, Jr.
US Army Public Health Center
Aberdeen Proving Ground
Aberdeen, MD, USA

LeeAnn Racz
ToxStrategies
Fort Walton Beach, FL, USA

Sammani Ramanayaka
Lancaster Environment Centre
Lancaster University
Lancaster, United Kingdom

Niall Ramsden
ENRg Consultants
Bucks, United Kingdom

Patrick Ransom
Abt Associates
Arlington, VA, USA

Stephen D. Richardson
GSI Environmental, Inc.
Austin, TX, USA

J. Alan Roberson
Association of State Drinking Water Administrators
Arlington, VA, USA

Ian Ross
Tetra Tech
Manchester, United Kingdom

Binoy Sarkar
Lancaster Environment Centre, Lancaster University
Lancaster, United Kingdom

Ted Schaefer
Energy and Resources Institute, Charles Darwin University
Darwin, Australia

Donald Schupp
Aptim Federal Services, LLC
Cincinnati, OH, USA

Tim Sidnell
University of Surrey
Surrey, United Kingdom

Raj Kamal Singh
Plasma Research Laboratory, Department of Chemical and Biomolecular Engineering, Clarkson University
Potsdam, NY, USA

Thomas F. Speth
US Environmental Protection Agency
Cincinnati, OH, USA

Peter Storch
Arcadis
Brisbane, Australia

Selma Mededovic Thagard
Department of Civil and Environmental Engineering, Clarkson University
Potsdam, NY, USA

Kyle Trewitz
Evoqua Water Technologies
Bellefonte, PA, USA

Meththika Vithanage
Ecosphere Resilience Research Center University of Sri Jayewardenepura
Nugegoda, Sri Lanka

David G. Wahman
US Environmental Protection Agency
Cincinnati, OH, USA

Hasintha Wijesekara
Department of Natural Resources, Sabaragamuwa University
Belihuloya, Sri Lanka

Marc A. Williams
US Army Public Health Center
Aberdeen Proving Ground
Aberdeen, MD, USA

Yubo Yan
School of Chemistry and Chemical Engineering, Nualyin Normal University
Beijing, China

Lawrence Zintek
US Environmental Protection Agency, Region 5
Chicago, IL, USA

Jeanmarie Zodrow
Geosyntec Consultants
Greenwood Village, CO, USA

Section I

Social Concerns

1 PFAS
Today and Tomorrow

LeeAnn Racz
Fort Walton Beach, Florida

David M. Kempisty
Colorado Springs, Colorado

CONTENTS

1.1 Introduction 3
1.2 PFAS Terminology 4
1.2.1 Linear and Branched Terminology 6
1.2.2 PFAS Families 6
1.3 PFAS Pathways 7
1.4 Policy 9
1.5 Remediation 10
1.6 Analytical Methods 12
1.7 Future Trends 14
References 16

1.1 INTRODUCTION

Per- and polyfluoroalkyl substances (PFAS) are found in a myriad of products in all locations across the globe. These compounds contain fully (*per*-) or partially (*poly*-) fluorinated carbon chains with various functional groups. They have low water surface tension, exhibit amphiphilic behavior, and have high thermal, chemical and biochemical stability. They have been used in applications such as electronics manufacturing, metal plating, textiles, lubricants, cookware, firefighting foams, carpeting, and food packaging materials all over the world since the 1950s. They survive in the environment a very long time, eventually reaching humans. These chemicals are linked to adverse health effects such as low birth weight; reduced immune responses; hepatic, reproductive, cardiovascular, and endocrine system effects; and cancer (IARC 2016, ATSDR 2018, Waterfield et al. 2020). PFAS also last in the human body for a long time, with half-lives ranging from 2.3 to 15.5 years (Li et al.

2018), and they are detectable in virtually the entire US population (Lewis et al. 2015, CDC 2019). Chapter 7 by Kato et al. discusses in detail biomonitoring insights in the United States via the National Health and Nutrition Examination Survey (NHANES). We also refer the reader to Chapter 8 by Zodrow et al. and Chapter E2 by Johnson et al. which cover risk assessments to humans and wildlife. (E-resource chapters are available at https://www.routledge.com/9780367456405.) Finally, Chapter 3 by Balan and Meng offers a nice summary of key PFAS exposure routes and determinants.

These hazards of substances were first brought to light with the litigation efforts of Robert Bilott in the 1990s and beyond (Bilott 2019). Since then, we continue to learn about the extent of the problem, with knowledge increasing on an exponential scale. Legacy PFAS, particularly long chain with six or more fluorinated carbons depending on the attached functional group, are arguably the most persistent and most toxic in this family. The most commonly recognized long-chain PFAS are perfluorooctanoic acid (PFOA) and perfluorooctanesulfonic acid (PFOS). With the addition of PFOS on the Stockholm Convention for Persistent Organic Pollutants in 2009 and PFOA in 2019, they are being phased out of production. However, short-chain (five or less fluorinated carbons for sulfonate-containing PFAS or six or less fluorinated carbons for carboxylic-containing PFAS) alternatives have been manufactured in their place. Some of these short-chain compounds are less persistent and less toxic than the long-chain homologues, but not all. Furthermore, more short-chain PFAS may be required per surface area to achieve the same desired effects as long-chain PFAS (Chu and Letcher 2014). There are over 6,330 PFAS with registered CAS numbers (EPA 2020a), and new compounds continue to be developed. Trade secrets and lack of regulations protect the identity of many of these novel PFAS, but environmental monitoring has revealed several classes of new compounds, including volatile ultra-short-chain (containing 2–3 carbons) PFAS, chlorinated polyfluorinated ether sulfonates (Cl-PFESA) and congeners, isomers and enantiomers of PFAS (Ateia et al. 2019, Zhang et al. 2020, Liu et al. 2020).

1.2 PFAS TERMINOLOGY

In order to provide a standard terminology for this large class of chemicals, in this section we summarize key points from the seminal work in Buck et al. (2011) on nomenclature and background, especially for legacy compounds. Other approaches included substances containing other perfluorocarbon moieties than those included by Buck et al. and promote a slightly different standardized nomenclature (OECD 2018; Interstate Technology Regulatory Council (ITRC) 2020). Chapter 3 by Balan and Meng provides additional discussion on PFAS categorization and an important discussion on PFAS terminology, namely the use of "PFAS" vs. "PFASs."

PFAS are aliphatic substances containing one or more carbon (C) atoms. The hydrogen (H) substituents have been replaced by fluorine (F) atoms. The difference between perfluoroalkyl and polyfluoroalkyl substances are that polyfluoroalkyl substances are aliphatic substances for which all H atoms attached to *at least one* (but not all) C atoms have been replaced by F atoms, whereas in perfluoroalkyl substances all H atoms in all C atoms have been replaced.

The broad term "fluorinated polymers" includes all polymers for which one or more of the monomer units contains the element F in the backbone and/or in side chains. Fluorinated polymers may or may not be PFAS, depending on whether they contain perfluoroalkyl moieties. Examples of nomenclature differences are illustrated by the two examples in Table 1.1.

TABLE 1.1
Examples of the correct and incorrect (or undesirable) uses of the nomenclature for perfluoroalkyl and polyfluoroalkyl substances (PFAS) (Adapted from Buck et al. (2011)).

	Example Statements	
Example	**Correct**	**Incorrect or Less Desirable**
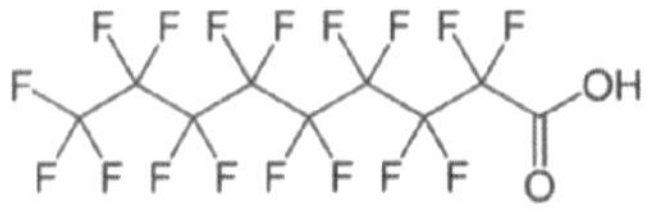 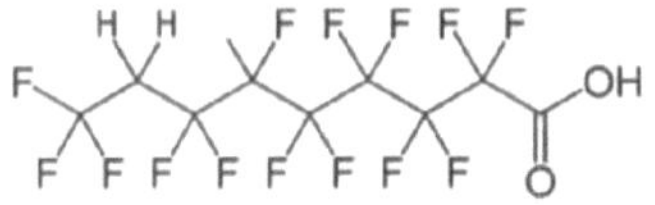	• Both are PFAS within the family of perfluoroalkyl and polyfluoroalkyl substances • Both are carboxylic acids	• Both are: - Perfluoroalkyl substances - Perfluorinated substances - Polyfluoroalkyl substances - Fluorocarbons - Perfluorocarbons - Fluorinated substances - Perfluorochemicals - Perfluorinated chemicals • Both contain fluorocarbons
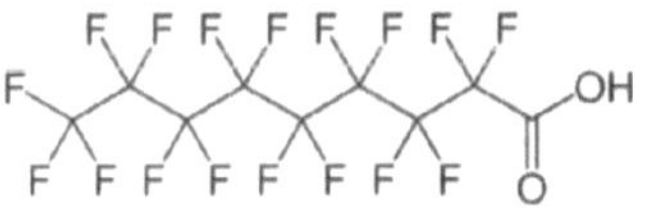	• All H atoms on all C atoms in the alkyl chain attached to the carboxylic acid function group are replaced by F • This is a: PFAS, perfluoroalkyl acid (PFAA), perfluoroalkyl carboxylic acid (PFCA) • Specifically, this is perfluorooctanoic acid, CAS number 335–67-1	• This is a: - Perfluorinated substance - Fluorinated substance - Fluorocarbon - Perfluorocarbon
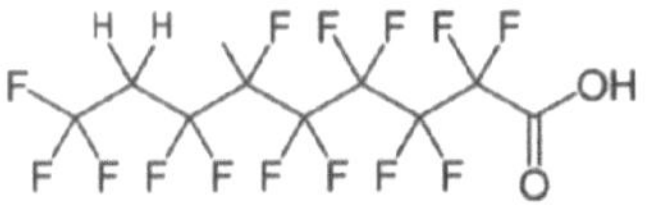	• The alkyl chain attached to the carboxylic acid functional group is polyfluorinated • This is a: PFAS, perfluoroalkyl acid, polyfluoroalkyl carboxylic acid • Specifically, this is 2,2,3, 3,4,4,5,5,6,6,8,8,8-trideca-fluorooctanoic acid	• This is a: - Polyfluorinated substance - Fluorinated substance - Perfluorinated substance • A portion of this compound is perfluorinated

1.2.1 Linear and Branched Terminology

The production of isomers varies by manufacturing process. Telomerization produces primarily linear PFAS, whereas the electrochemical fluorination process produces a mixture of branched and linear isomers. Branching of the main C backbone yields many PFAS families of isomers (Alsmeyer et al. 1994). Linear isomers are composed of carbons that are bonded to only one or two other C atoms. Branched isomers, on the other hand, are composed of C atoms that may be bound to more than two C atoms, resulting in a branching of the C backbone. For example, PFOS can be found as a mixture of the linear isomer and 10 branched isomers (Riddell et al. 2009), although 89 congeners are theoretically possible (Rayne et al. 2008).

1.2.2 PFAS Families

Table 1.2 lists families of some legacy PFAS. Perfluoroalkyl acids (PFAAs) includes perfluoroalkyl carboxylic, sulfonic, sulfinic, phosphonic, and phosphinic acids. PFAAs are highly persistent and are both directly discharged to the environment or are formed indirectly from the environmental degradation of other substances. PFAAs dissociate to their anions in aqueous conditions depending on their acid strength (pKa value). There is also a wide range of physiochemical properties between the protonated and anionic forms. However, it is not clear what is the environmentally relevant pKa for perfluoroalkyl carboxylic acids (PFCAs), also known as perfluorocarboxylic acids or perfluoroalkanoic acids (Burns et al. 2008; Goss 2008; Cheng et al. 2009; Rayne and Forest 2010). Perfluoroalkyl sulfonic acids (PFSAs),

TABLE 1.2
Classification hierarchy of environmentally relevant perfluoroalkyl and polyfluoroalkyl substances (PFAS) (Adapted from Buck et al. (2011)).

Non-Polymers	Polymers
Perfluoroalkyl Substances: Compounds for which all H on all C (except for C associated with functional groups) have been replaced by F - (Aliphatic) perfluorocarbons (PFCs) - Perfluoroalkyl acids - Perfluoroalkane sulfonyl fluorides - Perfluoroalkane sulfonamides - Perfluoroalkyl iodides - Perfluoroalkyl aldehydes	Fluoropolymers: C-only polymer backbone with F directly attached Perfluoropolyethers: C and O polymer backbone with F directly attached to C
Polyfluoroalkyl Substances: Compounds for which all H on at least one (but not all) C have been replaced by F - Perfluoroalkane sulfonamido derivatives - Fluorotelomer-based compounds - Semifluorinated n-alkanes and alkenes	Side-chain Fluorinated Polymers: Variable composition non-fluorinated polymer backbone with fluorinated side chains - Fluorinated acrylate and methacrylate polymers - Fluorinated urethane polymers - Fluorinated oxetane polymers

also known as perfluoroalkane sulfonic acid, includes perfluorooctane sulfonic acid (PFOS), the PFSA that has commanded the greatest attention.

1.3 PFAS PATHWAYS

Once they leave the manufacturing plant, PFAS materials on commercial products are much more likely to be transported and distributed throughout the environment rather than degrade. They have been detected on all parts of the globe, even the North Pole (Bossi et al. 2005). Eventually, they reach humans. Figure 1.1 gives a simplified illustration of the major PFAS pathways, although we continue to learn more about the fate and transport of these chemicals. Chapter E3 by Ross and Sidnell describes in detail the mechanism by which PFAS are transported throughout environmental media.

PFAS are used in a wide variety of products and industries Table 1.3 provides an example of the widespread use of these chemicals across different segments of industry.

Such diverse use complicates measures to effectively control release into the environment. Primary producers and users of these chemicals are a source of environmental release, as are the end products produced. Although products from these industrial sectors make our lives more convenient, and in many cases preserve life as with firefighting foam, they eventually reach the end of their useful service life. After discarding and even through normal use, PFAS containing materials degrade and release PFAS into the environment sometimes directly to humans. PFAS and their precursors are widely found in indoor dust (Zhang et al. 2020) and indoor air (Shoeib et al. 2005), a major pathway since people spend most of their time indoors. When they reach the end of their useful life, they may be dismantled or landfilled. Dismantling, such as in the case of recycling electronic waste, directly releases PFAS to the air and is entrained to dust. Engineered landfills are designed to prevent

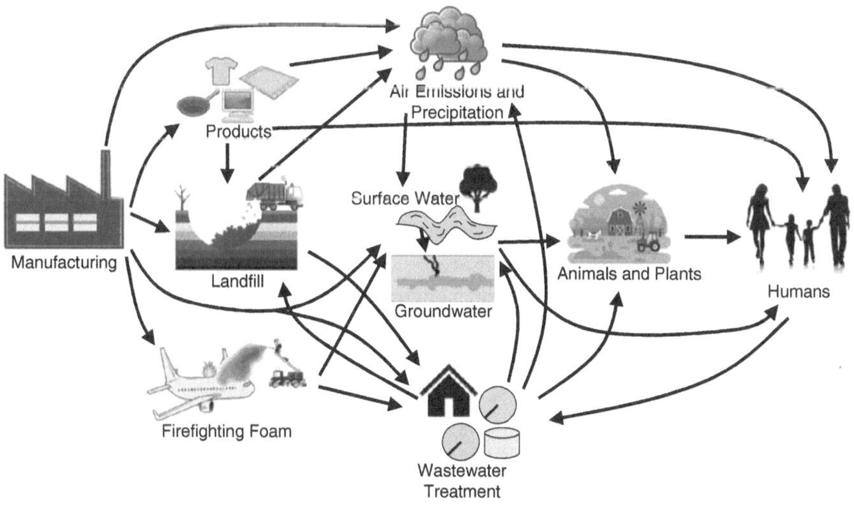

FIGURE 1.1 PFAS Pathways.

TABLE 1.3
Industrial Branches using PFAS (Adapted from Gluge et al. 2020).

Industrial Branches	
Aerospace	Mining
Biotechnology	Nuclear
Building and Construction	Oil and Gas
Chemical	Pharmaceutical
Electroplating	Photographic
Electronic	Production of Plastic and Rubber
Energy Sector	Semiconductor
Food Production	Textile Production
Machinery and Equipment	Watchmaking
Manufacture of Metal Products	Wood

groundwater contamination during the natural degradation process of discarded waste. Landfills produce leachate and air emissions, both of which contain elevated levels of PFAS (Hamid et al. 2018; Seo et al. 2019). In particular, the methanogenic stage of landfill biodegradation produces a higher pH (>7) of leachates which also yield a higher mobility of certain PFAS (Hamid et al. 2018).

Landfill leachate and manufacturing waste is typically sent to a wastewater treatment process sometimes with an industrial pre-treatment step and sometimes discharged directly to a municipal wastewater treatment plant. Humans eliminate PFAS primarily through urine, which also enters wastewater treatment (Jian et al. 2018). However, the highly hydrophobic and persistent nature of certain PFAS makes them difficult to treat (Phong Vo et al. 2020). This means that PFAS scan leave the treatment plant by multiple pathways: sorbed to wastewater sludge, via the treatment plant effluent water, or emitted to the air (Seo et al. 2019). Since wastewater sludge is recycled in the activated sludge process and is often disposed to landfills, PFAS can accumulate in landfills and wastewater treatment plants (Seo et al. 2019). Chapter 6 by Bolan et al. provides more discussion on landfills as a source of PFAS contamination.

Aqueous film-forming foam (AFFF), which has been an important firefighting agent at military bases and airports, is recognized as a major contributor of PFAS to the environment. Before the toxicity of PFAS was understood, AFFF was frequently released to the environment though fire-training exercises, equipment calibration testing, and emergency response operations. The releases impacted the soil and often reached surface water via runoff or groundwater via infiltration. Now that the hazards are more understood, use of AFFF in non-emergency response activities has been reduced and, when used, greater controls are in place. When used in emergency situations, activities post-response attempt to capture AFFF wastewater and implement remediation as necessary. Legacy formulations of AFFF are being replaced with less toxic alternatives and collected and sent to wastewater treatment plants or high-temperature waste incinerators. As discussed above, conventional treatment at the wastewater treatment plants is problematic and the ultimate fate of PFAS through incineration is not fully documented or understood (EPA 2020b). Chapter 2, by Ross et al., provides more discussion on alternative AFFF formulations and the status of fluorine-free foams.

Drinking water sources near contaminated sites have been found with PFAS concentrations over two orders of magnitude over the US Environmental Protection Agency's (EPA) health advisory level of 70 parts per trillion (ppt) and can account for the majority of PFAS exposure at such sites (Emmett et al. 2006; Landsteiner et al. 2014; Gyllenhammar et al. 2015; Worley et al. 2017). Based on the data collected during the Unregulated Contaminant Monitoring Rule 3 (UCMR3) effort, approximately 1.3% of the 36,972 public water systems sampled had PFOA and/or PFOS results above the health advisory level (EPA 2017). Since UCMR 3 was conducted in 2013–2015, analytical detection limits have decreased and interest in PFAS other than the six substances tested for has grown. Toxicological data on these other PFAS and understanding exposure routes and source contributions continues to improve, but is still incomplete. On the average, although there is a great deal of variability among individuals' consumption rates and toxicokinetics, most regulatory agencies have adopted a relative source contribution value of 20% for all PFAS in drinking water (DeWitt 2015, Hu et al. 2019). It is also of interest to note that the source studies for the development of drinking water standards mostly rely on animal models. To account for the translation of values from rodent to human, modifying and uncertainty factors ranging from 3–300 are applied, thereby reducing the derived allowable concentrations. Furthermore, toxicology studies among PFAS may not be applicable to different compounds. For example, recent work by Rice et al. (2020) concluded that toxicology data for perfluorohexanoic acid (PFHxA) was not suitable for assessing potential human health effects from exposure to other short-chain PFAS such as 6:2 fluorotelomer alcohol (6:2 FTOH) which had a more concerning toxicological profile.

Contaminated surface water and groundwater is also known to affect plants and animals through update and consumption, respectively. Fish, meat, egg, and dairy products are the primary food groups that contribute to human exposure (Ericson et al. 2008; Haug et al. 2010; Vestergren et al. 2012). Plants are known to uptake PFAS from contaminated surface or ground waters, precipitation, and wastewater treatment sludge when used as a soil conditioner (Sepulvado et al. 2011; Zareitalabad et al. 2013). Long-chain compounds (PFOA and PFOS) have been found in potatoes and cereal seeds, while short-chain versions can accumulate in leafy fruits and vegetables (Ghisi et al. 2019). PFAS bioaccumulate in the environment and in plants and animals as well (Buck et al. 2011).

1.4 POLICY

Several actions spurred changes in PFAS manufacturing and treatment. Between 2000 and 2002, 3 M, the primary manufacturer of PFOS, voluntarily phased out production of the chemical. In 2006, eight additional manufacturers similarly agreed to phase out global production of PFOA (Lu et al. 2020). In 2008, the European Food Safety Authority established tolerable daily intake values for PFOA and PFOS, and, in their 2018 report, dramatically decreased their assessment to a tolerable weekly intake of 13 ng/kg body weight for PFOS and 6 ng/kg body weight for PFOA. PFOS was listed in Annex B of the Stockholm Convention for Persistent Organic Pollutants in 2009, and PFOA was listed in Annex A in 2019. The European Union has limited the PFOS concentration in textiles to 10 mg/kg and in coating materials to 1 $\mu g/m^3$.

Between 2013 and 2015, six PFAS compounds were included in the EPA's UCMR3 which included quarterly samples at mostly large water systems. In 2016, the EPA published a health advisory level of 70 ppt of combined PFOS and PFOA in drinking water. However, this is only guidance. Since the Safe Drinking Water Act Amendments of 1996, no new contaminants have been formally adopted. Without clear, consistent, and enforceable regulation, clear actions for water systems to address contamination and communicate associated risks are lacking. The absence of federal standards leaves states to make their own difficult decisions. Having separate states regulate separately is inefficient and presents challenges. Each state's regulatory agency must fund, research, and determine an allowable PFAS limit – with a self-assembled dataset and varying levels of pressure applied (by either industry or a concerned public). Disparate regulations also increase regulatory burdens and leave water systems to explain why certain contaminants are regulated less (or more) strictly in other states. Chapter 4 by Roberson offers perspective on the process and implications of PFAS regulation. Although many US states have adopted the EPA level as a health advisory level, other states are taking a more stringent approach. The state-specific regulatory front is one that is changing rapidly. As of this writing, Michigan has promulgated the strictest standards in the nation with a list of seven PFAS compounds and maximum contaminant levels for PFOA and PFNA at 8 and 6 ppt, respectively. Other states monitor for different PFAS compounds and have different enforcement actions and triggers for notification.

Although regulations in the US and other parts of the world are ambiguous, the courts have recognized the clear and present danger PFAS pose to the environment and humans. Even though they ceased production of PFAS years ago, 3 M has agreed to a precedent-setting consent order to clean up contaminated sites in the US (Thusius 2020). Some state regulatory agencies are also requiring cleanup actions, such as New Jersey ordering five chemical manufacturing companies to pay for investigation and remediation efforts (McDaniel and McCrystal 2019). Several other class action lawsuits are also being pursued (Ellison 2019).

Despite the many measures across the world to limit PFAS use, there remain several exemptions for their production and use. Metal plating, firefighting foams, insect baits, photographic coating, gas filters, and medical devices are some of the applications that still allow PFAS use (Vierke et al. 2012; IISD/ENB 2019). In addition, PFAS production has shifted from North America, Europe and Japan to emerging Asian countries. China has become the only producer of PFOS and the largest producer of PFOA (Chen et al. 2009; Wang et al. 2020). Many more PFAS, which now number in the thousands, are being developed and produced to replace those that are phased out as well.

1.5 REMEDIATION

The toxicity and persistency of PFAS compounds demands ready and cost-effective cleanup methods. We need remediation techniques for a variety of matrices, including soil, sludge, sediment, drinking water, wastewater, and even air. Even then, most remediation approaches will only achieve transferring PFAS from one media to another.

Biotransformation by microbial actions is an attractive method. It has been more successful with short chains, but typically takes considerable time and may produce PFAS precursors. Benskin et al. (2013) had some success in marine sediments with low molecular weight compounds, but found higher molecular weight compounds could not be transformed. Certain PFAS found in AFFF have a half-life on the order of months in aerobic sludge, but little is known about their transformation products (D'Agostino and Mabury 2017). In aerobic forest soil, transformation produced more PFOA (Dasu et al. 2013). Researchers have demonstrated some success using fungus enzymes, but progress must be made in commercial applications (Tseng et al. 2014; Luo et al. 2015; Ross et al. 2018). Other work has focused on the lab scale to understand the enzymes responsible for defluorination using the Acidimicrobium bacteria (Huang 2019).

Phytoremediation has been explored. Lee et al. (2014) discovered that plants uptook short- and long-chain metabolites from wastewater treatment biosolids. Long chains are more likely to be accumulated in roots, and short chains in buds, leaves, and fruits (Gobelius et al. 2017). However, this approach has accounted for up to 50% removal at best.

Adsorption, in particular with activated carbon, is the most commonly used treatment method for contaminated water. This mature technology has been cited as Best Available Technology by the EPA for the removal of PFAS and has resulted in high removal efficiency (greater than 85%) of long-chain PFAS (Li et al. 2019). It is also the focus of point-of-use/point-of entry treatment, as explained in Chapter 10 by Patterson et al. However, activated carbon has not been as successful in removing short chains, and some have even had pronounced desorption behavior (Inyang and Dickenson 2017; Mccleaf et al. 2017). Furthermore, natural organic matter is likely to occupy activated carbon pore space or adsorption sites which can inhibit PFAS removal (Xiao et al. 2017). Ion-exchange (IEX) resins have been used to remove PFAS from water with good success (Fujii 2014). Resins have been able to remove short-chain compounds that can be problematic for GAC; however, resins also have their drawbacks. The ion concentrations (nitrate, chlorate, sulfate, etc.) in the background water matrix can compete with PFAS for exchange sites. Natural organic matter (NOM) can also adversely affect PFAS removal performance. Therefore, a reasonable approach may be to use ion-exchange resins as a polishing procedure following activated carbon treatment (Li et al. 2019). Other sorption techniques, such as using minerals, nanomaterials, polymers, and other mesoporous materials, are being investigated as well. Many of these approaches can perform well enough to meet target removal rates. Berretta et al. explore important considerations for the conventional technologies of GAC and IEX resins in Chapter 11. Ling et al. discusses a novel cyclodextrin polymer adsorbent in Chapter 14. An important discussion on modeling water treatment performance and costs for removing PFAS from drinking water is given in Chapter 15 by Burkhardt et al. Furthermore, Laramay et al. present a method for conducting a risk-based cost–benefit analysis of intervention alternatives in Chapter E4, and Emery et al. discuss a life cycle assessment for alternative treatment methods in Chapter E5.

The uncertainty of point-of-use/point-of-entry filters to remove PFAS prompted National Science Foundation (NSF) International to develop NSF P473, a protocol for reducing PFOA and PFOS from drinking water through carbon-based filtration products and reverse osmosis systems. Since then, NSF P473 has been incorporated into NSF/ANSI standards in order to enable ANSI oversight. We now have NSF/ANSI 53 for carbon-based products and NSF/ANSI 58 for reverse osmosis systems which will become the sole certification standards on June 1, 2021. However, it is important for consumers to understand that a filtration product is certified to remove specific compounds of concern rather than all contaminants.

Photocatalytic processes have shown some promising results. Ultraviolet/Fenton techniques have been used to degrade PFOA (Tang et al. 2012; Liu et al. 2013). Titanium dioxide (TiO_2) with ultraviolet light is a common approach for removing PFAS. Many researchers have experimented with modifying TiO_2 to improve its photocatalytic performance. However, full-scale applications require further treatment in order to recover TiO_2 powder. In_2O_3 and Ga_2O_3 have demonstrated even greater capacities to remove PFOA than TiO_2 (Xu et al. 2017). These and other photocatalytic agents continue to be investigated.

Although not yet used outside of the laboratory and consuming high amounts of energy, acoustochemical processes have shown some success. When traveling through water, sound waves oscillate resulting in cavitation and the production of bubbles. As these bubbles compress, their internal temperature can reach 5000 K, and they can release large amounts of energy upon collapse. Saturated fluorocarbon compounds tend to prefer the gas phase, while compounds with hydrophilic functional groups exist in the liquid phase. Therefore, PFAS accumulate at the interface between bubbles and water and mineralization is possible when subjected to these high-temperature conditions (Rodriguez-Freire et al. 2016). However, other volatile organic compounds can interfere with this process (Merino et al. 2016).

Many other novel approaches are being investigated. Coagulation and electrocoagulation have removed PFAS (Bao et al. 2014, Wang et al. 2016). Advanced reduction techniques have also shown some promise as the strong electronegativity of fluorine makes it vulnerable to reducing conditions. However, the variability in conditions, presence of interfering ions, and high energy requirements have made large-scale applications elusive. Lester outlines recent advances in oxidation and reduction processes in Chapter 13, and Chapter 12 by Singh et al. describes plasma treatment. A trending alternative is to use several techniques in tandem with treatment trains in order to leverage the benefits of multiple approaches (Lu et al. 2020). These and other methods continue to be researched.

1.6 ANALYTICAL METHODS

While increasingly strict target concentrations are noble and necessary, it has been difficult to develop analytical methods that are cost-effective and can apply a wide variety of PFAS (Nakayama et al. 2019). Wang et al. (2016) identified three primary barriers to comprehensive PFAS analysis:

1. There is a lack of information on mixture effects, total burden, individual hazards, and mechanisms of action among PFAS classes.

2. There are not yet effective detection methods for PFAS already in the environment and those still being discharged.
3. New PFAS compounds are being continuously developed and released into the environment.

PFAS in air are typically collected using passive samplers. However, there is no standardized methodology, making it difficult to compare global studies. In addition, there are a few analytical methods for anionic and neutral PFAS in air. One alternative is to use bio-indicators such as vegetation samples (e.g., tree leaves and bark) for estimating the airborne transport of PFAS (Jin et al. 2018, Barroso et al. 2018).

Aqueous samples have traditionally been analyzed using liquid–liquid extraction, ion-pair extraction or solid-phase extraction followed by high-pressure liquid chromatography/tandem mass spectrometry (HPLC/MS–MS), or, to a lesser extent, gas chromatography/mass spectrometry (GC/MS). While these fundamental approaches have changed little over the past decade, there have been advances in miniaturization of extraction procedures which have decreased the required sample volume and the amount of required extraction solvent (Concha-Grana et al. 2018, Tröger et al. 2018). The total oxidizable precursor method is another analytical method which is becoming increasingly popular. The idea is to identify the unknown PFAS components of water sample by oxidizing all the PFAS to stable end products. It is also worth noting that currently there are no EPA-approved laboratory methods to measure PFAS in matrices other than drinking water. EPA method 537v 1.1. and EPA method 533 are only approved for drinking water; to use these methods for other media requires a modification and therefore they are not EPA-approved. Work continues with EPA, ASTM and others to develop additional methodologies to address this gap. McKnight's Chapter 5 offers an explanation of the various methods available and currently under development.

Analyses of PFAS sorbed onto solids have similarly made few advancements over the past several years. Solid samples are typically freeze-, air- or vacuum-dried, and then analytes are extracted before analysis using HPLC/MS–MS or GC/MS (Nakayama et al. 2019). However, analytical methods for dust samples have not been thoroughly researched, though dust can be an important mechanism for human exposure.

Biological samples tend to pose the greatest challenges given their complex matrices (Sadia et al. 2020). Therefore, there has been more attention paid to sample preparation methods to biological samples than other types (van Leeuwen and de Boer 2007, Nakayama et al. 2019). Although plasma, serum, and breast milk have been the primary focus of human biomonitoring research, other studies have investigated non-invasive samples such as urine, hair, and nail, with nail being an ideal matrix (Li et al. 2013). Other advances have simplified pre-treatment steps and matured high-throughput analyses reducing the volume of required blood samples from milliliters to tens of microliters (Nakayama et al. 2019). Sample volume reduction has also yielded progress in minimizing matrix effects.

As novel PFAS compounds continue to be developed, a trending analytical approach is non-targeted analysis. Advancements in exact mass spectrometry instruments (e.g., high resolution mass spectrometry, fast atom bombardment/mass spectrometry, and quadrupole time-of-flight/mass spectrometry), including improved fragmentation techniques, increased instrument sensitivity and improved software have made such approaches possible. With the appropriate extraction steps, non-targeted approaches can be used for a variety of environmental matrices. However, the technology still requires maturation for widespread use. Standards do not yet exist for sample pretreatment methods and data analysis procedures, making this approach more suited for discovery rather than comprehensive analysis today (Nakayama et al. 2019). Furthermore, there are still emerging PFAS, and even non-target methods may underestimate the contamination (Yu et al. 2020).

1.7 FUTURE TRENDS

In November 2017, a group of more than 50 scientists and regulators from several nations met to identify the needs and goals by the scientific and policy communities. The result was a list of recommendations to coordinate the spectrum of research and regulations (Ritscher et al. 2018). The scope of the issues demands prioritized efforts in order to avoid duplication of efforts and use resources efficiently. Highlights of their recommendations are summarized below.

1. Create a centralized repository of PFAS knowledge.
2. Research should address groups of PFAS rather than individual compounds.
3. Regulations should account for high PFAS persistence in the environment.
4. Improve understanding of PFAS toxicity.
5. Continue monitoring environmental media and human tissues for emerging PFAS.
6. Improve analytical methods.
7. Conduct socioeconomic analyses on PFAS use, including the long-term societal costs associated with their persistency.
8. Reduce and eventually phase out nonessential PFAS use.
9. Increase public awareness of PFAS-related issues.
10. Maintain and strengthen the interface between science and policy.

These recommendations offer a roadmap for the future of PFAS policy and research. The Association of State Drinking Water Administrators has offered similar recommendations to the US EPA (ASDWA 2018). As investigations reveal additional knowledge gaps and needs, work will proceed accordingly. However, we are far more likely to approach our desired end state when efforts are coordinated and prioritized. The patchwork of state-based regulations is one example highlighting the need for coordinated efforts. In the absence of a Federal MCL for these contaminants, states have taken steps to produce their own values allowable in drinking water, as discussed by Roberson in Chapter 4. The disparate efforts behind the regulations result in different values leading the end consumers of the drinking water to wonder why one state would require <16 ppt for a contaminant and another that <40 ppt is a safe and allowable level for the same contaminant. Further complications arise when we consider the entire scope and number of PFAS. Although two PFAS (PFOA and PFOS) receive the most attention in both the media and in research, there are thousands of different PFAS. There have been adverse health effects associated with them and these compounds can degrade to more stable PFAS compounds (like PFOA and PFOS). Without EPA action on this complex issue and the promotion of a singular unity of effort, the states are being asked to answer these difficult questions multiple (50?) times over.

Related to the above discussion is the idea of adopting a group PFAS regulation. The time and effort required to understand each individual PFAS compound (and isomer and precursor) would be incredibly extensive, if not impossible with new PFAS continuing to enter the market. A more conservative approach would involve regulating PFAS as a group. The EPA drinking water strategy supports addressing contaminants as a group to achieve cost-effective drinking water protection. Group regulation has been applied before; contaminants such as trihalomethanes, polychlorinated biphenols and radionuclides are currently regulated in this manner. The total oxidizable precursor assay allows samples to be oxidized to their greatest extent and thereby determine a representative "total PFAS" concentration. A single EPA group regulation of PFAS would be more comprehensive than individual State efforts, streamline the regulatory process, reduce compliance burden, and be applied consistently across the country as a federal standard. States would have a comprehensive federal process and dataset to refer to for qualitative and quantitative defense of the established levels. States would still maintain their authority to create lower standards but would not be compelled to do so by their constituents because of a lack of singular consistent federal regulation.

Although the States are moving forward with their own regulatory approaches, the EPA has nevertheless published its PFAS Action Plan. Highlights of this plan include eventually establishing MCLs for PFOA and PFOS. It also aims to strengthen enforcement authorities and aid PFOS and PFOS contamination cleanup strategies. The EPA is using its authority under the Toxic Substance Control Act (TSCA) to impose restrictions on certain PFAS compounds. The 2020 National Defense Authorization Act (NDAA) requires certain PFAS to be reportable as part of the Toxics Release Inventory (TRI). There have also been efforts to use the Significant New Use Rule (SNUR) to further regulate importation and manufacture of PFAS. Furthermore, the EPA is considering undertaking additional PFAS drinking water monitoring, supporting research activities, and developing risk communication tools as part of this plan (Environmental Protection Agency, 2020c).

The 2020 NDAA also prohibits military use of fluorinated AFFF after October 1, 2024, although there are exceptions for shipboard and emergency use. This action is driving research and development of non-fluorinated firefighting alternatives. In fact, since 2017, the US Department of Defense (DoD) has committed $11 million to this cause. Some promising solutions are being found in nanotechnology, polymers, and more. These funds are also being used to develop forensic methods for source tracking and allocation, ecological risk characterization, ecotoxicity, and remediation (Paley 2019).

Future regulations are certain to be promulgated. Currently, the Comprehensive Environmental Response, Compensation, and Liability Act (CERCLA), commonly known as Superfund, does not list PFAS as hazardous substances. However, the EPA acknowledges that they could be regarded as CERCLA contaminants in some circumstances. Applying CERCLA to PFAS cleanup will be more straightforward if

PFAS, in particular PFOA and PFOS, are designated "hazardous substances," an action the EPA is pursuing. Many other actions are also being considered. One example requires the Department of Defense to achieve compliance with more stringent state standards (versus federal standards) when remediation is necessary. Another example would require regular blood testing for the health and safety of firefighters. A third example allows the use of Federal Emergency Management Agency's funds to cover cleanup costs. Whether or not these *specific* activities should, or will, pass is not the important point; the take-away is that we can count on technology to advance and policy to be debated and acted upon. The amount of attention being given to this class of persistent chemicals will continue for the foreseeable future.

REFERENCES

Agency for Toxic Substances and Disease Registry (ATSDR). 2018. Toxicological Profile for Perfluoroalkyls (Draft for Public Comment). *US Department of Health and Human Services, Public Health Service, Atlanta, GA*. Available at https://www.atsdr.cdc.gov/ToxProfiles/tp200-p.pdf (Accessed July 15, 2020).

Alsmeyer, Y.W., Childs, W.V., Flynn, R.M., Moore, G.G.I., Smeltzer, J.C. 1994. Electrochemical fluorination and its applications. In: Banks, R.E., Smart, B.E., Tatlow, J.C., editors. *Organofluorine chemistry: Principles and commercial applications*. New York, NY: Plenum. pp. 121–144.

Association of State Drinking Water Administrators (ASDWA). 2018. *State drinking water program recommendations to EPA and CDC on PFAS*, January 12, 2018. Available at https://www.asdwa.org/wp-content/uploads/2018/01/asdwa-pfas-letter-to-epa-and-cdc-final-011220181.pdf (Accessed July 17, 2020).

Ateia, M., Maroli, A., Nishanth, T., Karanfil, T. 2019. The Overlooked Short- and Ultrashort-Chain Poly- and Perfluorinated Substances: A Review. *Chemosphere*. 220, 866–882.

Bao, Y., Nia, J., Xu, Z., Gao, D., Shi, J., Sun, X., Huang, Q. 2014. Removal of perfluorooctane sulfonate (PFOS) and perfluorooctanoate (PFOA) from water by coagulation: Mechanisms and influencing factors. *Journal of Colloid Interface Science*, 434, 59–64.

Barroso, P.J., Martin, J., Santos, J.L., Aparicio, I., Alonso, E. 2018. Analytical method for the evaluation of the outdoor air contamination by emerging pollutants using tree leaves as bioindicators. *Analytical and Bioanalytical Chemistry*, 409, 1643–1655.

Benskin, J.P., Ikonomou, M.G., Gobas, F.A.P.C., Begley, T.H., Woudneh, M.B., Cosgrove, J.R. 2013. Biodegradation of N-ethyl perfluorooctane sulfonamido ethanol (EtFOSES) and EtFOSE-based phosphate diester (SAmPAP diester) in marine sediments. *Environmental Science and Technology*, 47, 1381–1389.

Bilott, R. 2019. *Exposure*. New York: Simon & Schuster.

Bossi, R., Rigel, F.F., Dietz, R., Sonne, C., Fauser, P., Dam, M., Vorkamp, K. 2005. Preliminary screening of perfluorooctane sulfonate (PFOS) and other fluorochemicals in fish, birds and marine mammals from Greenland and the Faroe Islands. *Environmental Pollution*, 136, 323–329.

Buck, R.C., Franklin, J., Berger, U., Conder, J.M., Cousins, I.T., de Voogt, P., Jensen, A.A., Kannan, K., Mabury, S.A., van Leeuwen, S.P. 2011. Perfluoroalkyl and polyfluoroalkyl substances in the environment: Terminology, classification, and origins. *Integrated Environmental Assessment Management*, 7, 513–541.

Burns, D.C., Ellis, D.A., Li, H., McMurdo, C.J., Webster, E. 2008. Experimental pKa determination for perfluorooctanoic acid (PFOA) and the potential impact of pKa concentration dependence on laboratory-measured partitioning phenomena and environmental modeling. *Environ Science and Technology*, 42, 9283–9288.

Centers for Disease Control and Prevention (CDC). 2019. "Fourth National Report on Human Exposure to Environmental Chemicals". *Centers for Disease Control and Prevention*, Department of Health and Human Services, Atlanta, GA. Available at https://www.cdc.gov/exposurereport/index.html (Accessed July 16, 2020).

Chen, C., Lu, Y., Zhang, X., Geng, J., Wang, T., Shi, Y., Hu, W., Li, J. 2009. A review of spatial and temporal assessment of PFOS and PFOA contamination in China. *Chemical Ecology*, 25, 163–177.

Cheng, J., Psillakis, E., Hoffmann, M.R., Colussi, A.J. 2009. Acid dissociation versus molecular association of perfluoroalkyl oxoacids: Environmental implications. *Journal of Physical Chemistry A*, 113, 8152–8156.

Chu, S. Letcher, R.J. 2014. In vitro metabolic formation of perfluoroalkyl sulfonamides from copolymer surfactants of pre- and post-2002 Scotchgard fabric protector products. *Environmental Science and Technology*, 48, 6184–6191.

Concha-Grana, E., Gernandez-Martinez, G., Lopez-Mahia, P., Prada-Rodriguez, D., Muniategui-Lorenzo, S. 2018. Fast and sensitive determination of per- and polyfluoroalkyl substances in seawater. *Journal of Chromatography A*, 1555, 62–73.

D'Agostino, L.A., Mabury, S. A. 2017. Aerobic biodegradation of 2 fluorotelomer sulfonamide-based aqueous film-forming foam components produces perfluoroalkyl carboxylates. *Environmental Toxicology Chemistry*, 35, 2012–2021.

Dasu, K., Lee, L.S. Turco, R.F., Nies, L.F. 2013. Aerobic biodegradation of 8:2 fluorotelomer stearate monoester and 8:2 fluorotelomer citrate trimester in forest soil. *Chemosphere*, 91, 399–405.

DeWitt, J.C. 2015. *Toxicological Effects of Perfluoroalkyl and Polyfluoroalkyl Substances*. New York: Human Press.

Ellison, G., 2019. Wolverine, 3 M class action lawsuit plaintiffs join forces. *MLive*, Available at https://www.mlive.com/news/2019/09/wolverine-3m-class-action-lawsuit-plaintiffs-join-forces.html (Accessed July 27, 2020).

Emmett, E.A., Shofer, F.S., Zhang, H., Freeman, D. Desai, C., Shaw, L.M. 2006. Community exposure to perfluorooctanoate: Relationships between serum concentrations and exposure sources. *Journal of Occupational Exposure Medicine*, 48, 759–770.

Environmental Protection Agency. 2017. *The third unregulated contaminant monitoring rule (UCMR 3): Data Summary*, January 2017. Available at: https://www.epa.gov/sites/production/files/2017-02/documents/ucmr3-data-summary-january-2017.pdf (Accessed September 12, 2020).

Environmental Protection Agency. 2020a. *PFAS master list of PFAS substances*. Available at: https://comptox.epa.gov/dashboard/chemical_lists/pfasmaster (Accessed September 11, 2020).

Environmental Protection Agency, Gullett, B., Gillespie, A. 2020. Technical brief - Innovative research for a sustainable future. *Environmental Protection Agency*. Available at https://www.epa.gov/sites/production/files/2019-09/documents/technical_brief_pfas_incineration_ioaa_approved_final_july_2019.pdf (Accessed August 1, 2020).

Environmental Protection Agency. 2020c. *EPA PFAS Action Plan: Program Update*. Available at https://www.epa.gov/sites/production/files/2020-01/documents/pfas_action_plan_feb2020.pdf (Accessed August 6, 2020).

Ericson, I., Marti-Cid, R., Nadal, M., van Bavel, B., Linstrom, G., Domingo, J.L. 2008. Human exposure to perfluorinated chemicals through the diet: Intake of perfluorinated compounds in foods from the Catalan (Spain) market. *Journal of Agricultural Food Chemistry*, 56, 1787–1794.

European Food Safety Authority. 2018. Risk to human health related to the presence of perfluorooctane sulfonic acid and perfluorooctanoic acid in food. *European Food Safety Authority Journal*, 16, 5194. Available at https://efsa.onlinelibrary.wiley.com/doi/epdf/10.2903/j.efsa.2018.5194 (Accessed July 16, 2020).

Fujii, S. 2014. Adsorption of perfluorooctanoic acid (PFOA) onto anion exchange resin, non-ion exchange resin, and granular-activated carbon by batch and column. *Desalination and Water Treatment*, 526, 542–6548.

Ghisi, R., Vamerali, T., Manzetti, S. 2019. Accumulation of perfluorinated alkyl substances (PFAS) in agricultural plans: A review. *Environmental Research*, 169, 326–341.

Gobelius, L., Lewis, J., Ahrens, L. 2017. Plant uptake of per- and polyfluoroalkyl substances at a contaminated fire training facility to evaluate the phytoremediation potential of various plant species. *Environmental Science and Technology*, 12, 602–610.

Goss, K.-U. 2008. The pKa values of PFOA and other highly fluorinated carboxylic acids. *Environmental Science and Technology*, 42, 456–458.

Glüge, J., Scheringer, M., Cousins, I., DeWitt, J. C., Goldenman, G., Herzke, D., Lohmann, R., Ng, C., Trier, X., Wang, Z. 2020. *An overview of the uses of per- and polyfluoroalkyl substances (PFAS).* Available at: https://doi.org/10.31224/osf.io/2eqac (Accessed September 13, 2020).

Gyllenhammar, I., Berger, U., Sundström, M., McCleaf, P., Euren, K., Eriksson, S., et al. 2015. Influence of contaminated drinking water on perfluoroalkyl acid levels in human serum – a case study from Uppsala, Sweden, *Environmental Research*, 140, 673–683.

Hamid, H., Li, L.Y., Grace, J.R. 2018. Review of the fate and transformation of per- and polyfluoroalkyl substances (PFASs) in landfills. *Environmental Pollution*, 235, 74–84.

Haug, I.S., Salihovic, S. Josten, I.E., Thomse, C., van Bavel, B. Linstrom, G., Becher, G. 2010. Levels in food and beverage and daily intake of perfluorinated compounds in Norway. *Chemosphere*, 80, 1137–1143.

Hu, X.C., Tokranov, A.K., Liddie, J., Zhang, X., Grandjean, P., Hart, J.E., Laden, F., Sun, Q., Yeung, L.W.Y., Sunderland, E.M. 2019. Tap water contributions to plasma concentrations of poly- and perfluoroalkyl substances (PFAS) in a nationwide prospective cohort of U.S. women. *Environmental Health Perspectives*, 127, 067006.

Huang, S., Jaffe, P.R. 2019. Defluorination of perfluorooctanoic acid (PFOA) and perfluorooctane sulfonate (PFOS) by Acidimicrobium Sp. Strain A6. *Environmental Science and Technology*, 53, 11410–11419.

International Agency for Research on Cancer (IARC). 2016. *IARC monographs on the evaluation of carcinogenic risks to humans 110.* Available at https://monographs.iarc.fr/wp-content/uploads/2018/06/mono110-01.pdf (Accessed July 15, 2020).

International Institute for Sustainable Development/Earth Negotiations Bulletin (IISD/ENB). 2019. *Summary of the meetings of the conferences of the parties to the Basel, Rotterdam and Stockholm Conventions*, vol 15. Available at https://enb.iisd.org/vol15/enb15269e.html (Accessed July 16, 2020).

Interstate Technology Regulatory Council (ITRC). 2020. *Naming conventions and physical and chemical properties of per- and polyfluoroalkyl substances (PFAS).* Available at: https://pfas-1.itrcweb.org/fact_sheets_page/pfas_fact_sheet_naming_conventions__4_8_20.pdf (Accessed September 12, 2020).

Inyang, M., Dickenson, E.R.V. 2017. The use of carbon adsorbents for the removal of perfluoroalkyl acids from potable resue systems. *Chemosphere*, 184, 168–175.

Jian, J-M., Chen, D., Han, F-J., Guo, Y., Zeng, L., Lu, X., Wang, F. 2018. A short review on human exposure to and tissue distribution of per- and polyfluoroalkyl substances (PFASs). *Science of the Total Environment*, 636, 1058–1069.

Jin, H., Shan, G., Zhu, L., Sukn, H., Luo, Y. 2018. Perfluoroalkyl acids including isomers in tree barks from a Chinese fluorochemical manufacturing park: Implication for airborne transportation. *Environmental Science and Technology*, 52, 2016–2024.

Landsteiner, A., Huset, C. Williams, A., Johnson, J. 2014. Biomonitoring for perfluorochemicals in a Minnesota community with known drinking water contamination. *Journal of Environmental Health*, 77, 14–19.

Lee, H., Tevlin, A.G., Mabury, S.A. 2014. Fate of polyfluoroalkyl phosphate diesters and their metabolites in biosolids-applied soil: Biodegradation and plant uptake in greenhouse and field experiments. *Environmental Science and Technology*, 48, 340–349.

Lewis, R.C., Johns, L.E., Meeker, J.D. 2015. Serum biomarkers of exposure to perfluoroalkyl substances in relation to serum testosterone and measures of thyroid function among adults and adolescents from NHANES 2011–2012. *International Journal of Environmental Research and Public Health*, 12, 6098–6114.

Li, J., Guo, F. Wang, Y., Zhang, J., Zhong, Y., Zhao, Y., Wu, Y. 2013. Can nail, hair and urine be used for biomonitoring of human exposure to perfluorooctane sulfonate and perfluorooctanoic acid? *Environment International*, 53, 47–52.

Li, Y., Fletcher, T., Mucs, D., Scott, K., Lindh, C.H., Tallving, P., Jakobsson, K. 2018. Half-lives of PFOS, PFHxS and PFOA after end of exposure to contaminated drinking water. *Occupational and Environmental Medicine*, 75, 46–51.

Li, P., Zhi, D., Zhang, X., Zhu, H., Li, Z., Peng, Y., He, Y., Luo, L., Rong, X., Zhou, Y. 2019. Research progress on the removal of hazardous perfluorochemicals: A review. *Journal of Environmental Management*, 250, 109488.

Liu, D., Xiu, Z., Liu, F., Wu, G., Adamson, D., Newell, C., Vikesland, P.L., Tsai, A.L., Alvaraez, P.J. 2013. Perfluorooctanoic acid degradation in the presence of Fe(III) under natural sunlight. *Journal of Hazardous Materials*, 262, 456–463.

Liu, Y., Li, A., Buchanan, S., Liu, W. 2020. Exposure characteristics for congeners, isomers, and enantiomers of perfluoroalkyl substances in mothers and infants. *Environment International*, 144, 106012.

Lu, D., Sha, S., Luo, J., Huang, Z., Zhang, S.J. 2020. Treatment train approaches for the remediation of per- and polyfluoroalkyl substances (PFAS): A critical review. *Journal of Hazardous Materials*, 386, 121963.

Luo, Q., Lu, J., Zhang, H., Wang, Z., Feng, M., Chiang, D., Woodward, D., Huang, Q., 2015. Laccase catalyzed degradation of perfluorooctanoic acid. *Envrionmental Science and Technology Letters*, 2, 198–203.

Mccleaf, P., England, S., Ostlund, A. Lindegren, K., Wiberg, K., Ahrens, L. 2017. Removal efficiency of multiple poly- and perfluoroalkyl substances (PFASs) in drinking water using granular activated carbon (GAC) and anion exchange (AE) column tests. *Water Research*, 120, 77–87.

McDaniel, J., McCrystal, L. 2019. New Jersey orders five companies to pay millions to fund PFAS cleanup. *The Philadelphia Inquirer*, Available at: https://www.inquirer.com/news/drinking-water-toxins-pfas-dow-dupont-cleanup-chemicals-20190325.html (Accessed July 27, 2020).

Merino, N., Qu, Y., Deeb, R.A., Hawley, E.L., Hoffmann, M.R., Mahendra, S. 2016. Degradation and removal methods for perfluoroalkyl and polyfluoroalkyl substances in water. *Environmental Engineering and Science*, 33, 615–649.

Nakayama, S.F., Yoshikane, M., Onoda, Y., Nishihama, Y., Iwai-Shimada, M., Takagi, M., Kobayashi, Y., Isobe, T. 2019. Worldwide trends in tracing poly- and perfluoroalkyl substances (PFAS) in the environment. *Trends in Analytical Chemistry*, 121, 115410.

OECD. 2018. Toward a new comprehensive global database of per- and polyfluoroalkyl substances (PFASs). *Series on Risk Management No. 39* Available at: https://www.oecd.org/officialdocuments/publicdisplaydocumentpdf/?cote=ENV-JM-MONO(2018)7&doclanguage=en (Accessed September 11, 2020).

Paley, M. 2019. *5 things to know about DoD's research on 'fluorine-free' firefighting foam.* Available at https://www.defense.gov/Explore/News/Article/Article/1953510/5-things-to-know-about-dods-research-on-fluorine-free-firefighting-foam/ (Accessed August 6, 2020).

Phong Vo, H.N., Ngo, H.H., Guo, W., Hong Nguyen, T.M., Li, J., Liang, H., Deng, L., Chen, Z., Hang Nguyen, T.A. 2020. Poly- and perfluoroalkyl substances in water and wastewater: A comprehensive review from sources to remediation. *Journal of Water Process Engineering*, 36, 101393.

Rayne, S., Forest, K., Friesen, K.J. 2008. Congener-specific numbering systems for the environmentally relevant C4 through C8 perfluorinated homologue groups of alkyl sulfonates, carboxylates, telomer alcohols, olefins, and acids, and their derivatives. *Journal of Environmental Science and Health A*, 43, 1391–1401.

Rayne, S., Forest, K. 2010. Theoretical studies on the pKa values of perfluoroalkyl carboxylic acids. *Journal of Molecular Structure THEOCHEM*, 949, 60–69.

Rice, P.A., Aungst, J., Cooper, J., Bandele, O., Kabadi, S.V. 2020. Comparative analysis of the toxicological databases for 6:2 fluorotelomer alcohol (6:2 FTOH) and perfluorohexanoic acid (PFHxA). *Food and Chemical Toxicology*, 138, 111210.

Riddell, N., Arsenault, G., Benskin, J.P., Chittim, B., Martin, J.W., McAlees, A., McCrindle, R. 2009. Branched perfluorooctane sulfonate isomer quantification and characterization in blood serum samples by HPLC/ESI-MS(/MS). *Environmental Science and Technology*, 43, 7902–7908.

Ritscher, A., Wang, Z., Scheringer, M., Boucher, J.M., Ahrens, L., Berger, U., Bintein, S., Bopp, S.K., Borg, D., Buser, A.M., Cousins, I., DeWitt, J., Fletcher, T., Green, C., Herzke, D., Higgins, C., Huang, J., Hung, H., Knepper, T., Lau, C.S., Leinala, E., Lindstrom, A.B., Liu, J., Miller, M., Ohno, K., Perkola, N., Shi, Y., Haug, L.S., Trier, X., Valsecchi, S., ver der Jagt, K., Vierke, L. 2018. Zürich statement on future actions on per- and polyfluoroalkyl substances (PFASs). *Environmental Health Perspectives*, 126, 084502.

Rodriguez-Freire, L., Abad-Fernandez, N., Sierra-Alvarez, R., Hoppe-Jones, C., Peng, H., Giesy, J.P., Snyder, S., Keswani, M. 2016. Sonochemical degradation of perfluorinated chemicals in aqueous film-forming foams. *Journal of Hazardous Materials*, 317, 275–283.

Ross, I., Mcdonough, J. Miles, J., Storch, P., Kochunarayanan, P.T., Kalve, E. Hurst, J., Dasgupta, S.S., Burdick, J. 2018. A review of emerging technologies for remediation of PFASs. *Remediation Journal*, 28, 101–126.

Sadia, M., Yeung, L.W.Y., Fiedler, H. 2020. Trace level analysis of selected perfluoroalkyl acids in food: Method development and data generation. *Environmental Pollution*, 263, 113721.

Seo, S-H., Son, M-H., Shin, E-S., Choi, S-D., Chang, Y-S. 2019. Matrix-specific distribution and compositional profiles of perfluoroalkyl substances (PFASs) in multimedia environments. *Journal of Hazardous Materials*, 364, 19–27.

Sepulvado, J.G., Blaine, A.C., Hundal, L.S., Higgins, C.P. 2011. Occurrence and fate of perfluorochemicals in soil following the land application of municipal biosolids. *Environmental Science and Technology*, 45, 8106–8112.

Shoeib, M., Harner, T. Wilford, B.H., Jones, K.C., Zhu, L.P. 2005. Perfluorinated sulfonamides in indoor and outdoor air and indoor dust: occurrence, partitioning, and human exposure. *Environmental Science and Technology*, 39, 6599–6606.

Tang, H., Ziang, Q., Lei, M., Yan, J., Zhu, L., Zou, J. 2012. Efficient degradation of perfluorooctanoic acid by UF-Fenton process. *Chemical Engineering Journal*, 184, 156–162.

Thusius, A. 2020. State, 3 M reach agreement for cleanup of contaminants at Decatur plan, other north Alabama sites, *WAAY31*. Available at https://www.waaytv.com/content/news/State-3M-reach-agreement-for-cleanup-of-contaminants-at-Decatur-plant-other-North-Alabama-sites--571893991.html (Accessed July 27, 2020).

Tröger, R., Klöckner, P., Ahrens, L., Wiberg, K. 2018. Micropollutants in drinking water from source to tap – method development and application of a multiresidue screening method. *Science of the Total Environment*, 627, 1404–1432.

Tseng, N., Wang, N., Szostek, B., Mahendra, S. 2014. Biotransformation of 6:2 fluorotelomer alcohol (6:2 FTOH) by a wood-rotting fungus. *Environmental Science and Technology*, 48, 4012–4020.

van Leeuwen, S.P.J., de Boer, J. 2007. Extraction and clean-up strategies for the analysis of poly- and perfluoroalkyl substances in environmental and human matrices. *Journal of Chromatography A*, 1153, 172–185.

Vestergren, R., Berger, U., Glynn, A., Cousins, L.T. 2012. Dietary exposure to perfluoroalkyl acids for the Swedish population in 1999, 2005 and 2010. *Environment International*, 49, 120–127.

Vierke, L., Staude, C., Biegle-Engler, A., Drost, W. Schulte, C. 2012. Perfluorooctanoic acid (PFOA) – main concerns and regulatory developments in Europe from an environmental point of view. *Environmental Sciences Europe*, 24, 16. Available at: https://doi.org/10.1186/2190-4715-24-16 (Accessed February 25, 2021).

Wang, Y., Lin, H., Jin, F., Kiu, J., Zhao, J., Bi, Y., Li, Y. 2016. Electrocoagulation mechanism of perfluorooctanoate (PFOA) on a zinc anode: influence of cathodes and anions. *Science of the Total Environment*, 49, 7770–7775.

Wang, Q., Ruan, Y., Lin, H., Lam, P.K.S. 2020. Review on perfluoroalkyl and polyfluoroalkyl substances (PFASs) in the Chinese atmospheric environment. *Science of the Total Environment*, 737, 139804.

Waterfield, G., Rogers, M., Grandjean, P. et al. 2020. Reducing exposure to high levels of perfluorinated compounds in drinking water improves reproductive outcomes: evidence from an intervention in Minnesota. *Environ Health*, 19, 42. Available at: https://doi.org/10.1186/s12940-020-00591-0 (Accessed September 13, 2020).

Worley, R.R., Moore, S.M., Tierney, B.C., Ye, X. Calafat, A.M., Campbell, S., et al. 2017. Per- and polyfluoroalkyl substances in human serum and urine samples from a residentially exposed community. *Environment International*, 106, 135–143.

Xiao, X., Ulrich, B.A., Chen, B., Higgins, C.P. 2017. Sorption of poly- and perfluoroalkyl substances (PFASs) relevant to aqueous film-forming foam (AFFF)-impacted groundwater by biochars and activated carbon. *Environmental Science and Technology*, 51, 6342–6351.

Xu, B., Ahmed, M.B., Zhou, J.L., Altaee, A., Wu, M., Xu, G. 2017. Photocatalytic removal of perfluoroalkyl substances from water and wastewater: Mechanism, kinetics and controlling factors. *Chemosphere*, 189, 717–729.

Yu, N., Wen, H., Wang, X., Yamazaki, E., Taniyasi, S., Yamashita, N., Yu, H., Wei, S. 2020. Nontarget discovery of per- and polyfluoroalkyl substances in atmospheric particulate matter and gaseous phase using cryogenic air sampler. *Environmental Science and Technology*, 54, 3103–3113.

Zareitalabad, P., Siemens, J., Harmer, M., Amelung, W. 2013. Perfluorooctanoic acid (PFOA) and perfluorooctanesulfonic acid (PFOS) in surface waters, sediments, soils, and wastewater – a review on concentrations and distribution coefficients. *Chemosphere*, 91, 725–732.

Zhang, B., He, Y., Huang, Y. Hong, D., Yao, Y., Wang, L., Sun, W., Yang, B., Huang, X., Song, S., Bai, X., Guo, Y., Zhang, T., Sun, H. 2020. Novel and legacy poly- and perfluoroalkyl substances (PFASs) in indoor dust from urban, industrial, and e-waste dismantling areas: The emergence of PFAS alternatives in China. *Environmental Pollution*, 263, 114461.

2 Fluorine Free Foams
Transitioning Guide

Ian Ross
Tetra Tech, United Kingdom

Peter Storch
Arcadis, Australia

Ted Schaefer
Energy and Resources Institute, Charles Darwin University, Australia

Niall Ramsden
ENRg Consultants, Monks Risborough, United Kingdom

CONTENTS

2.1 Introduction 24
2.2 The Evolution of F3 foams 26
  2.2.1 The Origins of Modern Fluorine Free Firefighting Foams 26
  2.2.2 The Genesis of Modern High-Performance F3 Foams 27
2.3 PFAS and Firefighting Foams 29
2.4 Evolving Concerns Regarding C6 PFAS Chemistries 30
2.5 Advancing Environmental Regulations 31
2.6 Regulations Focussed on Foam Usage 32
  2.6.1 Europe 32
  2.6.2 United States 32
    2.6.2.1 Washington 33
    2.6.2.2 California 33
    2.6.2.3 Colorado 33
    2.6.2.4 Arizona 33
    2.6.2.5 Virginia 33

2.6.2.6 Kentucky 33
2.6.2.7 Georgia 33
2.6.2.8 New York 33
2.6.2.9 Minnesota 34
2.6.2.10 Wisconsin 34
2.6.2.11 Michigan 34
2.6.3 Australia 34
2.6.4 Summary of Regulatory Trends 35
2.7 Availability and Users of Fluorine Free Firefighting Foam 36
2.8 Testing the Effectiveness of Fluorine Free Firefighting Foam 36
2.9 Water-Soluble Fuels 39
2.10 Environmental Certifications for F3 Foams 40
2.10.1 Harmonized Offshore Notification Format 40
2.10.2 GreenScreen Certification™ 40
2.11 AFFF Management and Transition 41
2.11.1 Foam Transition 42
2.11.2 Cleanout Challenges 42
2.11.3 Impact on Fire Water Systems 43
2.11.4 Equipment Replacement versus Cleanout and Retain 43
2.11.5 Decontamination Case Studies 43
2.12 Waste Management 45
2.13 Conclusions 46
References 46

2.1 INTRODUCTION

Increasing attention to the environmental and human health effects of per- and polyfluoroalkyl substances (PFASs) is leading to the development of increasingly conservative (low) regulatory levels for PFAS in drinking water (Hogue 2018; DeMeo and Brooks, 2019; Ray 2019). As an increasing number of PFASs, including both long-chain (C8) and shorter-chain (≤ C6), are regulated in drinking water, surface waters, soils and groundwater (Ross et al. 2018, 2019) and the uses of firefighting foams containing PFASs are being curtailed in multiple jurisdictions, so many foam users are transitioning to using fluorine free firefighting (F3) foams (Allcorn et al. 2018; Bluteau et al. 2019).

Firefighting foams that contain PFAS include not just aqueous film forming foams (AFFF) but also fluoroprotein (FP) foams and film-forming fluoroprotein (FFFP) foams. These are all used to extinguish Class B, flammable liquids fires, with their ongoing use coming under significant regulatory scrutiny (Ross and Storch 2020). Fluorosurfactants were introduced for use into firefighting foams in the early 1960s and have been used for decades at military bases, civil fire training facilities, airports, oil terminals and refineries for repeated fire training events. The historical regulatory focus has been on three individual "long-chain" highly bioaccumulative PFASs, namely perfluorooctane sulfonic acid (PFOS), perfluorohexane sulfonic acid (PFHxS) and perfluorooctanoic acid (PFOA), but an expanding range of PFASs are now regulated in many locations (Ross et al. 2019; Ross and Storch 2020; Dlugogorski and Schaefer 2021).

The effective use of F3 foams in most fire protection scenarios is supported by a growing body of independent testing data on increasingly larger-scale applications. For example, independent tests performed by the Large Atmospheric Storage Tank Fires (LASTFIRE) organization demonstrate comparable performance when used at standard NFPA application rates between F3 and PFAS-based firefighting foams such as the "C6-pure" alcohol-resistant AFFF (AR-AFFF) in some large diameter tank fire simulations (Ramsden 2018), and their work continues to include other hazards and scenarios. Equinor (formerly Statoil), a major Scandinavian petrochemical company, switched completely to F3 foams from 2013 for both its onshore and offshore (North Sea) operations, after having carried out extensive testing and due diligence on alternatives to fluorinated firefighting foams before changing over (Bluteau et al. 2019). Several major airports, including, for example, Heathrow and Schiphol, have transitioned to F3 foam based on in-house testing and evaluation, including recognised aviation related fire performance test standards (Ross and Storch 2020).

As firefighting activities represent one of the most environmentally emissive uses of these extremely persistent and highly mobile anthropogenic chemicals, through both training exercises and incident response, the uses of PFASs in firefighting are perceived to cause an increased potential for environmental impacts. Historical and ongoing use of AFFF and FP foams can generate long-term soil or concrete highly concentrated "source areas" of PFASs, which can potentially generate large plumes of PFASs dissolved in groundwater and/or impact surface waters meaning PFASs can travel well beyond the original source area, with some PFAS plumes impacting >250 km^2 of groundwater (Yingling 2015).

From a waste management perspective treating foam concentrates and spent firewater resulting from the use of AFFF or FP foams is not possible using biological treatment systems, such as via conventional wastewater treatment plants, as no PFASs are biodegradable. Discharge of these wastes to the sewer will therefore not result in treatment so should be avoided. In contrast some F3 foams have been identified as containing only biodegradable ingredients thus can potentially be treated biologically via conventional wastewater treatment plants. F3 foams could potentially be released to soil and groundwater in some circumstances without environmental harm. A GreenScreen Certified™ process is now available for F3 foams to mitigate concerns from end users regarding their environmental profile (Franjevic et al. 2020).

One growing concern is effective decontamination of fire suppression systems as fluorosurfactant PFASs adhere to surfaces and can form multiple bilayers. This means the interior of fire suppression systems can be coated with a significant mass of PFASs which cannot be removed with repeated water rinses. This behavior can result in significant rebound of PFAS into F3 foams, causing regulatory concerns and potentially negating the benefit of transitioning to PFAS-free foams. However, fortunately, the decontamination of fire suppression system infrastructure is achievable using some recently developed approaches meaning replacement of costly systems is not required.

As an accelerating number of foam transitions projects start, where AFFF and FP based foams are replaced by F3, there is a need for a competent and experienced team of fire and environmental engineers that can effectively decontaminate fire suppression systems, engineer any required minor suppression system modifications, ensure compliance with fire safety and insurance requirements, and manage regulatory expectations (Rice et al. 2020).

This chapter describes the evolution of F3 foams and the historic performance testing to demonstrate their effectiveness; it also provides an overview of environmental concerns relating to PFASs and the fast-evolving regulatory landscape. It then provides an overview of recent large-scale fire extinguishment tests, demonstrating the efficacy of F3 foams vs. AFFF and describing how successful foam transition and decontamination of fire suppression systems has been implemented. The challenges to overcome when transitioning away from AFFF, FP and FFFP are described to highlight that when changing out foams it is not as simple as removing a fluorosurfactant-based foam and replacing it with a F3 foam. However, successful foam transitions are possible using an experienced team with a combination of fire and environmental engineering skills. The lessons learned from many successful foam transition projects are described.

2.2 THE EVOLUTION OF F3 FOAMS

There is sometimes confusion regarding F3 foams with commentary that these are inferior products developed in the 1940s. To provide some context regarding the development of modern high-performance F3 foams, a brief history of the conception and evolution is provided.

2.2.1 The Origins of Modern Fluorine Free Firefighting Foams

In the 1960s when AFFF was invented and established as the operational firefighting foam of the future, it was necessary for firefighters to train so they could become proficient in the use of AFFF on flammable liquid fires. Hot fire training pads became the normal training methodology through the late 1960s and into the 1970s and 1980s utilizing a metal or concrete pan and using a water base while floating a hydrocarbon fuel on the water. The training exercises ranged from smaller pan sizes, using handlines, to larger configurations involving several crash rescue tenders using foam monitors and significant amounts of AFFF.

In 1985, one military aviation training school looked to resolve a problem restarting fires due to the amount of AFFF that remained on the surface of the fuel in the training ground, making it difficult to have more than two training sessions in a day. The productivity of the training college had dropped significantly. Most of the day was spent trying to re-ignite the test pans to allow more training sessions; however, the high usage of AFFF left significant residue in the sediment of the pan. The organization sought assistance in a new product designed for training Aircraft Rescue and Fire Fighting (ARFF) personnel. Dilution of the AFFF to half the concentration was previously tried; however, the burnback resistance of the AFFF continued to remain, as the residual fluorosurfactants would remain as they had accumulated in the soil and water of the training area.

The fire training facility wanted a product that produced a foam that looked real, yet had faster, and more controlled burn back. Achieving low surface tensions, and interfacial tensions was not possible, and the initial spreading coefficient could not be achieved to satisfy the mathematical model for a positive spreading coefficient. However, the scientific basis behind spreading was not considered absolute, as long as the firefighting foam was effective when applied.

If the standard AFFF was reformulated by just removing the fluorosurfactants, the resulting product was not effective for fire extinguishment. In 1985, alternative hydrocarbon surfactants were added to the AFFF base formulation (without fluorosurfactants) to boost the visual performance of this product, which came to be known as "Training Foam." This foam performed well when putting out a flammable liquid fire, such as kerosene, and had a safe and acceptable drainage rate, and vapor sealability to allow fire training without using any form of PFAS. Film formation was not considered essential for a controlled training situation. In fact, the use of training foams became more widely accepted as a solution on military flammable liquid training facilities. Training foams, which are devoid of fluorosurfactants, are also considerably less expensive than AFFF, FFFP, and FP, which was a further driver for their use. With the high prices for these fluorosurfactant-based foams the lower training foam costs foams made for an additional driver for their use (Pabon 2005). The use of "training foam" successfully grew from 1985 to today.

2.2.2 The Genesis of Modern High-Performance F3 Foams

The announcement in 2000 by 3M to discontinue PFOS-based products and all forms of 3M AFFF came as a shock to the firefighting industry. Three differing reports from environmental and animal studies were described to provide converging evidence about the adverse consequences of continued PFOS production and led 3M in consultation with the EPA to announce an immediate phaseout (Singh et al. 2005). At the time, PFOS chemistry was long described as being unreactive and inert, termed "chemical rocks". On that day in May 2000, it seemed that what had been described within the firefighting industry as environmentally inert was now termed environmentally persistent. The announcements were immediate and broadcasted globally; the implication being that 3M was leaving the AFFF market (Ross et al. 2018; Fire Fighting Foam Coalition 2019). At that time, one option was to continue using fluorosurfactant (PFAS)-based firefighting foams, which were mainly based on shorter-chain (C6) PFAS chemistries. The other option was to make a flammable liquid firefighting foam without the use of any fluorosurfactants, therefore "fluorosurfactant free" which was later termed as F3 foam. What already existed pre-2000 on the market that was described as fluorosurfactant free included high expansion foam, medium expansion foam, low expansion foams/wetting agents, training foams, and Class A wildfire foams.

In pursuing novel F3 foam concentrates for Class B flammable liquids, the product performance target must be specification-based such that a product is judged on fire performance and tolerates the use of non-fluorosurfactant raw materials without a fire extinguishment performance penalty. However, in some tests there were prescriptive specifications, such as the US Mil Spec, which stipulated that fluorosurfactants are required in the firefighting foam formulations. Alternatively, foam specifications like UK MOD/DEF(Aust), the International Civil Aviation Organization

(ICAO), EN 1568 and LASTFIRE were fire performance-based, without prescriptive requirements for specific chemicals such as fluorosurfactants. These tests then allowed for a choice where product fire performance can be combined with lower environmental impacts as foams are not stipulated to contain fluorosurfactants. Therefore, the decision regarding which foam to apply can be based on performance without a mandatory chemical content.

From the experience gained formulating training foams in the 1980s, it was clear that just removing the fluorosurfactants from conventional AFFF formulations would not create a functional product. From 2000, foam formulators embarked on a long journey trialing hundreds of ingredient to create functioning F3 foams. They made over 1200 experimental batches to create a viable F3 foam concentrate. Many of these early experimental formulations could not extinguish a test fire. Performance factors such as foam quality, measuring fire performance data primarily on kerosene and aviation gasoline, physical characteristics measurements (including surface and interfacial tension), vapor suppression of hydrocarbon spill simulations, vapor ignition suppression, and tests extinguishing large scale fires, including 100 and 150 m^2 fire pits showed that F3 formulations could be effective using standard foam branches and fog/ss nozzles and monitors.

In 2002, 3M had formulated a F3 foam that had comparable performance to two PFOS-based AFFF foams meeting the ICAO Level B specification (appropriate for AFFF and FFFP), which at that time was the highest performance level in the ICAO standard. The 3M F3 product was termed Rehealing Foam (RF) and was produced to be diluted to 3% and 6% (hence RF3 and RF6) (Schaefer 2003). The key foam characteristics were reported to include: no longer having aqueous film technology, a long drain time, to keep foam on fuel, good heat resistance, free-flowing like AFFF, self-repairing of punctures, and re-healing for burn back resistance. Fire performance tests comparing the ICAO specification criteria of the performance of two 3M AFFF formulations and the RF F3 foam were available from 3M in 2002 (see Table 2.1) (Schaefer 2002). Therefore, ICAO tests have demonstrated that F3 foams with superior performance to AFFF have been available for nearly 20 years. It was reported in 2005 that the RF6 product had been tested successfully for its ability to be used throughout an ARFF vehicle commonly used in Australia, and the proportioning system could accurately administer the foam (Schaefer et al. 2005).

TABLE 2.1
ICAO Level B Fire Performance (4.5 m^2 pan) Test Results for F3 Foam vs. AFFF in 2002 (Schaefer 2002).

	ICAO Level B Spec	3M™ Foam RF6 (F3)	3M™ FC-206CF (AFFF)	3M™ FC-3003 (AFFF)
Test witnessed by		DNV (Norway)	SP (Sweden)	ASA (Australia)
Solution Strength	3 or 6%	6	6	6
90% Control	-	30 s	38 s	-
Extinguishment	<60 s	46 s	46 s	50s
Burn Back Time	>5:00	>8:00	>8:00	7:06

There was no visible difference in the performance of 3M F3 Foam RF6 and the two PFOS-based AFFF foams. Subsequently, the RF6 formulation has been retested with similar results. A new higher performance level for ICAO has been established and named Level C. More recent free formulations RF products have been certified by a third-party laboratory to Level C performance rating.

There are a great number of F3 foam concentrates that are being developed and qualified to higher performance levels. That also includes listing with sprinklers and being qualified at both the "synthetic" application density, and at the "AFFF" application density. Improvements in F3 performance will continue; many more manufacturers are seeking environmentally improved options and removing PFAS from their formulations.

2.3 PFAS AND FIREFIGHTING FOAMS

Firefighting foams containing PFAS include all AFFF, FFFP and FP formulations. If a foam is classified as a FP or film forming it will contain PFAS. The only current exception is the recently developed fluorine free film forming firefighting (F5) foam by Angus termed Jetfoam ICAO-B (2020). Foams developed specifically for high expansion applications are not expected to contain PFAS, but an exhaustive survey of their chemical components has not yet been published.

PFASs are a broad group of some 4730 anthropogenic (human-made) xenobiotics (foreigners to the biosphere) (Wang et al. 2017; OECD 2018), that as bulk manufactured chemicals have been used in a wide array of commercial goods and products since the 1940s. None of these PFASs can biodegrade, some however can biotransform to create other PFASs.

The term PFASs includes many thousands of individual compounds each with their own acronyms, which can initially be very confusing. For clarity, it should be highlighted that PFAS refers to all the compounds in this class, i.e., those in both C8 and C6 firefighting foams (Buck 2011). The acronyms of importance considering firefighting foams are highlighted below and provide a brief overview of their importance when considering whether to transitioning between firefighting foams.

Perfluoroalkyl substances have previously been referred to as perfluorinated compounds (PFCs), but are now more commonly termed perfluoroalkyl acids (PFAAs). The PFAAs include PFOS, PFOA, (termed C8 PFAAs) and PFHxS or perfluorohexanoic acid (PFHxA) (termed C6 PFAAs). PFASs present in firefighting foams currently sold, principally comprise polyfluoroalkyl substances, which are termed fluorotelomers (Place and Field 2012; Backe et al. 2013). If a fluorotelomer based foam is termed as a C8 or C6 product, this means it comprises precursors to C8 or C6 PFAAs.

These polyfluorinated fluorotelomers in the foams, biotransform or abiotically transform in the environment (soil, sediment, groundwater, surface waters etc.) and may be partially metabolised in higher organisms (Vestergren et al. 2008) to create PFAAs (Ross et al., 2019; Ross and Storch 2020). Following foam release to ground the fluorotelomers can create PFAAs, so they are termed PFAA "precursors." The PFAAs are "dead end" terminal products of transformation, they persist indefinity in the environment and are currently far more commonly regulated in drinking water, soils and groundwater than their precursors.

The fluorotelomer precursors in the foams are proprietary molecules, for which no analytical standards are available, so they cannot be directly measured and quantified and can therefore go undetected until they transform into the PFAAs. More advanced chemical analytical tools such as the total oxidizable precursor (TOP) assay will reveal their presence of these PFAA precursors and is able to quantitatively estimate the concentration of PFASs containing a detectable perfluoroalkyl group (Houtz 2013; Houtz et al. 2013).

Recognizing the science behind this process, regulators in Australia have recently adopted this advanced analytical tool for sampling environmental matrices and compliance. The TOP assay is widely available commercially and may be considered best practice for assessment of PFASs in firefighting foams (Ross et al. 2018, 2019). However, recent analytical guidance in the US from the Fire Fighting Foam Coalition (FFFC) recommends the use of standard analyses (Fire Fighting Foam Coalition, 2019) to assess PFASs in firefighting foam concentrates, such that the vast majority of PFASs in the fluorotelomer based firefighting foams would not be detected (Ross et al. 2019).

Older formulations of AFFF (1964–2003) contained significant concentrations of PFOS (Dlugogorski and Schaefer 2021). The sale of firefighting foams containing PFOS ceased in 2003 (Santoro 2008), but many AFFFs marketed as "short-chain" (C6) firefighting foams still contained PFOA (C8) and its precursors (Bagenstose 2017). From the 1970s until 2015, there were also fluorotelomer-based AFFF formulations available that contained significant amounts of precursors to PFOA. In 2004 it was reported that AFFF was not a likely source of PFOA (Cortina 2004), but subsequent analysis of C6-based foams revealed that approximately 20% of the PFASs present were PFOA precursors and therefore had the potential to form PFOA in the environment, with the remainder being precursors to short-chain PFAAs (Backe et al. 2013; Houtz et al. 2013; D'Agostino and Mabury 2014).

As a result of the EPA PFOA stewardship program, between 2006 and 2015 (US EPA 2015), the amount of PFOA or PFOA-related substances (i.e., precursors) in firefighting foams was diminished to achieve a maximum of 50 mg/kg in C6-pure foams by 2015 (Mennie 2017). The current AFFF formulations, referred to as "C6-pure," mainly contain PFASs with six or fewer fluorinated carbons and thus will not result in the release of PFOS, but can form a myriad of short-chain PFASs, including PFHxA in the environment and should contain minor amounts of PFASs that are precursors to PFOA (i.e., <50 mg/kg PFOA-precursors).

2.4 EVOLVING CONCERNS REGARDING C6 PFAS CHEMISTRIES

Very little research has been done considering the toxicology of the unregulated fluorotelomer precursors in AFFF, FFFP and FP or their transient environmental transformation intermediates. However, studies have shown that some fluorotelomer precursors are 10,000 times more toxic than the PFAAs they transform into (Ritter 2010).

The risks posed by short-chain PFASs (i.e., fluorotelomers and their transient metabolites, not just PFAAs) are often mischaracterized by referral to select studies focused on PFHxA (Willson. 2020) failing to mention the toxicity or bioaccumulation potential of the parent fluorotelomer precursors and the myriad of intermediate

PFASs they can form. Intermediate transformation products include the 6:2 fluorotelomer sulphonate (6:2 FTS) and 5:3 acid which are described to bioaccumulate in marine invertebrates and rats, respectively, and the final short-chain PFAAs transformation products concentrate in the edible portion of crops (Ritter 2010; Blaine, Rich et al. 2013; Blaine, Rich et al. 2014; Blaine, Rich et al. 2014; Kabadi, Fisher et al. 2018; Langberg, Breedveld et al. 2019; Rice et al. 2020). Recent articles have highlighted a "*flawed focus on PFHxA as a reference chemical for short-chain PFAS class*" and highlighted scientific studies undertaken by independent scientists working for the US Food and Drug Administration (FDA) which highlighted the issue of the potential bioaccumulation of short-chain PFAS as being an "*elephant in the room*" (Neltner and Maffini 2019; Neltner 2020).

The focus of the criteria on which environmental regulations to classify chemical for restrictions have been traditionally based on persistence (P), bioaccumulation (B), and toxicity (T) (i.e., via Stockholm Convention PBT criteria). However, in Europe the adoption of mobility (M) criteria are being introduced as PBT criteria have been described to be only marginally effective for protection of drinking water supplies (Reemtsma et al. 2016), as the PBT criteria seemed aimed to protect the food chain. This means the shorter-chain (C6) replacement PFAS in C6 foams are increasingly being regulated based on PMT and very persistent very mobile (vPvM) criteria (Neumann et al. 2017, Vierke 2017). Being more mobile that the long-chain PFASs and significantly more difficult to remove from wastewater, the replacement short-chain PFAS will be more likely to be discovered in drinking water than long-chain PFAS as the replacements continue to be used commercially (Ross et al. 2019; Ross and Storch 2020).

2.5 ADVANCING ENVIRONMENTAL REGULATIONS

Globally, environmental regulations, which are generally focused on the PFAAs, are rapidly being proposed to very conservative (low) levels. After initial regulatory attention in the late 1990s environmental concerns considering PFASs from the US Environmental Protection Agency were acted on in 2000, when on May 16 it was announced that PFOS would be withdrawn from sale (US 2000). While PFOS sales continued until 2003 (Santoro 2008), it took until 2009 before PFOS production and uses were restricted by the Stockholm Convention which classed PFOS as PBT (UN Stockholm Convention 2009). The Stockholm Convention listed PFOA as a POP in 2019 (Williams 2019; UN Stockholm Convention 2019) with PFHxS also currently under consideration (2020). There has been a dramatically accelerating focus on PFASs by environmental regulators in the last five years. For example, in 2016, the detection of PFASs in drinking water in the USA, combined with the US Environmental Protection Agency (US EPA) issuing a long-term health advisory level of 70 ng/L (for combination of PFOS and/or PFOA) (US EPA 2016), led to 6.5 million people's drinking water being considered unsafe (Hu et al. 2016).

There have been calls for a complete ban of the whole class of PFASs (Van Hamme 2020; Simkins 2020), but so far this has only been instigated by Denmark in relation to food packaging (PFAS Central 2019), with Washington State also proposing the same approach (Chemical Watch 2020).

It was noted that the Stockholm Convention Persistent Organic Pollutants Review Committee recently recommended "*not to replace fire-fighting foam that contained or may contain PFOA, its salts and PFOA-related compounds with short-chain per- and polyfluoroalkyl substances (PFASs) due to their persistency and mobility as well as potential negative environmental, human health and socioeconomic impacts*" (UN Stockholm Convention on Persistent Organic Pollutants 2019).

Advances considering the regulation of firefighting foam use, which also place restrictions on PFASs via limiting use of AFFF, FFFP, and FP are also being proposed in multiple countries with regulations in specific US states (Denton et al. 2019) and Australia proposed or promulgated in the last few years.

2.6 REGULATIONS FOCUSSED ON FOAM USAGE

2.6.1 Europe

The European Chemicals Agency (ECHA) has recently proposed a European Union (EU)-wide restriction that will prevent and reduce the emissions of PFHxA, its salts and related substances within the EU (European Chemicals Agency 2019). As PFHxA (along with other short-chain PFAAs) is one of the terminal environmental transformation products of the fluorotelomer PFASs used in C6 firefighting foams this restriction has direct impact on the ongoing use of AFFF, FFFP, and FP foams. The rationale for the proposed restriction is that PFHxA fulfils the P-criterion and vP-criterion and has mobility and long-range transport potential, with unpredictable and irreversible adverse effects on the environment or human health over time. Exemptions are proposed over a five-year period for the emergency use of firefighting foams, but there is no exemption for fire testing and training, unless the whole release is contained. However, containment without appropriate treatment allowing discharge will be a significant challenge, as is mentioned in later sections, biological treatment is not effective for any PFASs. There is also a 12-year exemption in place for Class B firefighting foam user to protect storage tanks, with a surface area above 500 m^2. The proposed restriction was described as being practical because it is affordable, implementable, enforceable and manageable.

European regulations implementing restrictions of PFOA (its salts) at and PFOA-related substances in products will be applicable from 4 July 2020 (2017). PFOA and its salts shall not be used when the concentration of PFOA, including its salts, exceeds 25 parts per billion (ppb or mg/L) or the concentration of one or a combination of PFOA-related substances exceeds 1000 ppb (or mg/L) in the production of, or placed on the market in: another substance, as a constituent; a mixture; any part of an article. There is a question of whether C6-pure foams contain PFOA-related substances (i.e., precursors) below the 1000 ppb (1 mg/L) threshold since the C8 stewardship program limit for PFOA and its precursors is 50 mg/kg (Mennie 2017), which is significantly higher than restrictions imposed in Europe in 2020.

2.6.2 United States

In the US, regulations addressing the use of firefighting foams containing PFAS have so far been proposed or enacted in 11 States. There are some differences between

States, with the main elements of the restrictions in differing States outlined below (Denton et al. 2019).

2.6.2.1 Washington

A new bill was signed into law that expands existing restrictions on PFAS-containing foams. The measure eliminates exemptions included in the State's 2018 law that banned most applications of AFFF, FFFP, and FP (Chemical Watch 2020). The new law will phase out exceptions for the continued manufacture, sale, and distribution of the foams for use in chemical plants, oil refineries, and oil terminals in 2024, unless a waiver is secured. It would also eliminate exemptions for continued use of the products where required by federal law, effective two years after those federal requirements drop off.

2.6.2.2 California

Proposition 65 regulations (Safe Drinking Water and Toxic Enforcement Act) include three routes of exposure for PFAS: Occupational, Consumer and Environmental. With listing of PFOA as developmental toxicant in 2018, discharge prohibition came into effect in 2019.

2.6.2.3 Colorado

Law prohibits the use of Class B firefighting foam that contains PFAS for training purposes, and violations may result in the imposition of a civil penalty. There are limits on the sale of PFAS-containing foam, which requires manufacturers to notify their customers of this law.

2.6.2.4 Arizona

In 2019, the law prohibiting discharge or other testing or training uses of PFAS-containing Class B foam carves out usage "required by law or federal regulation."

2.6.2.5 Virginia

As of 2019, there are restrictions on the use of PFAS-containing foams for training and testing while allowing continued use for real-world fires.

2.6.2.6 Kentucky

As of 2020, there is a ban on uncontained release of PFAS-containing foams, except in an emergency.

2.6.2.7 Georgia

There is a 2019 house bill ruling to restrict the use of PFAS-containing foams for training and testing, while allowing continued use for real world fires.

2.6.2.8 New York

In 2019, a law banned the use of PFAS-containing Class B foams for training purposes and will prohibit their manufacture, sale, or distribution two years later. There are certain exemptions, such as when the products' use is required by federal law or to fight fires at oil refineries or chemical plants.

2.6.2.9 Minnesota

A 2019 law prohibits PFAS-containing Class B foam for testing or training unless required by federal law, but excludes from this ban the use of AFFF in emergency firefighting and fire prevention activities. Any release of PFAS-containing Class B foam must be reported within 24 hours.

2.6.2.10 Wisconsin

A 2020 law bans the use of PFAS-containing foam for training purposes. The foams will be allowed for use in emergency firefighting and testing purposes, although testing facilities must implement "appropriate containment, treatment, and disposal or storage measures to prevent discharges of the foam to the environment." Legislation took effect on September 1, 2020 (Mikkelson 2020).

2.6.2.11 Michigan

In 2019, a bill was proposed that would ban the use of PFAS-containing foam during firefighting training, beginning on December 31, 2023. Until that date, firefighters would need to be instructed on the proper use, handling, and storage of the foams, as well as on containment and proper disposal. Further bills outline the "best health practices" for using, handling, and storing the foam, including decontamination of a firefighter's body and equipment and would require that fire departments submit a written report to the State within 48 hours of using an AFFF.

2.6.3 Australia

In Australia, the key pieces of legislation guiding the management of firefighting foam include the PFAS National Environmental Management Plan (NEMP) and the Queensland Environmental Management of Firefighting Foam, Operational Policy. The regulation of PFAS in Australia is undertaken by the regulatory agencies in each state or territory. However, an intergovernmental agreement on a National Framework for Responding to PFAS Contamination has been signed by each Australian state and territory. A key action item of the agreement is the implementation of the PFAS NEMP which has been endorsed by the Heads of the Environmental Protection Agencies in Australia and New Zealand (HEPA). While this legislation does not provide specific guidance on foam transition or include recommendations for acceptable concentrations of PFAS in AFFF or other foams, it provides guidance on the assessment, management, and remediation of PFAS contaminated sites and materials and references key criteria protective of the human health and the environment which do apply to the management of foam fire suppression systems.

In 2016, the Queensland Department of Environment and Science (DES) introduced a policy for the management of firefighting foam, including the phase out of PFOA and PFOS-based AFFF. The compliance for the policy came due on July 7, 2019 in which foam users were required to comply with the policy or provide justification and a compliance plan to avoid non-compliance (Queensland Government Department of Environment and Heritage Protection 2016a).

For foam concentrates, the policy stipulates a limit on fluorinated organic compounds in concentrates equivalent to <10 mg/kg PFOS+PFHxS and < 50 mg/kg C7-C14 as fluorine, based on TOP Assay. For foam mixes and wastewater generated from foam system testing and decontamination, the Interim Contaminated Water (waste) criteria stipulates a limit on foam-water mixtures of 1.0 µg/L for C4-C14 perfluorinated carboxylates plus C4-C8 perfluorinated sulfonates following oxidation using the TOP Assay. A limit of 0.3 µg/L also applies to each of PFOS and PFOA measured in accordance with US EPA Method 537.1. These criteria apply to any liquids released from the foam system, including foam mix discharged during a test or fire event, fire water supply, and concentrate should there be loss of containment.

Given foams are typically proportioned with between 1–10% concentrate by volume, it is possible to use a concentrate that is compliant with the Queensland Foam Policy, but produces a foam with PFAS concentrations above the interim PFAS contaminated water criteria.

The policy provides also outlines further requirements for the off-site disposal of waste foam concentrate and contaminated water. Specifically, the policy states that, "*Notwithstanding that firefighting foams containing PFOS and PFOA must not be held or used, water contaminated by fluorinated organic compounds must not be released to the environment if the levels of fluorinated organics exceed 0.3 µg/L PFOS, 0.3 µg/L PFOA and 1.0 µg/L for sum of TOP Assay C4-C14 plus C4-C8 sulfonates.*"

In January 2018, the South Australian government amended the Environment Protection (Water Quality) Policy 2015 under the Environmental Protection Act 1993 to ban the use of potentially hazardous fluorinated firefighting foams. Under this legislation, all PFAS containing firefighting foams are banned and there is no maximum allowable concentration for PFAS in operational foams used in South Australia.

2.6.4 Summary of Regulatory Trends

As environmental regulation advance to include multiple PFASs, including both long- (C8) and short-chain (C6) varieties, it seems prudent that owners and operators of high-hazard facilities that require the use of PFAS-containing Class B foam products may need to consider the environmental liabilities associated with the continued or historical use of these foams. Additionally, training and use of some AFFF, FFFP and FP foams may still be an ongoing activity at high-hazard sites.

It may be wise to consider the future potential environmental liabilities of the historical and continued use of all PFAS-containing foams. This may be particularly important if the high-hazard site is situated on an aquifer used as a drinking water supply or used by a business that manufactures foodstuffs, such as a dairy. Over the last two years, there has been a rise in the number of firefighting foam-related sites that are being investigated to evaluate environmental and human health impacts, thus requiring potential future clean up via remediation. The opportunity to transition to F3 foams as soon as practicable appears to be a wise commercial decision considering the fast pace of advancing environmental regulations addressing PFASs and those restricting the use of AFFF, FFFP and FP.

2.7 AVAILABILITY AND USERS OF FLUORINE FREE FIREFIGHTING FOAM

Environmental and public health concerns regarding PFAS have been developing since 2000 and have stimulated significant innovation to create F3 foams. As a result, over the last 20 years firefighting foams and their delivery systems have evolved to be far more effective, without a need for PFAS in most circumstances (Ramsden 2018; LASTFIRE 2018). The new generation of F3 foams have been available since the early 2000s, and development and global use are increasing (Allcorn et al. 2018). The first F3 foams were successfully tested on a large scale in 2002 (Bavarian State Office for Environment 2017) with evolving improvements in performance.

Testing of new generation F3 foams is ongoing with an extensive number of foam manufacturing companies now providing F3 foams, the performance of which has steadily improved as they continually evolve and are optimized. Currently, the following foam vendors supply F3 foams for multiple commercial application systems for use on several different fire scenarios: Perimeter (Solberg/Auxquimia), National Foam, Angus, Dr. Sthamer, Bio-Ex, Novacool, Fomtech, 3F, Chemguard, Aberdeen Foam, FireRein, FireFreeze Worldwide, Orchidee Fire, Pyrocool Technologies, Verde Environmental Inc., Tyco, VS Focum and FoamFatale.

Fires need to be extinguished quickly to limit the damage and harm to people, environment and assets. For large Class-B fires there is the perception of uncertainty regarding the performance F3 foams; however there is also uncertainty regarding the performance of C6-pure AFFF as these foams are also untested for large-scale fire extinguishment. However, firefighters have been choosing foam containing PFAS, when both C6-pure foams and F3 foams offer comparable extinguishment performance.

2.8 TESTING THE EFFECTIVENESS OF FLUORINE FREE FIREFIGHTING FOAM

Carrying out large-scale fire testing is expensive and causes environmental issues in some locations. Consequently, it is preferable to develop testing that represents real-life situations but on a small scale. In effect, this is what "standard" tests try to do. Because different applications have different requirements, there are a number of standard small-scale test protocols that are used by industry to assess the performance of a firefighting foam for their particular unique circumstances. There is no "one size fits all" fire test.

No small-scale test will ever represent every incident situation, so it is still essential to validate small-scale testing with larger tests and incident experience whenever possible. This is particularly true given the current situation where different performance mechanisms are being used with F3 foams.

Examples of different standard fire performance tests include:

- CAP 168/ICAO – the European (and other areas) aviation sector standard for rapid rescue situations such as aircraft crash spill fires
- LASTFIRE – the oil and petrochemical sector standard for storage tank fire application

- MIL F24385 – USA Department of Defense and the Federal Aviation Authority for rescue situations and other defense related applications
- Underwriters Laboratories (UL) Standard UL162 – for general purpose foam application including different application methods
- EN1568 – for general purchase foam application.

Critical criteria for the ideal fire test when selecting the most appropriate for any particular application are:

- Relevance to and true simulation of the application – tank fire, aviation crash fire, shallow spill fire, etc.
- Reproducing "real-life" foam characteristics in terms of expansion, drainage time and shear.
- Critical test conditions that include a "safety factor" when compared to actual system design conditions (e.g., lower application rates).
- Validation against larger fires through testing and/or incident experience.
- Repeatability within defined test conditions.
- Performance ranking.
- Adaptability to other fuels or application techniques.

Currently, a number of organizations are carrying out small-scale evaluations. For example, the NFPA Research Foundation has published results for a series of tests based on UL protocols. With any published data it is important to analyze it in detail to assess its relevance, independence, interpretation of results and its validity against larger-scale work.

As an example, the industry group LASTFIRE, which is funded on a non-profit basis by fuel storage companies and develops best practice guidance for storage tank protection, and free from the often-commercial bias of foam manufacturers (LASTFIRE) has carried out an extensive series of tests using "new generation" foams of both C6 and F3 types. This series has included a number of both small scale and large-scale tests with different foams, different application techniques (including innovative application methods) and different preburn times, producing a large data base. A summary of the tests is given in Figure 2.1.

These tests were aimed at storage tank-related scenarios, but have relevance to all foam applications. Fire tests have involved a range of incident scenarios including tank fires, dike area fires and general spill fires as well as standard test protocols such as EN1568 and LASTFIRE. The LASTFIRE standard test protocol was developed initially by Mobil Research and Development Corporation, but finalized by LASTFIRE specifically to evaluate foam performance for the critical application of tank fires. LASTFIRE has also worked with other industry sectors such as aviation, carrying out large-scale tests involved 40 m x 7 m spill fires with JetA fuel in cooperation with the DFW Airport Research and Training Facility.

This test series is undoubtedly the most comprehensive work on this subject carried out by end users and independently managed for more than 50 years. LASTFIRE is keen to emphasize that the research is ongoing and further tests are planned. Conclusions regarding foam performance should not be generic, as there are

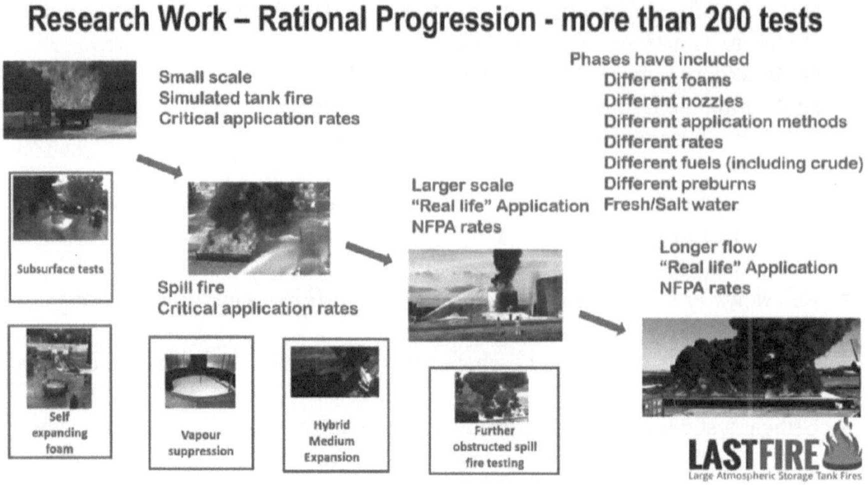

FIGURE 2.1 Summary of LASTFIRE Tests.

examples of varying levels of effectiveness and performance for each foam category on the market. The results show that F3 Foams can successfully extinguish both spill and tank fires using standard application rates – and indeed at more critical application rates too, given the correct application techniques and foam quality. Application techniques tested have included both forceful (monitor) and gentle (pourer) conventional proprietary aspirating and "non-aspirating" devices. An interesting part of the work is testing alternative application methods, including compressed air foam (CAF), "Hybrid Foam" and Self Expanding Foam. In a Dallas-Fort Worth airport test, for example, a 40 m long test pan with JetA fuel was efficiently and effectively extinguished using a CAF foam pourer at 50% of the application rate of a conventional aspirating foam pourer. Comparative testing with conventional application devices was carried out on tank scenarios as part of the overall series. It is recognized that there are still gaps in the testing – for example, a forceful monitor attack requiring over 30 m foam travel length and large-scale testing of water soluble fuels such as ethanol, but such work is planned by LASTFIRE.

The work of LASTFIRE is complemented by that of the Etank Project carried out by RISE in Stockholm which included small-scale testing of the use of F3 foam on water-soluble fuels and other work being carried by NFPA Research Foundation and the US Department of Defense. While each industry sector has specific critical issues, there are also many common learning points.

Based on this work, LASTFIRE is confident that they can show that F3 foams can be used with confidence. Their focus now is to optimize application methods to minimize any on-cost of system modification or increases in application rate. The work has clearly shown that the combination of foam type and application equipment to optimize bubble structure is of far greater importance than is recognized in design standards. This is clearly shown by the effectiveness of CAF which produces a more consistent bubble structure in comparison with that produced by conventional means.

FIGURE 2.2 Lage Scale Test Pit.

Current test plans include a series of small-scale tests changing bubble characteristics and with a wide range of different chemicals.

Also, a large (50 m × 6 m) test pit (Figure 2.2) has been built in cooperation with GESIP, the French group providing industrial fire and safety training. This will be used to test forceful foam application with a variety of application techniques and so will answer some of the more critical outstanding questions about foam effectiveness. When this is completed, F3 foam types will have been subjected to far greater levels of testing than previous types or new generation C6 types.

The LASTFIRE Group is now moving towards other aspects of the transition including monitoring developments of decontamination and disposal routes. Their assurance protocol provides best practices for all stages of foam procurement, application, management and ultimately disposal. This is aimed specifically at storage tank applications, but the principles and most of the detail are applicable to all foam applications including aviation.

Some LASTFIRE members have made the commitment to transition to F3 making sure that they adopt sustainable solutions for the long term and avoid "regret spend" by changing to a foam that may be subject to restrictions in the future.

2.9 WATER-SOLUBLE FUELS

There are now several F3 foams that that have "Approval Listings" for water-soluble fuels such as ethanol. However, larger-scale testing is required for validating these concentrates as well as other foam types such as new generation C6 concentrates.

Water-soluble fuels pose special problems for firefighters for large incidents such as storage tank fires where application tends to be forceful unless fixed systems are in place. All "alcohol-resistant" multi-purpose foams require relatively gentle application because they provide a polymer-based layer on the fuels surface when the first bubbles to land on the fuel are destroyed. The advantages that fluorosurfactants give in terms of oleophobicity and greater resistance to fuel contamination do not apply to the same degree as they do with hydrocarbon fuels. Industry best practice is to apply the foam as gently as possible by, for example, bouncing it off the inner surface of tank walls and causing more gentle application with movement of the foam around the tank. This might be much more difficult in practice than it is in theory. The problem applies to all foam types, yet somehow some end users are demanding that it must not be if F3 foam is used. In other words, some users demand greater performance from a F3 foam than is available from PFAS based types.

2.10 ENVIRONMENTAL CERTIFICATIONS FOR F3 FOAMS

It is important that when considering purchasing a F3 foam, that there is some confidence that ingredients of the F3 foam do not pose a potential future environmental liability.

Introducing another class of persistent compounds into foams and thus the environment would not be wise. As foams are discharged directly to a soil and groundwater environment, their biodegradability is important. Evaluating the biological oxygen demand (BOD) vs. chemical oxygen demand (COD) of a whole foam may not necessarily assist with determining if the foam contains only biodegradable ingredients. If one ingredient is relatively minor but extremely persistent, the BOD/COD ratio may not reflect this unless it is 1 (i.e., meaning 100% of the ingredients have biodegraded). Testing the BOD/COD ratio of each foam ingredient is a more robust method to determine if each organic component of an F3 foam is biodegradable.

The use of siloxanes in F3 foams appears to present a future environmental hazard as Environment Canada assessments concluded that cyclotetrasiloxane and cylcopentasiloxane, also known as D4 and D5, are toxic, persistent, and have the potential to bioaccumulate in aquatic organisms (Canada 2008). The use of certain polymer-based surfactants, ionic liquids, and oleophobic surfactants in F3 foams requires very careful consideration of their potential for biotransformation and biodegradation.

2.10.1 Harmonized Offshore Notification Format

In Europe the Harmonized Offshore Notification Format (HOCNF) has been applied to determine whether F3 foams pose a potential future environmental hazard. This must be completed on all chemicals to be registered offshore for use in the North Sea. A detailed list of all of the ingredients in the F3 foams are submitted to HOCNF under a non-disclosure agreement. The HOCNF system identifies chemicals by a colour coding system (Bergeson and Campbell 2014).

2.10.2 GreenScreen Certification™

Clean Product Action has recently developed a GreenScreen Certified™ standard for firefighting foams. Certification depends on foam manufacturers providing a full product inventory to Clean Product Action under a confidentiality agreement. Clean Product Action then reviews the relevant environmental and human health data. The data requirements vary by certification level and include: GreenScreen List Translator™ scores and GreenScreen Benchmark scores; product-level acute aquatic toxicity data for fish, aquatic invertebrates and algae; and ingredient-level aquatic toxicity and fate data for compliance with USEPA safer choice criteria (Master criteria or Direct Release criteria). The restricted ingredients include: organohalogens, PFAS, siloxanes, alkyl phenols and alkylphenol ethoxylates, and chemicals listed under the Manufacturing Restricted Substances List (MRSL). There are three levels of certification: Bronze, Silver and Gold. The first four companies with certified products are Angus Fire Ltd., BIOEX, National Foam Inc., and Verde Environmental, Inc.

To assist with ensuring that all brands of F3 foams do not pose a future environmental hazard, it is suggested that foam suppliers confirm that all organic components of the foams, irrespective of concentration, can be confirmed to be readily biodegradable as per Organization for Economic Cooperation and Development (OECD) guidelines (OECD 1992). Consideration of the very comprehensive Queensland Department of Heritage and Environmental Guidance Policy regarding Firefighting Foams (Queensland Government Department of Environment and Heritage Protection 2016a, b) could help guide policy in many jurisdictions.

To verify that F3 foams are fluorine free and do not contain PFAS, it is recommended that tests are performed with the TOP assay and Total Organic Fluorine (TOF) by Combustion Ion Chromatography (CIC) as both methods are now commercially available (ALS Global 2019).

2.11 AFFF MANAGEMENT AND TRANSITION

When considering foam transition, one of the first steps can be to inventory Class B firefighting foam stocks and classify them based on occurrence of PFOS, PFOA, 6:2FTS, PFOA and PFHxA -precursors (C6 foams) or dominant 6:2 FTS and PFHxA precursors (C6-pure foams). There may be a perceived priority to remove PFOS- and PFOA-containing AFFF from stock as C6-pure foams can sometimes pose less current liability than legacy AFFF stocks. However, there are concerns that the short chain PFAS and precursors present in these foams are increasingly being regulated. The current best practice includes, at a minimum, eliminating fire training and nozzle/coverage testing with AFFF, complete containment during proportioning testing, and proper clean-out of equipment to limit environmental releases of the new foam which contain short chain PFAS proposed for regulation.

Decontamination of firefighting and fire suppression equipment is essential to limit carryover of PFAS from old foam usage. Triple rinsing with water is not sufficient and leads to a significant volume of decontamination water that requires treatment. Arcadis recommends using specialized biodegradable cleaning agents such as V171 to effectively remove PFAS residuals from fire suppression systems to limit future liabilities and costs.

The transition to F3 foam will require foam supply system modifications, but more and more stakeholders are recognizing the benefits of eliminating PFAS from fire suppression and firefighting strategies.

2.11.1 Foam Transition

One of the first issues to address in planning for transition from AFFF and FP foams to F3 is bringing fire engineering, fire service contracting, and environmental engineering into the process from the beginning. Failing to bring a team with the required skill set could result in a project which delivers fire protection equipment that is incompatible with the new foam, a fire protection system that does not meet accreditation requirements for insurance and regulators, cross-contamination of the F3 foam with PFAS, and undocumented and irregular disposal of PFAS-impacted waste. Neglecting to assemble a comprehensive team is likely to result in a rework of the fire protection system at significant additional cost.

Successful foam transition projects often start with fire and environmental engineers working closely with the end users to select the best F3 foam to meet the project needs. Then, collaboration on a cost–benefit analysis that balances the feasibility to replace equipment with the viability to clean out and manage residual PFAS will result in a more cost-effective transition. Effective designs must address containment and management of foam mixtures to meet the fast-changing environmental regulations and reduce future environmental liability. Environmental policy and regulations are changing more rapidly than fire protection codes and standards. Therefore, collaboration between fire and environmental engineering from the beginning can help avoid lengthy delays in approvals of foam transition plans and the final commissioning of fire protection systems.

Minor modifications to existing fire protection systems are commonly required with F3 foam replacements to ensure appropriate standards for proportioning, flow requirements and foam quality are achieved, and often to maintain accreditation for insurance coverage. To ensure proper performance, it is now common practice for existing foam proportioners to be replaced with units tested and accredited with the replacement foam. For most cases, performance requirements cannot be achieved without the tested and certified proportioner. Foam application rate and discharge duration are often impacted by the differing physiochemical properties of the replacement foam. One clear example is kinematic viscosity, which will affect the performance of pumps and storage volumes of concentrate. The switch to aspirated discharge devices typically increases replacement foam performance and reduces the need for more extensive system modifications.

2.11.2 Cleanout Challenges

The PFAS used in firefighting foams are high-performance surfactants. They adhere to surfaces and are known to self-assemble in multiple layers to coat surfaces, forming a waterproof coating. Therefore, these layers of PFAS cannot be effectively removed from fire suppression systems by flushing with water. When replacement foam is added to a suppression system, the fluorosurfactants in the prior foams, which may have contained legacy PFAS such as PFOS, PFHxS and PFOA, can dissolve from the surfaces of the system and contaminate the new foam.

The common misconception has been that repeat washing of fire suppression systems with water can effectively remove PFAS, as for example after a few rinses, less PFOS may be detected in rinse water. However, PFAS can form a waterproof coating on surfaces, so PFAS concentration in the rinse water will not be representative of that still remaining entrained within fire suppression systems.

Arcadis recently conducted a comprehensive PFAS investigation of a foam supply and delivery system at an aircraft hangar in Australia approximately 20 months after transition to F3 foam. A dual water flush method had been used for the AFFF clean-out. The sampling and analysis of foam solutions throughout the system revealed levels of PFAS in replacement foam ranging up to 1.6 g/L Sum of PFAS (28 suite) by TOP Assay, as shown in Figure 2.3. The residual PFAS rebounded from the fire suppression system into the F3 foam replacement to such an extent that regulators no longer considered this system fluorine free. Clearly, the two water flushes conducted on piping had little effect in removing PFAS coated on surfaces. Difficulties were also encountered during flushing due to inadequate flow through laterals. The consequences for ineffective decontamination include non-compliance with environmental regulations, the need for expensive containment systems during discharge events, and the loss of the major benefit of F3 foam, reduced environmental liability.

Some techniques have been developed to address this issue. For example, Arcadis has developed methods for PFAS decontamination of piping and tank systems, including the use of a proprietary biodegradable cleaning agent, V171. These methods and the cleaning agent have been successfully applied in foam transition projects to remove PFAS from steel and PVC piping systems, stainless steel concentrate tanks, and underground wastewater tanks.

2.11.3 Impact on Fire Water Systems

The most significant factor in the contamination of fire water systems is inadequate backflow prevention on water lines upstream of the proportioners. The periodic testing of fire water pumps appears to further distribute PFAS into fire water storage tanks, creating risks of release of PFAS to the environment should loss of containment on fire water systems occur. Therefore, the investigation of fire water systems should be included as an important component of foam transition planning.

2.11.4 Equipment Replacement versus Cleanout and Retain

Underground or elevated piping conveying concentrate or foam mix can be very costly to remove and replace. Therefore, decontamination and reuse or abandonment in place can be the more cost-effective solution. From experience with multiple fire suppression systems, valves of all types can potentially be difficult to clean and therefore the replacement of such components as deluge valves, alarm check valves, and solenoid valves, can help avoid PFAS rebound. As mentioned previously, a thorough cost–benefit analysis by a team of fire and environmental engineers can identify those components best retained and those best replaced.

2.11.5 Decontamination Case Studies

Arcadis developed and then successfully applied cleaning agents to remove PFAS from surfaces at field scale in Australia in 2017. When a foam concentrate was

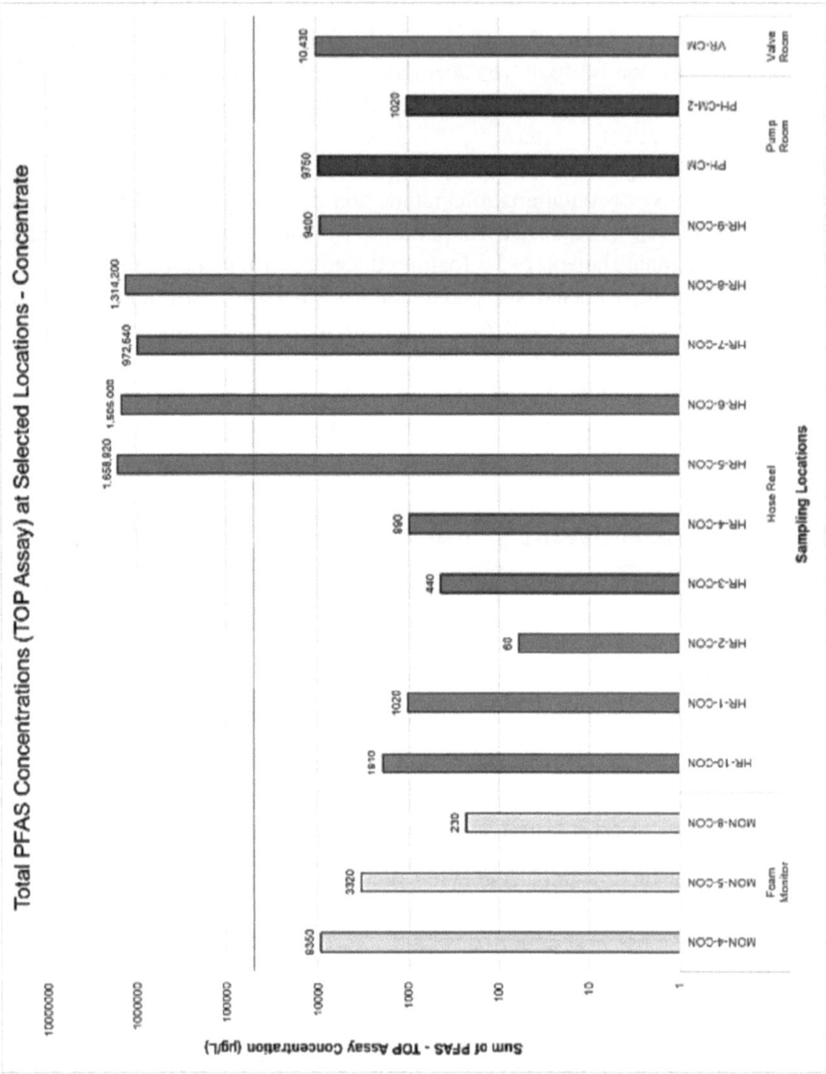

FIGURE 2.3 PFAS Detected in F3 Foam after an Ineffective Foam Cleanout Using Water.

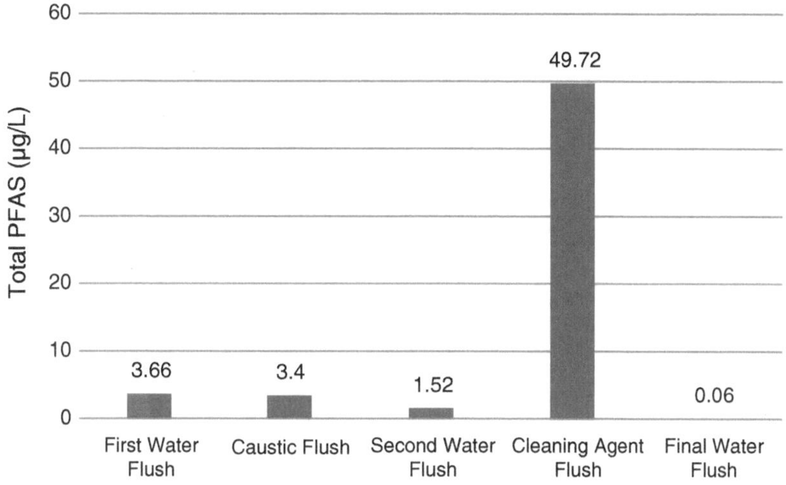

FIGURE 2.4 Sum of PFAS concentrations during cleaning agent application for AFFF impacted sewer pipe.

released to a sewer system, the complete network of piping, pump stations, and concrete wet wells was shut down, and then decontaminated to remove residual PFAS. Since then, several successful piping system decontamination projects have been implemented in Australia, Europe and the US.

Arcadis has used reagents to remove PFAS from the surface of PVC pipes, stainless steel tanks and hangar sprinkler piping and is currently progressing completion of a foam change out using the reagent at a large commercial hangar. The initial PFAS cleanout project in 2017 used a sequential series of aqueous rinses, high pH flushes and application of the cleaning reagent. The results showed that water and high pH are ineffective for the removal of PFAS from surfaces, as demonstrated by the relatively low concentrations of PFAS measured in these flush solutions. The elevated concentration in the cleaning agent indicated significantly greater PFAS mass removal even after multiple flushes of water and caustic solution.

Further work to clean PFAS out of a 20-m^3 foam concentrate tank was conducted and results are presented in Figure 2.4. This application demonstrated that soaking with the cleaning agent, followed by high-pressure washing, can be effective. During this work, TOP assay for PFAS analysis was essential as conventional analysis only detected approximately one percent of the PFAS identified using the TOP assay.

2.12 WASTE MANAGEMENT

From a waste treatment perspective, wastes generated during foam transition include AFFF, FFFP or FP concentrate, foam mix from proportioning test work, fire water drainage, decontamination solutions, and solid waste in the form of impacted piping, components, and tank bladders. If all the organic components of a firefighting foam are assessed to be readily biodegradable, they could be effectively managed using a biological wastewater treatment plant (WWTP). For example, discharge to a sewer system for a biodegradable foam mix generated from a proportioning test may be considered as an appropriate disposal method with regulatory and utility approval. The converse is true for any foam containing PFAS, such as AFFF, FFFP and FP, as they are extremely persistent organic compounds which do not biodegrade and cannot be treated using sewerage and conventional WWTPs. Alternative treatment systems, such as the use of granular activated carbon (GAC) for the long-chain PFAS and ion exchange resin (IX) for the short-chain PFAS, could be applied to treat firewater impacted with PFAS. However, the performance of GAC and resin treatment systems can be severely compromised by other organic compounds in the foam matrix (e.g., natural organic matter and glycols) which can coat adsorption sites and inorganics (e.g., natural anions) which compete with PFAS for binding capacity. Foam fractionation is one alternative treatment process that will not be impacted by co-contaminants.

The use of incineration has recently been questioned in the US as a result of the uncertainty relating to complete destruction of PFAS when incinerating liquids. There is some potential for incineration to release fluoroorganics such as tetrafluoromethane and perfluoroisobutylene. Other waste treatment alternatives such as cement kilns as demonstrated in Australia show promise as a practical destruction process because of their high temperatures, long residence time, and excess calcium that takes up fluorine. However, along with incineration, cement kilns may also require comprehensive assessment of organo-fluorine compounds in air emissions.

Arcadis has furthered the development of destructive methods for PFAS which operate effectively at ambient temperature using ultrasound. This approach, termed sonolysis, can destroy PFAS, mineralizing them to form fluoride and carbon dioxide at temperatures around 30 °C. This technology is being scaled up and adapted with pre-treatment technologies to address firefighting foam concentrates.

2.13 CONCLUSIONS

Due to reputational risks, regulatory changes, and potential litigation, many foam users are transitioning away foams containing PFAS, including both the C8 and C6 products. F3 foams are widely available and used across multiple sectors due to their effective fire protection in most situations. Increased confidence in F3 foam performance and developing environmental concerns regarding C8/C6 foams has accelerated foam transition. Balancing fire safety by using effective F3 foams to alleviate environmental liabilities is now generally perceived as essential as regulatory agencies begin to appreciate the risk and cost to society imposed by firefighting with PFAS.

In the next few years, the pace of transition to F3 foams will accelerate. Ensuring that F3 foams do not become contaminated with PFAS from residuals coating the interior of fire suppression systems will become increasingly important, especially taking into account the diminishing regulatory criteria for PFAS.

REFERENCES

Allcorn, M., Bluteau, T., Corfield, J., Day, G., Cornelsen, M., Holmes, N.J.C., Klein, R.A., McDowall, J.G. Olsen, Ramsden, N., Ross, I., Schaefer, T.H., Weber, R., Whitehead, K., 2018. *Fluorine-Free Firefighting Foams (3F) – Viable Alternatives to Fluorinated Aqueous Film-Forming Foams (AFFF). White Paper prepared for the IPEN by members of the IPEN*

F3 Panel and associates, POPRC-14, Rome 17–21 September 2018. Available from: https://ipen.org/documents/fluorine-free-firefighting-foams (accessed October 5, 2020).

ALS Global, 2019. *EnviroMail™ 125 - ALS Now Provides Testing Capabilities for Total Organic Fluorine (TOF)*. Available from: https://www.alsglobal.com/en-us/myals/news/2019/07/enviromail-125-als-now-provides-testing-capabilities-for-total-organic-fluorine (accessed October 5, 2020).

Angus Fire, 2010. *FF – Fluorine Free Foam*. Available from: https://angusfire.co.uk/products/foam-concentrates/product-range/ff/ (accessed October 5, 2020).

Backe, W.J., Day, T.C., Field, J.A., 2013. Zwitterionic, cationic, and anionic fluorinated chemicals in aqueous film forming foam formulations and groundwater from U.S. military bases by nonaqueous large-volume injection HPLC-MS/MS. *Environmental Science and Technology*, 47(10):5226–5234.

Bagenstose, K., 2017. Dangers of firefighting foam discussed in 2001, document shows. *The Intelligencer*. Available from: https://www.theintell.com/news/20170609/dangers-of-firefighting-foam-discussed-in-2001-document-shows (accessed October 5, 2020).

Bavarian State Office for Environment, 2017. *Preliminary Evaluation Guidelines of PFC impurities in water and soil*. Available from: https://www.lfu.bayern.de/analytik_stoffe/doc/leitlinien_vorlaufbewertung_pfc_verunreinigungen.pdf (accessed October 5, 2020).

Bergeson and Campbell 2014. *OSPAR Commission's Harmonized Offshore Chemical Notification Format*. Available at: https://www.jdsupra.com/legalnews/ospar-commissions-harmonized-offshore-c-32747/ (Accessed March 5 2021).

Blaine, A. C., Rich, C. D., Hundal, L. S., Lau, C., Mills, M. A., Harris, K. M., Higgins, C. P. 2013. Uptake of perfluoroalkyl acids into edible crops via land applied biosolids: field and greenhouse studies. *Environmental Science & Technology*, 47 (24), 14062–14069.

Blaine, A. C., Rich, C. D., Sedlacko, E. M., Hundal, L. S., Kumar, K., Lau, C., Mills, M. A., Harris, K. M., Higgins, C. P. 2014a. Perfluoroalkyl acid distribution in various plant compartments of edible crops grown in biosolids-amended soils. *Environmental Science & Technology*, 48 (14), 7858–7865.

Blaine, A. C., Rich, C. D., Sedlacko, E. M., Hyland, K. C., Stushnoff, C., Dickenson, E. R., Higgins, C. P. 2014b. Perfluoroalkyl acid uptake in lettuce (Lactuca sativa) and strawberry (Fragaria ananassa) irrigated with reclaimed water. *Environmental Science & Technology*, 48 (24), 14361–14368.

Bluteau, T., Cornelsen, M., Day, G., Holmes, N.J.C., Klein, R.A., Olsen, K.T., McDowall, J.G., and R. Stewart, Tisbury, M., Webb, S., Whitehead, K., Ystanes, L., 2019. The Global PFAS Problem: Fluorine-free Alternatives as Solutions to Firefighting Foams and Other Sources – Going Fluorine-Free. *IPEN 2019/Stockholm Convention COP-9 White Paper 2019*. Available from: https://ipen.org/sites/default/files/documents/the_global_pfas_problem-v1_5_final_18_april.pdf (accessed October 5, 2020).

Buck, R.C., Franklin, J., Berger, U., Conder, J.M., Cousins, I.T., de Voogt, P., Jensen, A.A., Kannan, K., Mabury, S.A., van Leeuwen, S.P., 2011. Perfluoroalkyl and polyfluoroalkyl substances in the environment: Terminology, classification, and origins. *Integrated Environmental Assessment Management*, 7, 513–541.

Chemical Watch, 2020. *Washington legislature advances bill to expand PFAS restrictions*. Available from: https://chemicalwatch.com/98804/washington-legislature-advances-bill-to-expand-pfas-restrictions#overlay-strip (accessed October 5, 2020).

Commission Regulation (EU) 2017. 1000 of 13 June 2017 amending Annex XVII to Regulation (EC) No 1907/2006 of the European Parliament and of the Council concerning the Registration, Evaluation, Authorisation and Restriction of Chemicals (REACH) as regards perfluorooctanoic acid (PFOA), its salts and PFOA-related substances. Available at: https://www.legislation.gov.uk/eur/2017/1000/adopted (accessed October 5, 2020).

Cortina, T., 2004. EPA Workgroup determines AFFF not a likely source of PFOA. Industrial Fire Journal, June 2004: 14.

D'Agostino, L.A., Mabury, S.A., 2014. Identification of novel fluorinated surfactants in aqueous film forming foams and commercial surfactant concentrates. *Environmental Science and Technology*, 48(1):121–129.

DeMeo, R.A., Brooks, L.D., 2019. *What in the PFAS is Going On?*, Available at: https://www.youtube.com/watch?v=Ga4N6Y0qhlg (accessed October 14, 2020).

Denton, C., Helminski, T., Longsworth, J., 2019. *Expert Focus: US states outpace EPA on PFAS firefighting foam laws*. Available at: https://chemicalwatch.com/78075/expert-focus-us-states-outpace-epa-on-pfas-firefighting-foam-laws#overlay-strip (accessed October 14, 2020).

Dlugogorski, B.Z., Schaefer, T.H., 2021. Compatibility of aqueous film-forming foams (AFFF) with sea water. *Fire Safety Journal*, 120:103288.

European Chemicals Agency, 2019. *Annex XV Restriction Report*. Available at: https://echa.europa.eu/documents/10162/c4e04484-c989-733d-33ed-0f023e2a200e (accessed October 5, 2020).

Fire Fighting Foam Coalition, 2019. *Analytical Method for Measuring PFAS in AFFF Firefighting Foam Concentrate*. Available from: https://docs.wixstatic.com/ugd/331cad_4eb7dca6a804487cb21943d26f27bf34.pdf (accessed October 5, 2020).

Franjevic, S., Schreder, E., Plant, D., Krause, R., 2020. *GreenScreen Certified™ Standard for Firefighting Foam*. Available from: https://www.greenscreenchemicals.org/resources/entry/webinar-gsc-fff (accessed October 5, 2020).

Hogue, C., 2018. *U.S. class-action case targets nine PFAS makers*. Available at: https://cen.acs.org/policy/litigation/US-class-action-case-targets/96/i41 (accessed October 14, 2020).

Houtz, E.F., Sedlak, D.L., 2012. Oxidative conversion as a means of detecting precursors to perfluoroalkyl acids in urban runoff. *Environmental Science and Technology*, 46(17):9342–9349.

Houtz, E., 2013. "Oxidative Measurement of Perfluoroalkyl Acid Precursors: Implications for urban runoff management and remediation of AFFF-contaminated groundwater and soil". Doctor of Philosophy, University of California, Berkeley.

Houtz, E.F., Higgins, C.P., Field, J.A., Sedlak, D.L., 2013. Persistence of perfluoroalkyl acid precursors in AFFF-impacted groundwater and soil. *Environmental Science and Technology*, 47(15):8187–8195.

Hu, X.C., Andrews, D.Q., Lindstrom, A.B., Bruton, T.A., Schaider, L.A., Grandjean, P., Lohmann, R., Carignan, C.C., Blum, A., Balan, S.A., Higgins, C.P., Sunderland, E.M., 2016. Detection of Poly- and Perfluoroalkyl Substances (PFASs) in U.S. Drinking Water Linked to Industrial Sites, Military Fire Training Areas, and Wastewater Treatment Plants. *Environmental Science and Technology Letters*, 3(10):344–350.

Kabadi, S. V., Fisher, J., Aungst, J., Rice, P. 2018. Internal exposure-based pharmacokinetic evaluation of potential for biopersistence of 6:2 fluorotelomer alcohol (FTOH) and its metabolites. *Food Chem Toxicol*, 112, 375–382.

Langberg, H. A., Breedveld, G. D., Gronning, H. M., Kvennas, M., Jenssen, B. M., Hale, S. E. 2019. Bioaccumulation of Fluorotelomer Sulfonates and Perfluoroalkyl Acids in Marine Organisms Living in Aqueous Film-Forming Foam Impacted Waters. *Environ Sci Technol*.

LASTFIRE, 2018. *LASTFIRE Foam Application Tests, Dallas Fort Worth Airport*. Available from: http://www.lastfire.co.uk/default.aspx?ReturnUrl=%2f. (accessed October 5, 2020).

Mennie, C., 2017. *Draft amendment to ban the use of fire-fighting foams containing perfluorinated alkylated substances (PFAS) – Consultation report*. Available at: https://www.epa.sa.gov.au/files/13280_pfas_consultation_report.pdf (accessed October 5, 2020).

Mikkelson, M., 2020. Evers Signs Bill Restricting Use of Firefighting Foam Containing PFAS. *WUWM 89.7*. Available from: https://www.wuwm.com/post/evers-signs-bill-restricting-use-firefighting-foam-containing-pfas#stream/0 (accessed October 5, 2020).

Neltner, T., Maffini, M., 2019. *The elephant in the room: potential biopersistence of short-chain PFAS*. Available at: http://blogs.edf.org/health/2019/02/20/potential-biopersistence-short-chain-pfas/ (accessed October 14, 2020).

Neltner, T., 2020. *FDA scientists push back on an industry-funded analysis about bioaccumulation and toxicity of short-chain PFAS*. Available at: http://blogs.edf.org/

health/2020/04/01/fda-scientists-push-back-industry-analysis-pfas/ (accessed October 14, 2020).

Neumann, M., Sättler, D., Vierke, L., Schliebner, I.,. 2017. *A proposal for criteria and an assessment procedure to identify Persistent, Mobile and Toxic (PM or PMT) substances registered under REACH*. Available at: https://www.mn.uio.no/kjemi/english/research/projects/ICCE2017/monday-19.06/helga-eng-auditorium-u35/hr.-13:45/1445-michael-neumann-20170616_neumann_pmt_concept.pdf (accessed October 14, 2020).

OECD, 1992. *OECD Guideline for Testing of Chemicals Ready Biodegradability*. Available from: https://www.oecd.org/chemicalsafety/risk-assessment/1948209.pdf (accessed October 5, 2020).

OECD 2018. *Portal on Per and Poly Fluorinated Chemicals*. Available at: http://www.oecd.org/chemicalsafety/portal-perfluorinated-chemicals/. Accessed March 5 2021.

Pabon, M., 2005. High Performance Firefighting Foam. *Industrial Fire Journal*, April 20.

PFAS Central, 2019. *Denmark becomes first country to ban all PFAS from food contact materials*. Available at: https://pfascentral.org/policy/denmark-becomes-first-country-to-ban-all-pfas-from-food-contact-materials. (accessed October 14, 2020).

Place, B.J., Field, J.A., 2012. Identification of novel fluorochemicals in aqueous film-forming foams used by the US military. *Environmental Science and Technology*, 46(13):7120–7127.

Queensland Government Department of Environment and Heritage Protection, 2016a. *Operational Policy Environmental Management of Firefighting Foam*. Available from: https://www.ehp.qld.gov.au/assets/documents/regulation/firefighting-foam-policy.pdf (accessed October 5, 2020).

Queensland Government Department of Environment and Heritage Protection, 2016b. *Environmental Management of Firefighting Foam Policy Explanatory Notes Revision 2*. Available from: https://environment.des.qld.gov.au/assets/documents/regulation/firefighting-foam-policy-notes.pdf (accessed October 5, 2020).

Ramsden, N., 2018. Foam Testing, in Petroleum Review. August 2018, Energy Institute, 32–33.

Ray, J. P., 2019. *PFAS Litigation: Just Getting Started?*, Available at: https://www.americanbar.org/groups/litigation/committees/environmental-energy/articles/2019/winter2019-pfas-litigation-just-getting-started/ (accessed October 14, 2020).

Reemtsma, T., Berger, U., Arp, H.P., Gallard, H., Knepper, T.P., Neumann, M., Quintana, J.B., Voogt, P., 2016. Mind the Gap: Persistent and Mobile Organic Compounds-Water Contaminants That Slip Through. *Environmental Science and Technology* 50(19):10308–10315.

Rice, P.A., Aungst, J., Cooper, J., Bandele, O, Kabadi, S.V., 2020. Comparative analysis of the toxicological databases for 6:2 fluorotelomer alcohol (6:2 FTOH) and perfluorohexanoic acid (PFHxA). *Food and Chemical Toxicology*. 138: 111210.

Ritter, S.K. 2021. Fluorochemicals go Short. *Chemical and Engineering News*. Available at https://cen.acs.org/articles/88/i5/Fluorochemicals-Short.html (Accessed March 5 2021).

Ross, I., Houtz, E., Kalve, E., McDonough, J., Hurst, J. and Miles, J.A.L., 2018. PFASs and the TOP Assay. *Industrial Fire Journal*, First Quarter, 26.

Ross, I., Kalve, E., McDonough, J., Hurst, J., Miles, J., Pancras, T., 2019. Per- and Polyfluoroalkyl Substances, in *Emerging Contaminants Handbook*, M.G. Caitlin Bell, Erica Kalve, Ian Ross, John Horst, Suthan Suthersan, Editors. CRC Press: Boca Raton, FL, USA. p. 85–257.

Ross, I., Storch, P., 2020. *Foam Transition: Is it as simple as "foam out / foam in?"*. Catalyst, JOIFF:1–19.

Santoro, M.A., 2008. *Brief History of Perfluorochemical Brief History History of Perfluorochemical Perfluorochemical Production, Products and Environmental Presence. ASTSWMO Mid-Year Meeting, Director Regulatory Affairs Affairs 3M Company*, St. Paul, MN.

Schaefer, T. H., 2002. *Class B Foams… Is It Time To Innovate? Australasian Fire Authorities Conference 2002 Conference Proceedings CD Abstracts* p. 48 (September).

Schaefer, T.H., 2003. *Performance Challenges of Fluoro Free Foams*. Available from: http://www.firedirect.net/_pdfs/_technical/tech_1205_0013.pdf (accessed October 5, 2020).
Schaefer, T.H., Dlugogorski, B.S., Kennedy, E.M., 2005. *New Foam Technology, New Found Benefits. International Aviation Fire Protection Association Conference Proceedings, Sydney, 2005*. Available from: https://www.solbergfoam.com/getattachment/c5bef149-b850-48df-81a8-19b977c6daed/New-Foam-Technology,-New-Found-Results.aspx (accessed October 5, 2020).
Simkins, G. 2020. *'Ban forever chemicals now', says PFAS lawyer*. Available at: https://www.endsreport.com/article/1673359/ban-forever-chemicals-now-says-pfas-lawyer (accessed October 14, 2020).
Singh, J., Jayanti, R.K., Kilgore, J.E., Agarwal, K., Gandarvakottai, R.R. 2005. What Goes around Comes Around: Understanding Trust–Value Dilemmas of Market Relationships. *Journal of Public Policy & Marketing*, 24(1):38–62. doi: 10.1509/jppm.24.1.38.63882
UN Stockholm Convention, 2009. *PFOS, its Salts and PFOSF*. Available at: http://chm.pops.int/Implementation/IndustrialPOPs/PFOS/Overview/tabid/5221/Default.aspx (accessed October 14, 2020).
UN Stockholm Convention, 2019. *The New POPs Under the Stockholm Convention*. Available at: http://chm.pops.int/TheConvention/ThePOPs/TheNewPOPs/tabid/2511/Default.aspx (accessed October 14, 2020).
UN Stockholm Convention on Persistent Organic Pollutants, 2019. *Recommendation by the Persistent Organic Pollutants Review Committee to list perfluorooctanoic acid (PFOA), its salts and PFOA-related compounds in Annex A to the Convention and draft text of the proposed amendment*. Available at: http://chm.pops.int/TheConvention/ConferenceoftheParties/Meetings/COP9/tabid/7521/ctl/Download/mid/20312/Default.aspx?id=67&ObjID=26478. (accessed October 14, 2020).
US EPA, 2015. *Fact Sheet: 2010/2015 PFOA Stewardship Program*. Available from: https://www.epa.gov/assessing-and-managing-chemicals-under-tsca/fact-sheet-20102015-pfoa-stewardship-program (accessed October 5, 2020).
US EPA, 2016. *Drinking Water Health Advisories for PFOA and PFOS*. Available at: https://www.epa.gov/ground-water-and-drinking-water/drinking-water-health-advisories-pfoa-and-pfos. (accessed October 14, 2020).
US EPA, 2000. *EPA and 3M Announce Phase Out of PFOS*. Available at: https://archive.epa.gov/epapages/newsroom_archive/newsreleases/33aa946e6cb11f35852568e1005246b4.html (accessed October 14, 2020).
Van Hamme, J. 2020. *A blanket ban on toxic 'forever chemicals' is good for people and animals*. Available at: https://theconversation.com/a-blanket-ban-on-toxic-forever-chemicals-is-good-for-people-and-animals-127879 (accessed October 14, 2020).
Vestergren, R., Cousins, I.T., Trudel, D., Wormuth, M., Scheringer, M., 2008. Estimating the contribution of precursor compounds in consumer exposure to PFOS and PFOA. *Chemosphere*, 73(10):1617–1624.
Vierke, L. 2017. *Regulation needs support from research: Short-chain PFASs under REACH*. Available at: https://www.mn.uio.no/kjemi/english/research/projects/ICCE2017/wednesday-21.06/helga-eng-auditorium-1/hr.-16:30/1715-vierke.pdf (accessed October 14, 2020).
Williams, D., 2019. *PFOA Added to Stockholm Convention POP list*. Available at: https://www.chemanager-online.com/en/news-opinions/headlines/pfoa-added-stockholm-convention-pop-list (accessed October 14, 2020).
Yingling, V., 2015. Karst Influence in the Creation of a PFC Megaplume. *Proceedings from National Cave and Karst Research Institute Symposium* 5, 319–326.Bergeson and Campbell 2014. *OSPAR Commission's Harmonized Offshore Chemical Notification Format*. Available at: https://www.jdsupra.com/legalnews/ospar-commissions-harmonized-offshore-c-32747/ (Accessed March 5, 2021).

3 PFASs in Consumer Products

Exposures and Regulatory Approaches

Simona Andreea Bălan and Qingyu Meng
California Department of Toxic Substances Control,
Safer Consumer Products Program

CONTENTS

3.1 Introduction ... 52
3.2 A Note on Nomenclature ... 52
3.3 Which PFASs are Used in Consumer Products? ... 53
3.3.1 Perfluoroalkyl Acids (PFAAs) ... 53
3.3.2 PFAA Precursors ... 54
3.3.3 Fluoropolymers ... 55
3.3.4 Perfluoropolyethers (PFPEs) ... 56
3.3.5 Trends in PFAS Use in Consumer Products ... 57
3.4 From Consumer Products to Human Exposure ... 59
3.4.1 PFASs in Indoor Air ... 60
3.4.2 PFASs in Indoor Dust ... 61
3.4.3 PFASs in Food ... 64
3.4.4 PFASs in Drinking Water ... 66
3.5 PFAS Exposure Determinants ... 67
3.5.1 Exposure Quantification and Exposure Profiles ... 67
3.5.2 Exposure Determinants ... 68
3.6 Regulatory Approaches ... 70
3.6.1 Single Chemical Approach ... 71
3.6.2 Mixture Approach ... 71
3.6.3 Arrowhead or Subclass Approach ... 72
3.6.4 Chemical Class Approach ... 73
3.7 Conclusions ... 75
Acknowledgements ... 75
References ... 75

3.1 INTRODUCTION

Since the late 1940s, perfluoroalkyl and polyfluoroalkyl substances (PFASs) have been used in a wide range of industrial and consumer product applications, including in the aerospace, apparel, automotive, building and construction, food packaging, chemicals and pharmaceuticals, electronics and semiconductors, alternative energy, energy storage, oil and gas exploration, mining, first responder safety, and healthcare sectors (ACC 2021, OECD 2013, Posner et al. 2013). PFAS use in such a diverse range of applications can be ascribed to their unique physicochemical characteristics: they confer stability in extreme temperatures, have antistatic properties, and are resistant to wettability, staining, and corrosion (Buck et al. 2011, 2012; OECD 2013). According to their manufacturers, PFASs are critical to modern life (Bowman, 2015).

However, the widespread use of PFASs in industry and consumer products has also led to widespread environmental contamination and human and ecological exposures, resulting in adverse impacts to individuals, communities, and ecosystems across the globe. Researchers, policy experts, and community activists have been questioning whether many of the uses of PFASs are truly essential. As a result, manufacturers and retailers have started removing PFASs from certain products and a growing number of jurisdictions have begun restricting the uses of PFASs to reduce exposures. This chapter discusses some of the common consumer product uses of PFASs, the associated PFAS exposure pathways, and several regulatory approaches to reduce these exposures.

3.2 A NOTE ON NOMENCLATURE

The term perfluoroalkyl and polyfluoroalkyl substances, abbreviated as PFASs, was first introduced by Buck et al. (2011). Subsequently, the U.S. Environmental Protection Agency (U.S. EPA) and others have dropped the extra "s" at the end of the class name. In one of its latest fact sheets, the Interstate Technology & Regulatory Council (ITRC) argues that "no single chemical within the PFAS class can be both polyfluorinated and perfluorinated, so by definition "PFAS" is plural and a small "s" is not needed" (ITRC 2020). "PFAS" has now become the standard nomenclature in most of the U.S. However, the Organisation for Economic Cooperation and Development (OECD) and most scientists and regulators outside the U.S. continue to use the original nomenclature of PFASs (OECD 2018). When referring to a single PFAS, they change the conjunction from "and" to "or," i.e., one perfluoroalkyl or polyfluoroalkyl substance (PFAS).

Unlike many other chapters in this book, this chapter uses the abbreviation "PFASs," since this is the nomenclature used by us in the California Department of Toxic Substances Control's Safer Consumer Products Program. We have decided to retain the extra "s" to emphasize that PFASs are a class of chemicals, as opposed to a single compound, and to remain consistent with the nomenclature used for other chemical classes (e.g., PBDEs for polybrominated diphenyl ethers, PCBs for polychlorinated biphenyls, PAHs for polycyclic aromatic hydrocarbons, CFCs for chlorofluorocarbons). In our experience, removing the "s" from the acronym confuses communication with the general public. For instance, we've seen multiple public media articles use singular verbs when referring to "PFAS," giving the reader the impression that "PFAS" is a single chemical instead of a class of thousands of compounds. In order to clarify, policymakers and regulators who use the acronym "PFAS" often resort to expressions such as "PFAS chemical(s)," which technically reads as "perfluoroalkyl and polyfluoroalkyl substances chemical(s)." Retaining the extra "s" when referring to multiple PFASs facilitates communication, and changing the conjunction from "and" to "or" (i.e., perfluoroalkyl or polyfluoroalkyl) enhances the clarity of regulatory and legislative language.

3.3 WHICH PFASs ARE USED IN CONSUMER PRODUCTS?

As described in earlier chapters, the universe of PFASs is broad and has not been fully characterized. In 2019, the U.S. EPA assembled a master list of 6330 Chemical Abstract Services (CAS) numbers related to PFASs (U.S. EPA 2020b). The total number of members of this class may be even larger, since some PFASs lack CAS numbers. However, not all PFASs are intentionally used in consumer products. The U.S. Toxic Substances Control Act (TSCA) Chemical Substance Inventory lists just over one thousand PFASs, approximately half of which are known to have been in active use during the past decade (U.S. EPA 2019).

Most PFASs can be categorized into four main subclasses: perfluoroalkyl acids (PFAAs), PFAA precursors, fluoropolymers, and perfluoropolyethers (PFPEs) (Wang et al. 2017). As explained below, all these PFAS subclasses are relevant to consumer product applications and related exposures.

3.3.1 Perfluoroalkyl Acids (PFAAs)

PFAAs are perfluorinated substances in which fluorine atoms have replaced all hydrogen atoms attached to carbon atoms (except for those associated with functional groups). Due to the strength of the carbon-fluorine bond, these compounds are recalcitrant and extremely persistent in the environment. Examples include perfluoroalkyl carboxylic acids (PFCAs) such as perfluorooctanoate (PFOA), perfluoroalkyl sulfonic acids (PFSAs) such as perfluorooctane sulfonate (PFOS), perfluoroether carboxylic acids (PFECAs) such as GenX, and perfluoroether sulfonic acids (PFESAs) such as ADONA. PFAAs and some of their precursors are frequently subdivided into longer- and shorter-chain PFASs. The longer-chain PFSAs have six or more perfluorinated carbons; longer-chain PFCAs have seven or more perfluorinated carbons (OECD 2018).

While PFAAs are a small subset of PFASs (OECD 2018), they are terminal degradation products, manufacturing aids/feedstocks, or impurities of other PFAS class members (Figure 3.1). Some perfluoroether acids may be used in certain cosmetics and varnishes (Nordic Council of Ministers et al. 2020), but most PFAAs are not currently used intentionally in consumer products in the U.S. and EU. Nevertheless, low levels of PFAA degradants or manufacturing impurities are commonly found in a variety of consumer products sampled around the world, including clothing, furnishings, and cosmetics (Borg and Ivarsson 2017, KEMI 2015, Kotthoff et al. 2015, Posner et al. 2013, Ye et al. 2015).

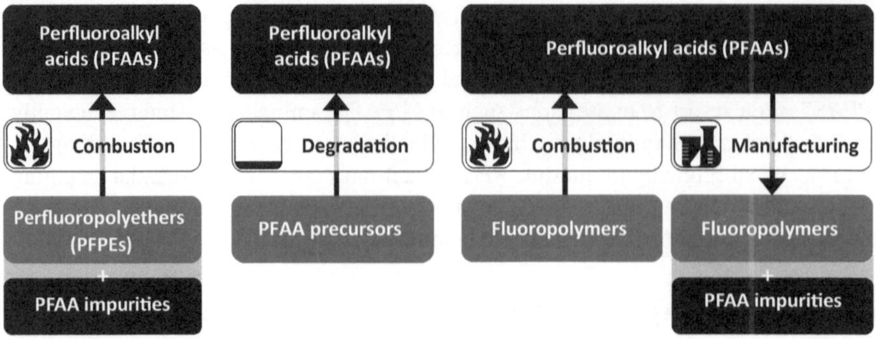

FIGURE 3.1 Nearly all PFASs can lead to exposure to PFAAs because they degrade to form PFAAs, release PFAAs if incinerated, require PFAAs for their manufacture, or contain PFAAs as impurities.

3.3.2 PFAA Precursors

The PFAA precursors, as the name implies, can degrade or metabolize into PFAAs. They are mostly polyfluoroalkyl substances, meaning fluorine atoms have replaced all hydrogen atoms attached to at least one (but not all) of their carbon atoms. More than 85% of all PFASs are PFAA precursors (OECD 2018), including the PFASs used as raw materials and as surface treatments, as well as the intermediate environmental transformation products of other PFAA precursors. Some examples are listed in Figure 3.2.

Typical consumer product applications of PFAA precursors include liquid formulations (e.g., treatment solutions or sprays, cleaning products, paints, coatings, varnishes, lubricants, cosmetics and personal care products) and surface coatings for consumer products made of textile, leather, or plant fiber materials (Bečanová et al. 2016, Borg and Ivarsson 2017, Fiedler et al. 2010, Herzke et al. 2012, KEMI 2015, Kotthoff et al. 2015, Liu et al. 2015, Posner et al. 2013, Tokranov et al. 2019, Vestergren et al. 2015, Ye et al. 2015). Non-polymeric PFAA precursors are also used as biocides, including in ant baits and other pesticide formulations (OECD 2013).

Non-polymeric PFAA precursors such as n:2 polyfluoroalkyl phosphoric acid esters (PAPs) and fluorotelomer ethoxylates (FTEOs) were widely used in the past. (Note that "n" refers to the number of perfluorinated carbon atoms; the number after the colon refers to the number of nonfluorinated carbon atoms in the molecule.) However, since the early 2000s manufacturers have transitioned to side-chain fluorinated polymers (e.g., fluorinated acrylate, methacrylate, urethane, and oxetane polymers). Side-chain fluorinated polymers are now widely used as repellents and soil-release agents in textile and food packaging products, among others. The large molecular sizes of these polymers make them less bioavailable, thus less likely to cause toxic effects. However, because their side chains are not part of the polymer backbone, side-chain fluorinated polymers break down into non-polymeric PFASs, with PFAAs as their end products. In other words, they are also PFAA precursors. Their final degradation to PFAAs under environmental conditions might take decades or centuries (Russell et al. 2008, 2010; Washington et al. 2009, 2015, 2019), which

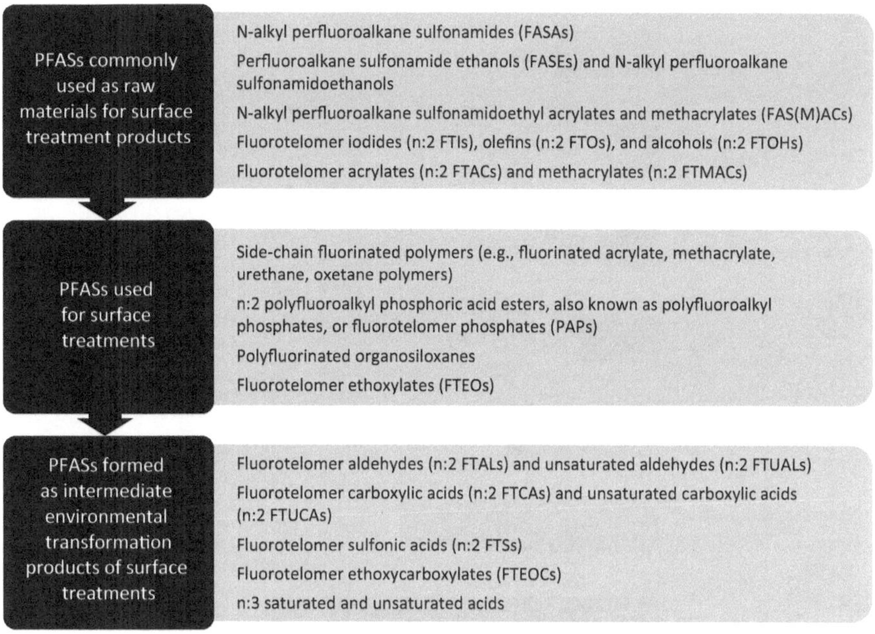

FIGURE 3.2 Examples of PFAA precursors.

makes them long-term sources of PFAAs to the environment even after the treated consumer products reach the ends of their lives.

3.3.3 Fluoropolymers

Fluoropolymers are large molecules with fluorinated carbon atoms as part of the polymeric backbone. They are very durable materials that cannot degrade to PFAAs under typical environmental conditions, but certain PFAAs have been used as processing aids in their manufacturing and can occur as impurities. Moreover, fluoropolymers may release PFCAs, including PFOA, during combustion at temperatures between 180 and 800 °C (Feng et al. 2015, Schlummer et al. 2015). Examples of fluoropolymers include polytetrafluoroethylene (PTFE), polyvinylidene fluoride (PVDF), and polyvinyl fluoride (PVF). Some fluoropolymers, such as perfluoroalkoxy alkanes (PFAs) and fluoroelastomers, are made from a mix of perfluoroether monomers and tetrafluoroethylene (TFE) or hexafluoropropylene (HFP) (Nordic Council of Ministers et al. 2020). They are still classified as fluoropolymers rather than PFPEs because the ether linkages occur in the side chains, not in the polymer backbone.

Fluoropolymers are used as high-performance materials in a wide range of applications that require chemical resistance, heat stability, good electrical properties, or resistance to water, grease, or oil. Examples include outdoor apparel, electronic parts, solar cells, semiconductors, and cookware (Chemours 2020c, 2020b; Daikin Global 2020c, 2020b; OECD 2013). Fluoroelastomers are commonly used in

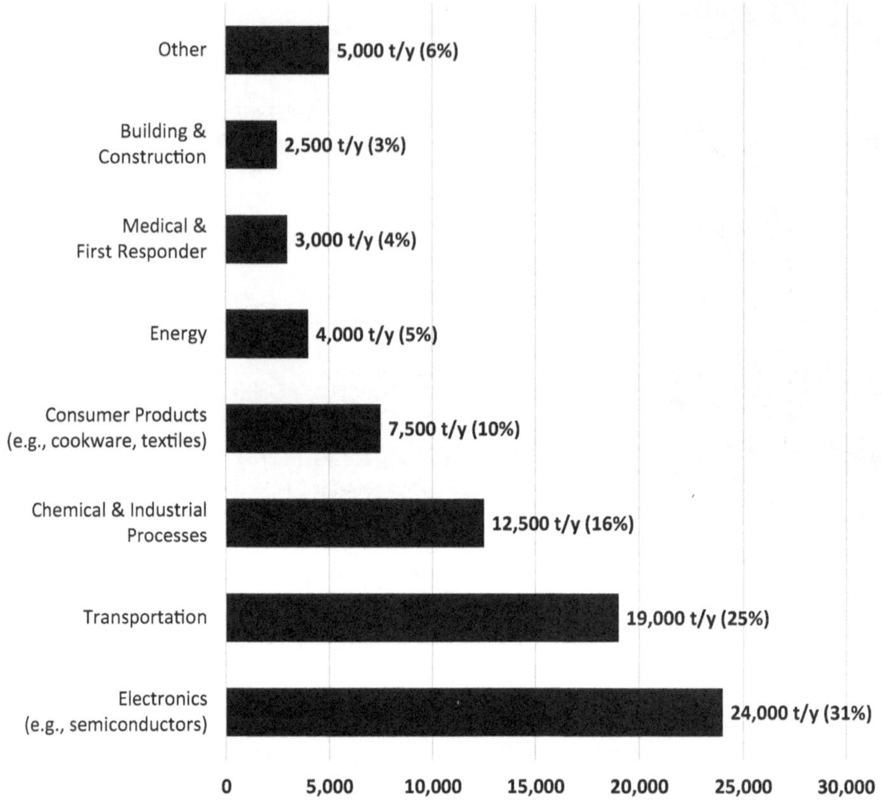

FIGURE 3.3 Annual use in 2018 of fluoropolymers in the U.S. by industry sector, in tons. Data from Wood Environment & Infrastructure Solutions UK Limited (2020).

consumer wearables such as fitness bands (Chemours 2020a, Daikin Global 2020a), as well as in seals, valves, gaskets, lubricants, and molded parts for food processing equipment (Nordic Council of Ministers et al., 2020). In the U.S., consumer product applications (including electronics and building and construction products) account for approximately 44% of the 77,500 tons of fluoropolymers used annually (see Figure 3.3) and 40% of the $2110 M annual revenue (Wood Environment and Infrastructure Solutions UK Limited 2020).

3.3.4 Perfluoropolyethers (PFPEs)

PFPEs are perfluoroalkyl substances with large molecular size (oligomers, polymers, and copolymers), and ether linkages. They are highly persistent and unlikely to degrade into PFAAs under typical environmental conditions but may contain PFAA impurities and release hazardous products during combustion (Nordic Council of Ministers et al. 2020).

PFPEs are used in some of the same applications as side-chain fluorinated polymers and other PFAA precursors, such as in cosmetics and personal care products, food packaging, and stain-resistant coatings for carpets and rugs. They can also be used in some of the same applications as fluoropolymers, such as lubricants for food processing equipment (Nordic Council of Ministers et al. 2020).

3.3.5 Trends in PFAS Use in Consumer Products

Tests in the 1940s found that fluorochemicals performed exceptionally well in imparting resistance to the spreading of liquids on solid surfaces. This led 3M to develop Scotchgard™ Fabric Protector in the mid-1950s, the first brand of stain and soil repellents for textiles (Audenaert et al. 1999). The first commercial Scotchgard™ rain and stain repellent, Scotchgard™ FC-208, was released in 1960. This was followed in 1963 by DuPont's Zepel, which was developed for use on clothing. In 1972, 3M produced Scotchgard™ Carpet Protector, the first product specifically marketed for use on carpets (Maitland 1982). Subsequent formulations with increased ability to withstand laundering fell under three major categories: esters, polyurethanes, and acrylate copolymers (Audenaert et al. 1999).

Since the early 1980s, PFASs have been widely used to impart stain-, soil-, and grease-resistance to textiles (Dusaj 1988, U.S. EPA 2012). In 2009, U.S. EPA estimated that, globally, 50% of fluorotelomers and associated side-chain fluorinated polymers produced were used in the treatment of textiles and apparel, including carpets and rugs (U.S. EPA 2009a). Despite the high performance and convenience provided by these coatings, mounting evidence of adverse health impacts has changed the types of PFASs produced and used in many countries, including in the U.S. and most of the EU (Table 3.1). In 2000, 3M announced that it was discontinuing the manufacture and sale of Scotchgard™, which at that time was based on C8 chemistry. The company reintroduced a reformulated version 16 months later (Singh et al. 2005) based on C4 chemistry (Renner 2006). The current Scotchgard™ formulation contains a trade secret fluorochemical urethane (3M 2018), which is a short chain PFAA precursor. Since 2002, most other global manufacturers have gradually replaced long-chain PFAAs and their precursors with shorter-chain PFASs or with non-PFAS compounds (OECD 2013).

Similar to textile manufacturers, paper and food packaging producers apply PFASs to a wide array of plant fiber-based packaging products to impart moisture, oil, and grease resistance (FDA 2020a, RIVM 2019). PFASs may also be used as an agent to ease the release of plant fiber-based and plastic food packaging products from their manufacturing molds (CEH 2018). Historically, paper and paperboard products were treated with non-polymeric long-chain PFASs, including PFOS precursors (Trier et al. 2011a, 2011c). Since 3M phased out PFOS and its precursors, most PFAS formulations used in food packaging products are now polymers, such as fluorotelomer-based acrylate polymers or PFPEs (Schultes et al. 2019). For instance, Solvay markets Solvera® PFPE for use in pizza boxes, multiwall paper sacks, fast food packaging, microwave popcorn packaging, and molded paper applications (Solvay 2020).

These changes in the types of PFASs used in consumer products were voluntary actions by industry, sometimes prompted by government-industry partnerships.

For instance, in 2006, the U.S. EPA developed a 2010/2015 Stewardship Program for reducing emissions of PFOA, its precursors, and related higher homologues (U.S. EPA 2020a). Participating companies included Arkema, Asahi, BASF, Clariant, Daikin, 3M/Dyneon, DuPont, and Solvay Solexis. These voluntary actions on longer-chain PFAAs and their precursors were followed by regulatory action in some countries, as well as at the UN level through the Stockholm Convention (OECD 2013). However, these phased-out PFASs were replaced with other, less well-studied, members of the class. As more studies have become available, these newer PFASs are also showing potential for adverse human and ecological impacts and other issues (Table 3.1).

TABLE 3.1
Recent changes in the types of PFASs used in consumer products.

Change	Rationale	Remaining Issues
Replacement of non-polymeric PFASs with side-chain fluorinated polymers and copolymers	Large molecules are less bioavailable, thus less likely to cause toxic effects.	Side-chain fluorinated polymers break down into non-polymeric PFASs and contain non-polymeric PFAS impurities. While exposures are reduced compared with non-polymeric PFASs, they are not eliminated. Not all countries have adopted this trend, particularly not all developing countries.
Replacement of longer-chain PFAA precursors with shorter-chain PFAA precursors	Shorter-chain PFAAs are less persistent inside human and animal bodies and do not bioaccumulate inside humans and animals. Because of this, they were expected to be less toxic.	Shorter-chain PFAAs are extremely persistent in the environment, just like the longer-chain PFAAs. They are more mobile, more difficult to remove from drinking water sources, and less well-studied than the longer-chains. Additionally, like some longer-chain PFAAs, they accumulate in plants, including those used for human and animal food. Recent research found evidence of toxicity for shorter-chain PFAAs and some of their precursors. Longer-chain PFAAs and their precursors continue to be manufactured or used in some countries, including China, India, Russia and other developing countries. They are also still used in the U.S. in some applications (Alliance for Automotive Innovation, 2020).
Replacement of some longer-chain PFASs with fluorinated ether compounds	The addition of one or more oxygen atoms to the PFAS molecule was thought to make these compounds more readily degradable in the environment.	These compounds (e.g., PFPEs and PFECAs) are also extremely persistent in the environment. They are less well-studied and more difficult to remove from drinking water sources than the compounds they replaced. Some show similar evidence of toxicity (e.g., GenX).

3.4 FROM CONSUMER PRODUCTS TO HUMAN EXPOSURE

PFASs can enter environmental media from consumer products during any life cycle stage: manufacturing, use, and end-of-life. These releases result in a continuous exchange of ever-increasing amounts of PFASs between environmental compartments, with no apparent loss from the environmental system. As a result, there is continuous, if not growing multimedia PFAS exposure to biological receptors (Figure 3.4).

Exposure media are environmental media through which people are exposed to a contaminant. They include indoor and outdoor air, dust, soil, and drinking water. Data on PFAS levels in various exposure media are summarized below, with a focus on cases where use of a consumer product has a direct impact on personal exposure

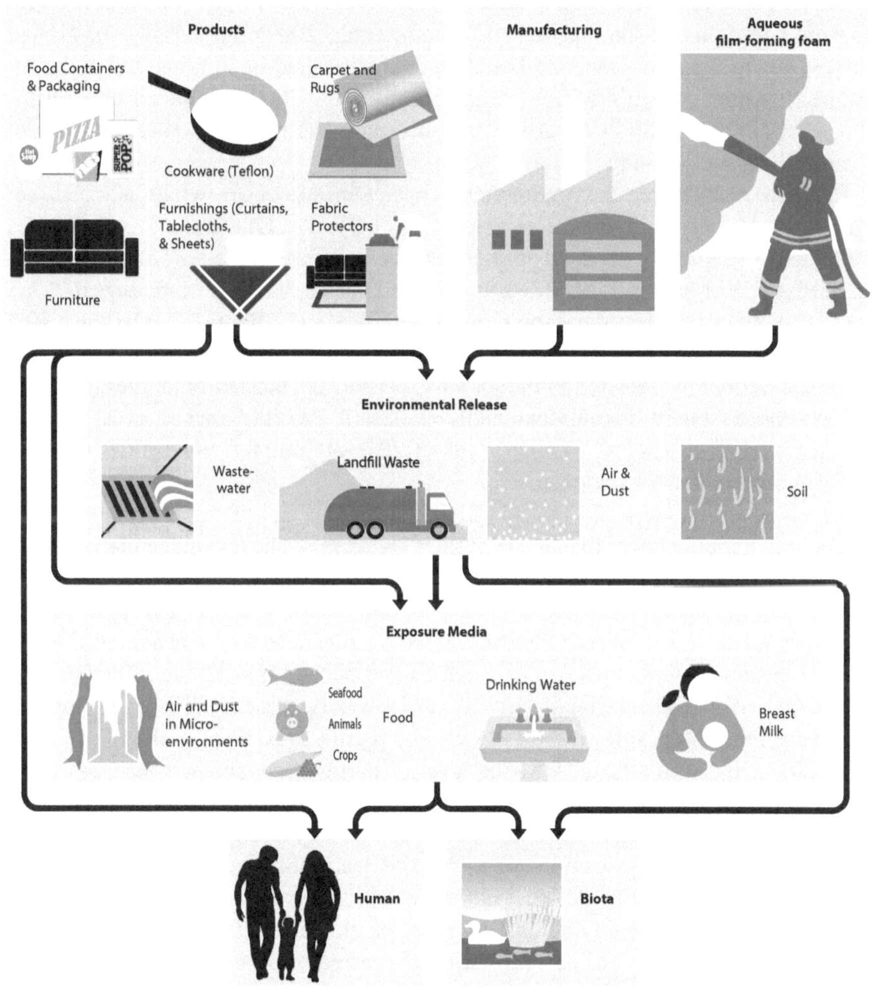

FIGURE 3.4 Main PFAS exposure pathways for humans and biota.

(for example, release of fluorotelomer alcohols (FTOHs) into indoor air from stain repellents or migration of PFASs from food packaging materials into food). The summary below is by no means exhaustive; it is intended to illustrate the potential for PFASs in consumer products to contribute to human exposure.

3.4.1 PFASs in Indoor Air

The level of an indoor air pollutant is a function of outdoor air concentrations, indoor sources, ventilation, and transformation and other decay mechanisms, which are in turn related to housing structure, building age, and personal activities. A few studies have reported elevated indoor air PFAS concentrations compared with outdoor air levels, indicating the direct impact of consumer products on indoor air quality (Goosey and Harrad 2012, Yao et al. 2018).

Indoor air studies have been conducted since the 2000s. Many of the earlier studies were reviewed and summarized by Fromme et al. (2009) and Goosey and Harrad (2012). Figure 3.5 was generated based on data published or summarized in the literature (Fromme et al. 2009, Goosey and Harrad 2012, Padilla-Sánchez et al. 2017, Shoeib et al. 2011, Yao et al. 2018) and illustrates the range of concentrations of several PFAS species in the air.

PFAA precursors are the predominant species in indoor air, which is consistent with their relatively high volatility compared with PFAAs. Of the precursors, FTOHs are the most frequently detected species and account for a large portion of the measured PFASs. For example, Shoeib et al. (2011) reported that 92% of measured PFAA precursors were FTOHs; the geometric means of 8:2 FTOH, 6:2 FTOH, and 10:2 FTOH were 2900 pg/m^3, 980 pg/m^3, and 950 pg/m^3, respectively. Another subclass of PFASs frequently detected in indoor air is perfluorooctanesulfonamides (FOSAs) and perfluorooctanesulfonamidoethanols (FOSEs). Padilla-Sánchez et al. (2017) detected MeFOSE (methyl-FOSE) in about 70% of samples and EtFOSE (ethyl-FOSE) in about 49% of samples.

Levels of PFAAs (e.g., PFOA and PFOS) in indoor air are generally one to two orders of magnitude lower than those of their precursors. However, the relative abundance of PFAAs varies across different cities and microenvironments, reflecting the changes in PFAS manufacturing and use patterns. Shoeib et al. (2011) reported that PFOA dominates the measured PFAAs (28 pg/m^3) in Vancouver, Canada, followed by perfluorohexanoic acid (PFHxA) (9.7 pg/m^3) and perfluoroheptanoic acid (PFHpA) (5.1 pg/m^3). However, in Tianjin, China, perfluorobutyric acid (PFBA) (271 pg/m^3) dominates the profile, followed by PFOS (185 pg/m^3) and PFOA (106 pg/m^3) (Yao et al. 2018). In Birmingham, UK, the level of perfluorohexane sulfonate (PFHxS) (94 pg/m^3) was the highest among measured ionic PFASs in offices, while PFOA (52 pg/m^3) was the highest in residential homes (Goosey and Harrad, 2012).

In general, the concentrations of PFAS species within the same subclass (i.e., FTOH, FOSE/FOSA, PFCA, etc.) correlate with each other, i.e., the concentrations of two PFAS species fluctuate together or increase/decrease in parallel. Yao et al. (2018) reported a strong correlation (Spearman $r_s = 0.90$) between 8:2 and 10:2 FTOHs. Similarly, Shoeib et al. (2011) reported strong correlations among 6:2 FTOH, 8:2 FTOH, and 10:2 FTOH. Spearman correlation coefficients among

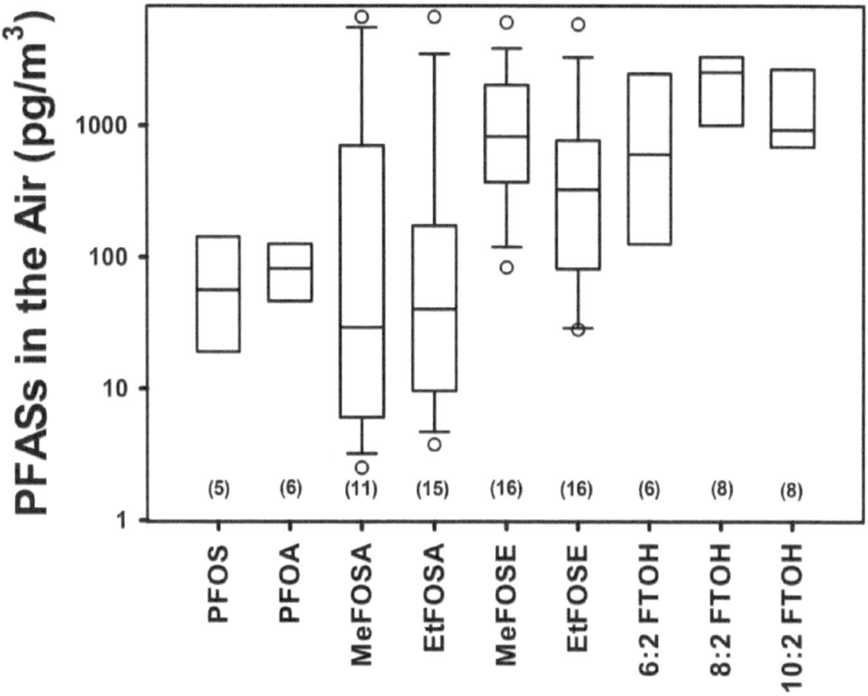

FIGURE 3.5 PFAS concentrations in the air, sampled across the world, compiled from several sources (Fromme et al. 2009, Goosey and Harrad 2012, Padilla-Sánchez et al. 2017, Shoeib et al. 2011, Yao et al. 2018). The data used in this figure reflect the central tendency of PFAS concentrations measured in each study. The box represents the 25th, 50th, and 75th percentile, the whiskers extend to the 95th percentile and to the 5th percentile, and the data points above the 95th percentile or below the 5th percentile are indicated by circles. The number in parentheses under each box is the number of data points used to generate the boxplot.

long-chain C8 – C11 PFCAs ranged from 0.82 to 0.92, 0.76 between PFHxS and PFOS, and 0.77 between 6:2 di-PAPs (perfluoroalkyl phosphate diester) and 8:2 di-PAPs (Yao et al. 2018). There are a number of reasons why PFAS species could be correlated with each other: 1) they originate from the same source(s), 2) they are present in the same environmental media due to environmental transport, and 3) they are related through transformation process (i.e., one chemical is the precursor of another one). Therefore, examining the associations among PFASs could provide useful insights into their source contribution and environmental behavior.

3.4.2 PFASs in Indoor Dust

U.S. EPA researchers studied the PFASs found in 13 consumer product categories and 116 household products (U.S. EPA 2009b). They concluded that many of the PFASs detected in household dust can be attributed to the use of treatment products

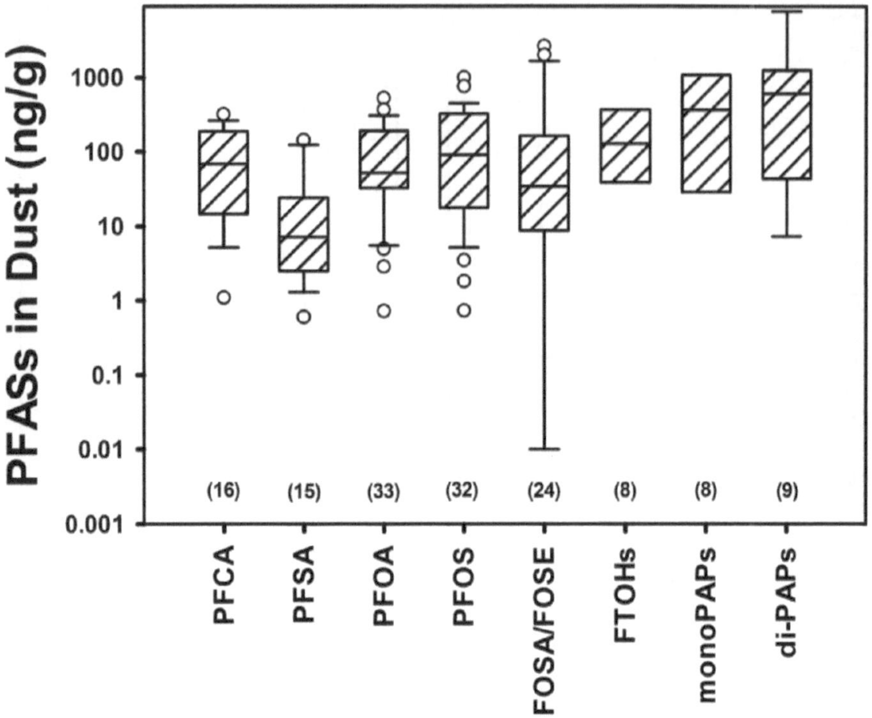

FIGURE 3.6 PFAS concentrations by various subclass, in indoor dust measured across the world, extracted from several studies (de la Torre et al. 2019, Eriksson and Kärrman 2015, Nadal and Domingo 2014, Tian et al. 2016). The data used in this figure reflect the central tendency of PFAS concentrations in indoor dust reported in each study. The box represents the 25th, 50th, and 75th percentile, the whiskers extend to the 95th percentile and to the 5th percentile, and the data points above the 95th percentile or below the 5th percentile are indicated by circles. The number in parentheses under each box is the number of data points used to generate the boxplot.

applied to textile and leather products such as carpeting, apparel, and upholstery. Carpet and fabric care liquids were among the greatest sources of exposure. It is estimated that the average household has 1 kg of household carpet or fabric care liquids for spot treatment, which can introduce up to 950 μg of PFCAs during application. Furthermore, a single professional treatment uses about 6 kg of commercial-grade carpet/fabric care liquids, resulting in the addition of up to 71,800 μg of PFCAs to the home (U.S. EPA 2009b). Figure 3.6 illustrates the concentrations of PFAS species in dust published in major studies and summarized by de la Torre et al. (2019), Tian et al. (2016), Eriksson and Kärrman (2015), and Nadal and Domingo (2014).

PFOS and PFOA are among the first detected PFASs in indoor air dust in the early 2000s (Miralles-Marco and Harrad, 2015). Since then, PFOS and PFOA levels in indoor dust have been extensively studied across the world. Table 3.2 summarizes worldwide indoor dust levels of PFOS and PFOA.

TABLE 3.2
PFOS and PFOA in indoor dust measured in different countries (ng/g).

Country	Citation	PFOS (ng/g)	PFOA (ng/g)
Canada	Nadal and Domingo (2014)	37–443.7	33–270
China	Zhang et al. (2010) Yao et al. (2018)	1.11–10.7	1.4–718
Germany	Nadal and Domingo (2014)	91.7–310	114–290
Japan	Nadal and Domingo (2014)	11–2500	42.3–3700
Spain	Nadal and Domingo (2014)	1.1–12	1.5–36
Sweden	Nadal and Domingo (2014)	31–110	41–93
U.S.	Nadal and Domingo (2014)	26.9–420	23.7–370

Note: Data presented in the table are the range of the mean concentrations in multiple studies for each country.

Other PFCAs and PFSAs frequently detected in indoor dust include perfluorononanoic acid (PFNA), PFHxA, perfluorobutane sulfonate (PFBS), PFHxS, among others (de la Torre et al. 2019, Eriksson and Kärrman 2015, Nadal and Domingo 2014, Shoeib et al. 2011, Winkens et al. 2018, Yao et al. 2018). The most frequently reported fluorotelomer sulfonates (FTSAs) in indoor dust are 6:2 FTSA and 8:2 FTSA. In a worldwide indoor dust study, Eriksson and Kärrman (2015) reported that 6:2 FTSA levels ranged from nondetectable (Nepal) to 11 ng/g (Faroe Islands). Zheng et al. (2020) reported that the median levels of 6:2 FTSA and 8:2 FTSA across childcare facilities in the U.S. were 12 ng/g and 5.8 ng/g, respectively.

Multiple studies report concentrations of FTOHs, FOSA/FOSEs, and PAPs in indoor dust (Padilla-Sánchez et al. 2017, Shoeib et al. 2011, Tian et al. 2016, Winkens et al. 2018, Yao et al. 2018, Zheng et al. 2020). Winkens et al. (2018) analyzed dust samples from 65 children's bedrooms in Finland collected during 2014 and 2015; 8:2 FTOH was detected in 80% of the samples, followed by 6:2 FTOH (72%) and 10:2 FTOH (65%). Eriksson and Kärrman (2015) analyzed and compared PFASs in indoor dust samples collected from eight countries and regions in the world. MeFOSE was the most frequently detected FOSE/FOSA with a detection rate of 33%, followed by EtFOSE (8%), MeFOSA (3%) and EtFOSA (3%). FOSE/FOSE concentrations varied widely by region with a mean concentration of 745 ng/g in Canada and 0.1 ng/g in Nepal. PAPs are also frequently detected in indoor dust. Eriksson and Kärrman (2015) observed a large variation of mono- and di-PAPs between different parts of the world: the lowest measured concentrations of mono- and di-PAPs were 6.9 ng/g and 7.3 ng/g, respectively, in Nepal, and the highest were 1790 ng/g and 1415 ng/g, respectively, in Japan.

Similar to PFASs in indoor air, PFAS species belonging to the same subclass usually significantly correlate with each other (i.e., concentrations of species within a subclass increase or decrease in parallel). de la Torre et al. (2019) reported that long-chain PFCAs ($C \geq 8$) are moderately correlated with each other, with a Spearman correlation coefficient above 0.487. Strong and positive correlations have also been reported among FTOHs ($r \geq 0.90$) (Shoeib et al. 2011, Winkens et al. 2018, Yao et al.

2018) and among PAPs (generally $r \geq 0.65$) (Eriksson and Kärrman 2015, Winkens et al. 2018).

The observed PFAS species and their concentrations in indoor dust are determined by several factors, including geographical location (de la Torre et al. 2019, Eriksson and Kärrman 2015), the presence of indoor sources (de la Torre et al. 2019, Eriksson and Kärrman 2015, Zheng et al. 2020), the extent of biotic or abiotic transformation (Eriksson and Kärrman 2015), and some behavioral and microenvironmental factors (de la Torre et al. 2019, Karásková et al. 2016). PFAS levels in air and dust are also expected to be coupled because PFASs are often released to both media from the same source; can partition between air and dust; and can be transformed in the environment (Shoeib et al. 2011, Winkens et al. 2018, Yao et al. 2018).

3.4.3 PFASs in Food

Numerous studies have reported the presence of PFASs in food items, including fish, vegetables, crops, dairy products, eggs and meat products (Berger et al. 2009, Bhavsar et al. 2014, Chen et al. 2018, Domingo 2012, Fair et al. 2019, Herzke et al. 2013, Hung et al. 2018, Jian et al. 2017, Kotthoff and Bücking 2018, Papadopoulou et al. 2017, Schaider et al. 2017, Scher et al. 2018, Sunderland et al. 2019, Tittlemier et al. 2007). Jian et al. (2017) ranked PFAS levels in various types of food items as follows: fish and shellfish > meat and egg > milk products > vegetables.

Due to their high bioaccumulation potential, the most abundant PFAS species detected in fish are long-chain PFCAs and PFSAs, including PFOS, PFOA, PFNA, perfluorodecanoic acid (PFDA), perfluoroundecanoic acid (PFUnDA), and PFHxS; they may be present at concentrations up to a few ng/g (Hung et al. 2018, Jian et al. 2017). Which PFAS species are present in fish and shellfish, as well as their levels, depends on the concentration of PFASs in water habitat, which is in turn affected by nearby industrial and municipal releases (Berger et al. 2009, Fair et al. 2019). Some PFAS species can accumulate along the food chain, with higher concentrations in higher trophic levels (Jian et al. 2017). The level of total PFASs also varies in different fish tissues: blood (112 ng/mL) > liver (44.6 ng/g) > egg (32.6 ng/g) > muscle (8.42 ng/g) (Hung et al. 2018), indicating differential levels of PFAS exposure depending on how people consume fish (whole fish vs. fillet). PFASs cannot be effectively removed by cooking (e.g., baking, broiling, or frying) to reduce exposure (Bhavsar et al. 2014).

In meat and meat products, the dominant PFAS species are PFOA and PFOS, followed by PFNA, PFDA, PFUnDA, PFHxS (up to a few hundreds of pg/g) (Jian et al. 2017). Similar PFAS species (e.g., PFOS, PFNA, PFDA, PFUnDA, PFHxS, PFOS) were observed in eggs at levels ranging from tens to hundreds of pg/g. PFOS levels in eggs decreased from 1999 to 2010, consistent with the phaseout of PFOS starting in 2002 (Jian et al. 2017).

Shorter-chain PFCAs (e.g., PFHxA, PFHpA) and PFOA are the predominant PFAS species seen in vegetables, at levels of up to tens of ng/kg (Herzke et al. 2013, Jian et al. 2017). Shorter-chain PFCAs are particularly mobile in water and can be retained in leaves, whereas longer-chain PFCAs accumulate in roots (Herzke et al.

2013, Jian et al. 2017). The species and levels of PFASs in vegetables depend on the PFAS levels in the irrigation water, soil, and biosolids used as fertilizer, as well as on the soil characteristics, and the type of vegetable (Ghisi et al. 2019).

PFAS impurities and degradation products present in the polymeric formulations currently used in food-contact materials can migrate into food items (RIVM et al. 2019). A recent report (Nordic Council of Ministers 2017) summarizes the studies showing migration of such non-polymeric PFASs from paper and paperboard used in food packaging applications. The authors conclude that:

- PFASs that are not covalently bound to the paper (such as the PFAA and FTOH impurities) are more easily released;
- PFASs migrate out of paper and board packaging via hydrolysis, which is accelerated by heating, moisture, and the presence of emulsifiers such as alcohol – this was demonstrated for PAPs, but is expected to also be true for fluoroacrylate ester bound coatings;
- the types of PFASs in the coating matters: what determines the transfer of perfluorinated PFASs is not necessarily true for the polyfluorinated ones;
- other factors that impact migration include the food composition, the presence of salts, microwaving, the total surface area, and the surface energy of the surfaces.

Other studies have also found correlations between factors such as high heat, acidity, and the presence of emulsifiers, and the migration of PFASs from food packaging. For instance, Still et al. (2013) found that increasing the time that butter (an emulsified food) is stored in fluorinated polymer-coated packaging increased its levels of PFAAs and FTOHs.

Note that most of the available studies, including most of those reviewed by the Nordic Council of Ministers (2017), were done on older PFASs formulations that may no longer be used in food packaging. Specifically, many describe the migration potential of PFOA, PFOS, and other longer-chain PFASs that have been phased out and are no longer authorized by the U.S. Food and Drug Administration (FDA). As noted above, shorter-chain PFASs tend to be more mobile and migrate out of food packaging materials more readily than their longer chain counterparts (Nordic Council of Ministers 2017). Some newer studies indicate that the intermediate degradation products of these PFAS impurities present in FDA-approved shorter-chain PFAS formulations currently used in food packaging are biopersistent and more toxic than previously thought (Kabadi et al. 2018, 2020; Rice et al. 2020). The best source of information regarding currently used formulations are the Food Contact Notification (FCN) documents submitted to FDA which list numerous PFASs that are expected to migrate out of packaging into food from formulations currently approved for use in the U.S. However, most of the names and CASRNs of these PFASs are redacted from the publicly-available versions of these documents. (FDA 2020a, Nelter 2018).

It has been estimated that a typical paper mill produces "825 tons of PFAS-coated paper per day and discharges 26 million gallons of water per day." This wastewater contains between 95 and 225 pounds of PFASs, based on an estimated concentration range between 43,000 to 103,000 parts per trillion (ppt) (CELA 2019). The PFASs used in food packaging degrade to PFAAs, which are taken up by plants grown in compost-treated soil (Gredelj et al. 2020, Trier et al. 2011b, Zhang et al. 2020, Zhou et al. 2019). Compost samples collected from California and four other U.S. states that included food service packaging had significantly higher PFAA levels than compost that did not include these items, with the shorter-chain PFAAs (PFBA, perfluoro-n-pentanoic acid [PFPeA], and PFHxA) being most prevalent (Lee and Trim 2018). These shorter-chain PFASs are extremely persistent in the environment, highly mobile in water, and preferentially taken up by plants (Blaine et al. 2013), including food crops (Wang et al. 2015).

3.4.4 PFASs in Drinking Water

Drinking water is the most studied environmental medium for PFAS exposures. Historically, PFOA and PFOS have been the most commonly measured PFAS species in drinking water. Away from large point sources, the reported PFOA and PFOS levels range from a few to tens of ng/L (Fromme et al. 2009, Miralles-Marco and Harrad 2015). However, drinking water sources near PFAS manufacturing facilities, military firefighter training sites, landfills, or wastewater treatment plants can have elevated PFAS levels (Dauchy 2019). Shifts in the types of PFASs produced and used in recent years have led to corresponding changes in the PFAS species seen in drinking water, with more shorter-chain and other emerging PFASs (e.g., GenX) being detected (Hopkins et al. 2018).

Release of PFASs from industrial manufacturing facilities and consumer products can contaminate drinking water. Source apportionment studies are much needed to quantify the impact of consumer product use on the drinking water PFAS load, however these types of studies are lacking. Indirect evidence, such as measurements of PFAS levels in groundwater under municipal landfills and in the influent and effluent of municipal wastewater treatment plants, could be used to semi-quantitatively assess the impact of consumer product use on the types and levels of PFASs in drinking water. However, the data associated with indirect evidence are also extremely limited.

A study by Hu et al. (2016) illustrates the impact of industrial sites, military fire training areas, and wastewater treatment plants on PFASs in drinking water. Using U.S. EPA's Unregulated Contaminant Monitoring Rule (UCMR3) data, Hu et al. (2016) calculated that 6 million U.S. residents were consuming drinking water containing PFOS and PFOA at levels greater than 70 ng/L, and found that the number of wastewater treatment plants was positively associated with PFAS detection frequencies and concentrations in drinking water. More studies of this type are needed to evaluate the impact of consumer products use on PFASs in drinking water.

PFASs can also be introduced into previously uncontaminated drinking water when it is bottled or treated. Jian et al. (2017) reported the potential migration of PFHpA from drinking water bottles to bottled water, and the potential migration of PFASs, including PFOA and PFOS, from water purification systems to drinking water.

3.5 PFAS EXPOSURE DETERMINANTS

3.5.1 Exposure Quantification and Exposure Profiles

Human exposure to PFASs can be characterized either by measuring external exposures or by measuring biomarkers of exposure. External exposure measurements involve determining the PFAS concentrations in various exposure media, including air, dust, food, and drinking water, as well as people's personal activity patterns (e.g., time spent in various microenvironments) and their frequency of use of consumer products. Biomarkers of exposure can be used to determine a person's aggregate exposure to PFASs by characterizing and quantifying PFAS species in human serum, whole blood, urine, breast milk, hair, nails, and other tissues (Colles et al. 2020, Hu et al. 2018, Jian et al. 2018, Miralles-Marco and Harrad 2015, Winkens et al. 2017). Both methods are equally important and which of them is appropriate in a given situation depends on the research question and the study design. To interpret the findings from either approach requires a detailed understanding of personal activity patterns, product use patterns, the levels of PFASs in exposure media, and the pharmacokinetics or absorption, distribution, metabolism, and excretion (ADME) of each PFAS species (Lorber and Egeghy, 2011).

Differential exposure profiles, i.e., the relative abundance of exposure to multiple PFAS species (or the "fingerprint" of PFAS exposure that is characteristic of a particular source), have been reported in multiple studies for both external and internal exposure measurements. Differential exposure profiles could provide information on major PFAS sources, their fate and transport, and their pharmacokinetic properties (Hu et al. 2018). The most comprehensive biomonitoring data in the U.S. are from the National Health and Nutrition Examination Survey (NHANES). Egeghy and Lorber (2011) reported differential PFAS exposures according to age, gender, and ethnic group, using NHANES data (i.e., biomarkers of exposure). Chen et al. (2018) summarized geographical differences in PFAS external exposure estimates and reported that people in Taiwan are exposed to elevated levels of PFHxA (11.2 ng/kg/day), PFOA (85.1 ng/kg/day), PFDA (44.2 ng/kg/day), and PFUnDA (4.45 ng/kg/day) compared with those in western countries, due to dietary habits (higher consumption of rice and pork liver, etc.). The relative contributions of different exposure routes and pathways to aggregate PFAS exposures were the subject of a few multi-pathway studies summarized in Figure 3.7.

Dietary intake is consistently reported as the biggest contributor to the overall PFAA intake, and dermal absorption as the smallest one (Ao et al. 2019, Egeghy and Lorber 2011, Poothong et al. 2020, Sunderland et al. 2019). For PFAA precursors such as FTOHs, dermal uptake might contribute a significant portion of the overall exposure, but more research is needed to quantify dermal exposures (Birnbaum 2018). However, the relative contribution from different exposure pathways varies across populations, including across age groups and geographical areas. For example, dust ingestion is as important a pathway as dietary exposure for children because of their special activity patterns (Egeghy and Lorber 2011). For people living close to PFAS contaminated sites, drinking water ingestion could be the dominant exposure pathway (representing up to 75% of aggregate PFAS exposure) (Sunderland et al.

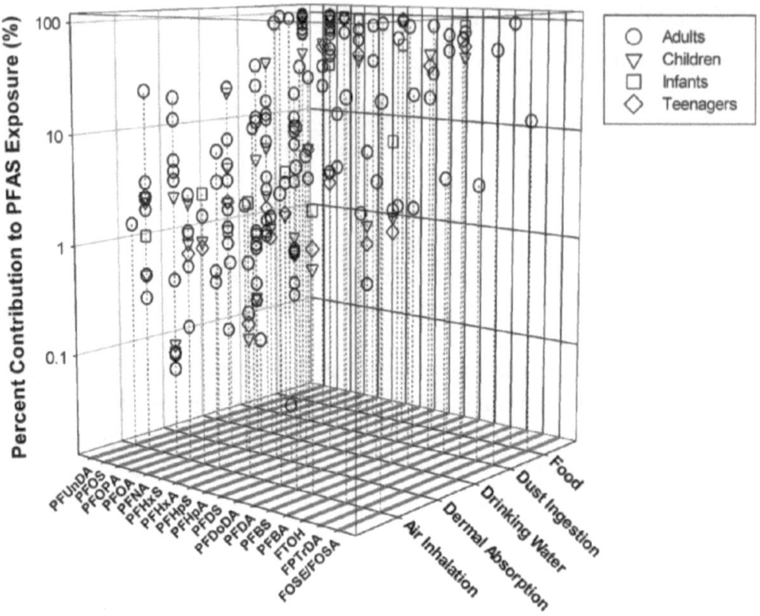

FIGURE 3.7 Relative contributions of food consumption, drinking water ingestion, dust ingestion, air inhalation, and dermal absorption to human PFAS exposure for various age groups.

2019). Poothong et al. (2020) also reported a large variation in the relative contributions from each exposure pathway: the contribution of food to overall PFAA intake ranged from 4% to 98%, with a median of 91%; the contribution of inhalation of indoor air and ingestion of indoor dust ranged from <1% to 95%, with a median of 3%. Exposure determinants need to be studied to explain this variability. In addition, prospective cohort studies need to be conducted to track the exposure trends for different PFAS species, including emerging PFASs.

3.5.2 Exposure Determinants

The differential exposure profile and the large variation in population exposures illustrated above can be interpreted through the lens of exposure determinants. These determinants account for differences in exposure sources, personal activity patterns, microenvironmental characteristics, and factors affecting ADME. These factors are briefly summarized below, with a focus on those factors related to consumer product use.

The impact of **consumer product use** on human PFAS exposures has been reported in a few studies (Boronow et al. 2019, Colles et al. 2020, de la Torre et al., 2019, Harris et al. 2017, Kingsley et al. 2018, Kotthoff et al. 2015, Park et al. 2019, Poothong et al. 2019, Ramli et al. 2020, Sunderland et al. 2019, Tian et al. 2018, Winkens et al. 2018). Kotthoff et al. (2015) summarized the PFAS levels in various consumer products, including textiles, carpets, cleaning and impregnating agents, leather samples, baking and sandwich papers, paper baking forms, and ski waxes, based on a random sampling of 115 consumer products in Germany. Many of these products are used indoors, where people can be exposed to the PFASs they contain through inhalation of indoor air and ingestion of indoor dust. Poothong et al. (2019) detected the same PFCAs, PFSAs, and precursors in indoor dust as in handwipe samples taken from building occupants, substantiating that dust ingestion is a significant PFAS exposure pathway. In a study of pregnant Canadian women, levels of PFAA precursors in indoor air and dust correlated with participants' PFOA, PFNA, and PFOS serum levels (Makey et al. 2017).

Harris et al. (2017) reported that the presence of carpeting or a rug in the bedrooms of 6–10-year-old U.S. children is a significant predictor for PFAS levels in their plasma. Kingsley et al. (2018) observed a 16% and 33% increase in serum PFOA and PFHxS, respectively, in people using stain repellents, compared to those who have never used these products. A similar association was observed by Boronow et al. (2019).

The associations between consumer product use and higher blood PFAS concentrations have also been reported for personal care products. Colles et al. (2020) reported that people using more than seven personal care products had 32% higher serum PFNA levels compared with people using fewer than seven personal care products. Boronow et al. (2019) observed higher PFHxS serum levels in Glide dental floss users. Ramli et al. (2020) similarly reported a 39.4% increase in serum PFDA for dental floss users.

Food packaging materials and cookware also contribute to human PFAS exposures through migration into food. The Nordic Council of Ministers (2017) summarized the levels of PFCA, PFSA, FTOH, PTFE, and PAPs in food packaging materials. Park et al. (2019) reported a positive association between consumption of salty snacks including popcorn, potato chips, and corn chips, and serum levels of PFOA, PFOS, and 2-(N-ethyl-perfluorooctane sulfonamido) acetic acid in women in the U.S. In an analysis of NHANES 2003–2014 data, Susmann et al. (2019) found that consumption of microwave popcorn is associated with significantly higher serum levels of PFOA, PFNA, PFDA, and PFOS, likely as a result of PFAS-treated microwave popcorn packaging. Similarly, Kingsley et al. (2018) reported 22% lower serum PFOA levels in people who had not consumed microwave popcorn in the past year. Boronow et al. (2019) observed that nonstick cookware use was positively (but not statistically significantly) associated with an increase in serum PFOA and Me-PFOSA-AcOH (2-(N-methyl-perfluorooctane sulfonamido) acetic acid). Ramli et al. (2020) found a positive and significant association between nonstick cookware use and serum PFNA level: people who used nonstick cookware had higher serum PFNA (89.6%) compared with those who did not.

Estimating the contribution of consumer products to human PFAS exposures requires the comprehensive speciation of PFASs in biomonitoring and in environmental monitoring studies. However, shorter-chain PFAAs and many of the precursors are not commonly measured in either serum or exposure media, leading to an underestimating of consumer products' contribution to human PFAS exposure.

Other exposure determinants include:

- **Age:** Studies have consistently reported a positive association between age and serum PFAS levels, presumably due to the persistence and bioaccumulation of

PFASs. (Colles et al. 2020, Hu et al. 2018, Jian et al. 2017, Nøst et al. 2014, Poothong et al. 2020, Sunderland et al. 2019).

- **Maternal PFAS burden:** Infants' blood PFAS levels are highly affected by their mothers' PFAS levels in breast milk (Kingsley et al. 2018, VanNoy et al. 2018).
- **Gender**: Women have consistently been reported to have lower levels of serum PFASs than men due to lactation and menstrual bleeding elimination (Colles et al. 2020, Glynn et al. 2020, Jian et al. 2018, Jusko et al. 2016, Kingsley et al. 2018, Park et al. 2019, Ramli et al. 2020, Sagiv et al. 2015).
- **Geographical location:** People living closer to PFAS contaminated sites or PFAS manufacturing facilities have higher levels of serum PFASs (Bao et al. 2017, Barton et al. 2020, Byrne et al. 2017, Denys et al. 2014, Egeghy and Lorber 2011, Fromme et al. 2009, Glynn et al. 2020, Jian et al. 2018, Manzano-Salgado et al. 2016, Olsen et al. 2017, Sunderland et al. 2019).
- **PFAS levels in exposure media:** Positive associations have been reported between PFAS levels in food and drinking water, as well as the frequency of consumption, and serum PFAS levels (Christensen et al. 2017, Colles et al. 2020, Egeghy and Lorber 2011, Glynn et al. 2020, Jusko et al. 2016, Papadopoulou et al. 2017, Ramli et al. 2020, Shu et al. 2018, Sunderland et al. 2019).

3.6 REGULATORY APPROACHES

Several approaches have been used to regulate chemical and environmental pollutants. These include, in order of increasing scope and breadth, single chemical, mixture, arrowhead, and chemical class approaches (Figure 3.8).

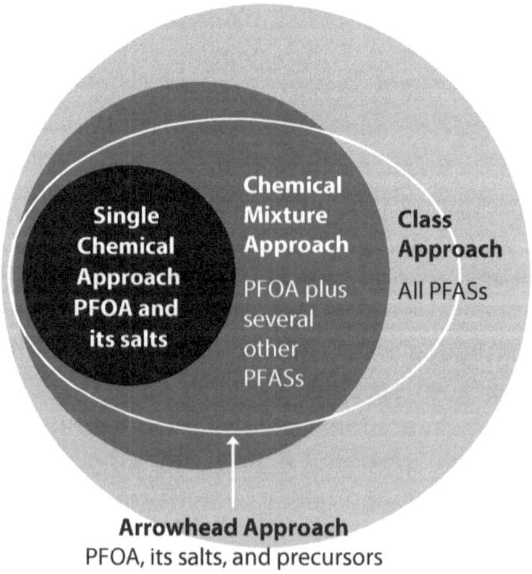

FIGURE 3.8 The four main types of regulatory approaches, with examples of application to PFASs.

3.6.1 Single Chemical Approach

The single chemical approach is the most traditional one in environmental regulations, including, for instance, the U.S. Clean Air Act and Clean Water Act. It is commonly used for regulating PFASs in drinking water sources. A recent example is the 2009 listing of PFOS, its salts, and perfluorooctane sulfonyl fluoride in Annex B to the Stockholm Convention (UNEP, 2020). In another example, several individual PFASs (PFBS, PFHxS, PFNA, PFDA, and a GenX precursor) and their salts were added to the REACH Candidate List of Substances of Very High Concern (SVHCs) for authorization (ECHA 2020a). The U.S. EPA also takes a single chemical approach under Section 5 of TSCA which authorizes the agency to issue Significant New Use Rules (SNURs) for new chemicals or existing chemicals used in a significant new way. A SNUR requires companies to notify U.S. EPA at least 90 days prior to manufacturing, importing, or processing substances for a significant new use, and to submit a notification that includes information about the chemical's identity, physical characteristics, processing and use, and available toxicity data. To date, the U.S. EPA has issued three SNURs covering over two hundred individual PFASs (U.S. EPA 2020c).

The single chemical approach relies on the availability of toxicity and exposure information for individual chemicals to estimate the risks to human and environmental health associated with each chemical. Therefore, it cannot address exposures to PFASs that lack toxicity or exposure data. Also, cumulative impacts, meaning impacts from exposures to multiple chemicals with similar hazard traits or toxicological endpoints, are typically not considered.

3.6.2 Mixture Approach

The mixture approach has also been used to develop environmental standards. Unlike the single chemical approach, it considers cumulative impacts to some extent. For instance, the concept of toxic equivalency (TEQ) was developed for dioxins and dioxin-like compounds, as well as for polycyclic aromatic hydrocarbons (PAHs). The TEQ is a weighted quantity based on the relative toxicity of each member of the group of chemicals (dioxins or PAHs) compared to the most toxic group member.

The Dutch National Institute for Public Health and the Environment (RIVM) developed a similar method, called Relative Potency Factor (RPF), which allows for calculating the cumulative impacts of exposures to a PFAS mixture (RIVM et al. 2018). They calculated RPFs for several PFAAs based on semi-chronic liver toxicity in male rats and proposed to use this approach to regulate multiple chemicals in drinking water. This method is currently limited by the lack of toxicological data for most PFASs, as well as by the incomplete understanding of toxicity mechanisms and modes of action.

Although individual PFASs rarely occur in isolation, it is uncommon for regulatory agencies to take a mixture approach and consider the cumulative impacts of exposures to multiple PFASs. One regulatory agency that does take such an approach is the FDA, which regulates chemicals in food packaging primarily through the Food Contact Notification (FCN) process. Under this program, the manufacturer of a food contact substance must submit a safety summary and comprehensive toxicological profile to FDA for review, including chemical, toxicological, and exposure information and the identity of any degradation products expected to migrate from packaging into food (FDA 2020b). When evaluating the FCNs, FDA considers cumulative impacts and aggregate exposures, but only from other dietary exposure routes that it regulates. It does not consider all cumulative impacts, aggregate exposures, and end-of-life impacts, and therefore does not address all potential exposures and adverse impacts associated with the use of PFASs in food packaging.

While more comprehensive than the single chemical approach, the mixture approach also requires detailed toxicity and exposure data for each chemical in the mixture. This means that chemicals can continue to be used in consumer products until sufficient evidence of their toxicity mounts. For instance, up until a decade ago, FDA permitted the use of long-chain PFASs in food contact materials. Due to concern about longer-chain (C8) compounds, FDA worked with manufacturers to voluntarily stop distributing C8-containing products as of October 1, 2011, though products existing in the marketplace could still be used until they ran out (FDA 2018). The FDA subsequently approved FCNs for 14 side-chain fluorinated polymers and three PFPEs for use in plant fiber-based food packaging applications, despite the fact that some of them reported PFAA impurities, including PFOA (FDA 2020a). However, a series of peer-reviewed studies by FDA scientists raised major concerns about the biopersistence and potential toxicity of a common impurity found in these side-chain fluorinated polymers, 6:2 FTOH and its metabolite 5:3 acid (Kabadi et al. 2018, 2020; Rice et al. 2020). On July 31, 2020, FDA announced that it had reached an agreement with three manufacturers to voluntarily discontinue their 6:2 FTOH-containing formulations over a three-year period starting January 2021, while a fourth manufacturer said it is no longer producing food packaging treatments (Hahn 2020). This phaseout impacts 15 FCNs for 11 distinct PFAS formulations (FDA 2020c). Two additional side-chain fluorinated polymers produced by a manufacturer that did not agree to participate in the phaseout will remain in use.

3.6.3 Arrowhead or Subclass Approach

Similar to the mixture approach, the arrowhead approach regulates multiple chemicals. This approach sets standards for an arrowhead chemical that is well characterized and can be considered representative for a group of related chemicals. The key assumption underlying this approach is that by setting standards to regulate the levels of the arrowhead chemical the levels of the other chemicals in the group will be also controlled. This approach has its roots in traditional environmental regulations. For example, under the Clean Air Act, the U.S. EPA sets standards for six criteria pollutants: particulate matter, ozone, nitrogen dioxide, sulfur dioxide, carbon monoxide, and lead. Ozone, nitrogen dioxide, and sulfur dioxide can all be considered arrowhead chemicals. Although the National Ambient Air Quality Standards are set only for those chemicals, these standards are also intended to control related photochemical oxidants and other oxides of nitrogen and sulfur.

The arrowhead approach has been commonly applied to regulating PFASs in consumer products, targeting a specific PFAS (typically a PFAA) and all its salts and precursors (Cousins et al. 2020). One example is the listing of PFOA, its salts, and PFOA-related compounds in Annex A to the Stockholm Convention (COP 2019). PFHxS, its salts, and PFHxS-related compounds are also being considered for listing in Annexes A, B, and/or C to the Convention (POPRC 2020). ECHA also uses the arrowhead approach in its restriction of PFOA, its salts, and PFOA-related compounds under REACH, and has proposed similar restrictions for several other PFAAs, along with their salts and precursors (ECHA 2020b).

While the arrowhead approach is broader than the single chemical or mixture approaches, covering potentially hundreds, or even thousands of compounds, it only considers the hazards and risks posed by the arrowhead substance. In the case of PFASs, this means that the basis for the regulation is the toxicity of the PFAA arrowhead chemical. The related risk management approaches fail to consider the hazard traits of the PFAA salts and precursor compounds, some of which can be more toxic (Rice et al. 2020).

3.6.4 Chemical Class Approach

The chemical class approach considers the aggregate exposures and cumulative impacts of a whole class of chemicals throughout their life cycles, including those of known and unknown environmental transformation products. For the class of PFASs, which comprises thousands of chemicals, it is neither feasible nor efficient to study the toxicity and exposure of each member of the class. Due to their characteristic carbon-fluorine bond, all PFASs or their terminal degradation products are extremely persistent in the environment. Persistence has been proposed as a sufficient basis for chemical regulation because, if adverse impacts are identified, contamination cannot be reversed at scale within a reasonable timeframe (Cousins et al. 2019a).

The California Department of Toxic Substances Control (DTSC) is among the growing number of regulatory agencies approaching PFASs as a class (Bălan et al. 2021). DTSC's Safer Consumer Products program identifies product-chemical combinations of concern and regulates them as Priority Products. Entities that wish to sell Priority Products in California must submit a comprehensive Alternatives Analysis to DTSC that investigates whether the chemical of concern is necessary in the product and whether safer alternatives are available. Based on the results of the Alternatives Analysis, DTSC issues a Regulatory Response that can include sale restrictions or prohibitions, product labeling, engineering or administrative controls to reduce exposure, product end-of-life management, and investment in green chemistry and engineering to develop safer alternatives. DTSC is proposing to regulate as Priority Products several consumer products that contain any member of the class of PFASs: carpets and rugs; treatments for converted textiles and leathers; and plant fiber-based food packaging products (DTSC 2019, 2020a, 2020b). DTSC is taking a class approach because all PFASs or their degradation, reaction, or metabolism products display at least one hazard trait according to the California Code of Regulations, and because these consumer products contribute to widespread human and ecological exposures to mixtures of PFASs, with potential for significant and widespread adverse impacts. DTSC's decision to take a class approach to regulating PFASs has been criticized by some industry stakeholders because not all PFASs are used, nor could they be used, in these products. These stakeholders have argued that DTSC should focus instead on regulating the specific PFASs that are used in these products. However, for most consumer products, such as carpets and rugs and the treatments for converted textiles or leathers, the chemical identity of the PFASs used is trade secret and has not yet been provided to DTSC. While not all PFASs are used in these products, any PFASs used in these products fall under the scope of DTSC's proposed regulations. The chemical class approach is most effective in this case and can help prevent the regrettable substitution of one PFAS with another that is poorly characterized but may eventually be shown to also possess significant hazard traits.

Washington state has also begun taking a class approach to regulating PFASs. On July 28, 2019, the state enacted Substitute Senate Bill 5135, which authorizes its departments of Ecology and Health to regulate classes of chemicals in consumer products (Washington State Legislature 2019). The law identifies PFASs as one of five initial priority chemical classes. Previously, in March 2018 two other PFAS-related bills were signed into law in Washington state: Engrossed Substitute Senate Bill 6413 and Engrossed Substitute House Bill 2658. The former prohibits the use of PFAS-containing firefighting foam for training purposes as of July 1, 2018, and the sale of firefighting foams with intentionally-added PFASs, with some exemptions, as of July 1, 2020 (Engrossed Substitute Senate Bill 6413, 2018 Regular Session, 65th Legislature 2018). Engrossed Substitute House Bill 2658 prohibits the sale of food packaging containing PFASs as of January 1, 2022, provided the state Department of Ecology identifies safer alternatives (House Bill 2658, 2018 Regular Session, 65th Legislature 2018).

Several other U.S. states have proposed legislation limiting the use of all PFASs in specific consumer products. Vermont bill S 295 would restrict the use of PFASs in food packaging, carpets and rugs, and class B firefighting foam (Vermont Legislature 2020). North Carolina house bill 1109 would ban the manufacture, use, and distribution of all PFASs and PFAS-containing products in the state (General Assembly of North Carolina 2020).

A few EU countries are similarly taking a chemical class approach to regulating PFASs. As of July 2020, Denmark banned the intentional use of all PFASs in cardboard and paper used as food contact materials (Danish Ministry of Environment and Food 2019, Keller and Heckman 2020). On May 11, 2020, the national authorities of Germany, the Netherlands, Norway, Sweden, and Denmark initiated a call for evidence to inform a joint REACH restriction proposal to limit the human health and environmental risks associated with the manufacture and use of PFASs. The proposal is expected to be submitted to ECHA in 2022, with restrictions possibly becoming effective as of 2025 (ECHA 2020c).

The scientific basis for managing PFASs as a chemical class was described in detail by Kwiatkowski et al. (2020). Furthermore, the Global PFAS Science Panel developed an approach to apply the concept of essential use to the class of PFASs (Cousins et al. 2019b). This concept can help determine which PFAS applications are truly essential to society and which can be regulated using the class approach.

3.7 CONCLUSIONS

PFASs have made many consumer products more convenient and more profitable but at the expense of widespread human and ecological exposures to these chemicals, with potential adverse impacts for generations to come. Although monitoring data are available for only a small subset of PFASs, it is clear that virtually every living creature on Earth is exposed to multiple PFASs from a variety of exposure media, including indoor and outdoor air, indoor dust, food, and drinking water. The manufacturing, use, and end-of-life management of consumer products contribute to the presence of PFASs in these media. All PFASs are extremely persistent or degrade to other PFASs that are extremely persistent. Therefore, the levels of these substances in the environment will continue to rise as long as they continue to be used in consumer products and other applications.

Given the large number of chemicals in this class, much more research is needed to fully understand the environmental fates, exposure potentials, toxicological profiles, and cumulative impacts of the PFASs associated with consumer products—especially those used in emerging technologies. However, gathering such data for every member of this class of thousands of compounds is cost- and time-prohibitive. Because of this, traditional regulatory approaches such as the single chemical, mixture-based, and arrowhead approaches are ineffective for regulating PFASs. Given the lack of toxicity data for most members of the PFAS class, regulators are left with two choices: assume that class members lacking such data are safe or take a precautionary approach that avoids unnecessary uses of any members of the class. Based on the known toxic effects of a few members of this class and the extreme persistence of all PFASs or their degradation products, the second option is the most responsible one for regulators charged with protecting public health and the environment. Regulating PFASs as a class is the most effective way to address the nonessential uses of PFASs in consumer products and drive the adoption of safer alternatives.

ACKNOWLEDGEMENTS

The authors wish to acknowledge David Grealish for the artwork; André Algazi, Topher Buck, and Jeff Wong for their helpful feedback on the manuscript; and the entire Safer Consumer Products team for their research and support.

REFERENCES

3M. (2018). *3M Safety Data Sheet: ScotchgardTM Fabric Protector (Cat. No. 4101, 4106).* Available at: https://multimedia.3m.com/mws/mediawebserver?mwsId=SSSSSuUn_zu8l00xMx_14x2GOv70k17zHvu9lxtD7SSSSSS (Accessed August 29 2020).

ACC. (2021). *Fluorotechnology makes important products for vital industries possible.* American Chemistry Council (ACC). Available at: https://www.americanchemistry.com/Fluorotechnology-Makes-Important-Products-for-Vital-Industries-Possible.html (Accessed April 9 2021).

Alliance for Automotive Innovation. (2020, April 28). *Comment submitted by Julia M. Rege, Vice President, Energy & Environment, Alliance for Automotive Innovation.* Available at: https://www.regulations.gov/document?D=EPA-HQ-OPPT-2013-0225-0220 (Accessed August 29 2020).

Ao, J., Yuan, T., Xia, H., Ma, Y., Shen, Z., Shi, R., Tian, Y., Zhang, J., Ding, W., Gao, L., Zhao, X., & Yu, X. (2019). Characteristic and human exposure risk assessment of per- and polyfluoroalkyl substances: A study based on indoor dust and drinking water in China. *Environmental Pollution*, 254, 112873. doi: 10.1016/j.envpol.2019.07.041.

Audenaert, F., Lens, H., Rolly, D., & Vander Elst, P. (1999). Fluorochemical textile repellents—synthesis and applications: A 3 M perspective. *Journal of the Textile Institute*, *90*(3), 76–94. doi: 10.1080/00405009908659480

Bălan, S. A., Mathrani, V. C., Guo, D. F., & Algazi, A. M. (2021). Regulating PFAS as a chemical class under the California Safer Consumer Products program. *Environmental Health Perspectives, 129(2)*. doi: 10.1289/EHP7431.

Bao, J., Liu, L., Wang, X., Jin, Y.-H., & Dong, G.-H. (2017). Human exposure to perfluoroalkyl substances near a fluorochemical industrial park in China. *Environmental Science and Pollution Research*, 24(10), 9194–9201. doi: 10.1007/s11356-017-8620-6.

Barton, K. E., Starling, A. P., Higgins, C. P., McDonough, C. A., Calafat, A. M., & Adgate, J. L. (2020). Sociodemographic and behavioral determinants of serum concentrations of per- and polyfluoroalkyl substances in a community highly exposed to aqueous film-forming foam contaminants in drinking water. *International Journal of Hygiene and Environmental Health*, 223(1), 256–266. doi: 10.1016/j.ijheh.2019.07.012

Bečanová, J., Melymuk, L., Vojta, Š., Komprdová, K., & Klánová, J. (2016). Screening for perfluoroalkyl acids in consumer products, building materials and wastes. *Chemosphere*, 164, 322–329. doi: 10.1016/j.chemosphere.2016.08.112

Berger, U., Glynn, A., Holmström, K. E., Berglund, M., Ankarberg, E. H., & Törnkvist, A. (2009). Fish consumption as a source of human exposure to perfluorinated alkyl substances in Sweden – Analysis of edible fish from Lake Vättern and the Baltic Sea. *Chemosphere*, 76(6), 799–804. doi: 10.1016/j.chemosphere.2009.04.044

Bhavsar, S. P., Zhang, X., Guo, R., Braekevelt, E., Petro, S., Gandhi, N., Reiner, E. J., Lee, H., Bronson, R., & Tittlemier, S. A. (2014). Cooking fish is not effective in reducing exposure to perfluoroalkyl and polyfluoroalkyl substances. *Environment International*, 66, 107–114. doi: 10.1016/j.envint.2014.01.024

Birnbaum, L. (2018). *Testimony before the Senate Committee on Homeland Security and Governmental Affairs Subcommittee on Federal Spending Oversight and Emergency Management*. Available at: https://www.hsgac.senate.gov/imo/media/doc/Birnbaum%20 Testimony.pdf (Accessed August 29 2020).

Blaine, A. C., Rich, C. D., Hundal, L. S., Lau, C., Mills, M. A., Harris, K. M., & Higgins, C. P. (2013). Uptake of perfluoroalkyl acids into edible crops via land applied biosolids: Field and greenhouse studies. *Environmental Science & Technology*, 47(24), 14062–14069. doi: 10.1021/es403094q

Borg, D., & Ivarsson, J. (2017). Analysis of PFASs and TOF in products (TemaNord 2017:543). *Nordic Council of Ministers*. Available at: https://norden.diva-portal.org/smash/get/diva2:1118439/FULLTEXT01.pdf (Accessed August 29 2020).

Boronow, K. E., Brody, J. G., Schaider, L. A., Peaslee, G. F., Havas, L., & Cohn, B. A. (2019). Serum concentrations of PFASs and exposure-related behaviors in African American and non-Hispanic white women. *Journal of Exposure Science & Environmental Epidemiology*, 29(2), 206–217. doi: 10.1038/s41370-018-0109-y

Bowman, J. S. (2015). Fluorotechnology is critical to modern life: The FluoroCouncil counterpoint to the Madrid Statement. *Environmental Health Perspectives*, 123(5). doi: 10.1289/ehp.1509910

Buck, R. C., Franklin, J., Berger, U., Conder, J. M., Cousins, I. T., de Voogt, P., Jensen, A. A., Kannan, K., Mabury, S. A., & van Leeuwen, S. P. (2011). Perfluoroalkyl and polyfluoroalkyl substances in the environment: Terminology, classification, and origins. *Integrated Environmental Assessment and Management*, 7(4), 513–541. doi: 10.1002/ieam.258

Buck, R. C., Murphy, P. M., & Pabon, M. (2012). Chemistry, properties, and uses of commercial fluorinated surfactants. In T. P. Knepper & F. T. Lange (Eds.), *Handbook of Environmental Chemistry, Volume 17, Polyfluorinated Chemicals and Transformation Products* (Vol. 17, pp. 1–24). Springer. doi: 10.1007/978-3-642-21,872-9_1

Byrne, S., Seguinot-Medina, S., Miller, P., Waghiyi, V., von Hippel, F. A., Buck, C. L., & Carpenter, D. O. (2017). Exposure to polybrominated diphenyl ethers and perfluoroalkyl substances in a remote population of Alaska Natives. *Environmental Pollution*, 231, 387–395. doi: 10.1016/j.envpol.2017.08.020

CEH. (2018). Avoiding hidden hazards: A purchaser's guide to safer foodware. *Center for Environmental Health (CEH)*. Available at: https://www.ceh.org/wp-content/uploads/CEH-Disposable-Foodware-Report-final-1.31.pdf (Accessed August 29 2020).

CELA. (2019). Scoping per- and polyfluoroalkly substances releases from the recycling of paper and textiles and their implications for the Great Lakes-St Lawrence River ecosystem. *Canadian Environmental Law Association (CELA)*. Available at: https://cela.ca/wp-content/uploads/2019/10/Report-PFAS-Sept-2019.pdf (Accessed August 29 2020).

Chemours. (2020a). *Fitness Bands with Viton™ Fluoroelastomers*. Available at: https://www.chemours.com:443/en/chemistry-in-action/chemistry-delight/fitness-bands (Accessed August 29 2020).

Chemours. (2020b). *Teflon™ and Tefzel™ Fluoropolymer Films*. Available at: https://www.teflon.com:443/en/products/films (Accessed August 29 2020).

Chemours. (2020c). *Teflon™ Industrial Coatings*. Available at: https://www.teflon.com:443/en/products/coatings (Accessed August 29 2020).

Chen, W.-L., Bai, F.-Y., Chang, Y.-C., Chen, P.-C., & Chen, C.-Y. (2018). Concentrations of perfluoroalkyl substances in foods and the dietary exposure among Taiwan general population and pregnant women. *Journal of Food and Drug Analysis*, 26(3), 994–1004. doi: 10.1016/j.jfda.2017.12.011

Christensen, K. Y., Raymond, M., Blackowicz, M., Liu, Y., Thompson, B. A., Anderson, H. A., & Turyk, M. (2017). Perfluoroalkyl substances and fish consumption. *Environmental Research*, 154, 145–151. doi: 10.1016/j.envres.2016.12.032

Colles, A., Bruckers, L., Den Hond, E., Govarts, E., Morrens, B., Schettgen, T., Buekers, J., Coertjens, D., Nawrot, T., Loots, I., Nelen, V., De Henauw, S., Schoeters, G., Baeyens, W., & van Larebeke, N. (2020). Perfluorinated substances in the Flemish population (Belgium): Levels and determinants of variability in exposure. *Chemosphere*, 242, 125250. doi: 10.1016/j.chemosphere.2019.125250

COP. (2019, May). *Conference of the Parties (COP): Ninth Meeting of the Conference of the Parties to the Stockholm Convention*. Available at: http://chm.pops.int/TheConvention/ConferenceoftheParties/Meetings/COP9/tabid/7521/Default.aspx (Accessed August 29 2020).

Cousins, I. T., Ng, C. A., Wang, Z., & Scheringer, M. (2019a). Why is high persistence alone a major cause of concern? *Environmental Science: Processes & Impacts*, 21(5), 781–792. doi: 10.1039/C8EM00515J

Cousins, I. T., Goldenman, G., Herzke, D., Lohmann, R., Miller, M., Patton, S., Scheringer, M., Trier, X., Vierke, L., Wang, Z., & DeWitt, C. (2019b). The concept of essential use for determining when uses of PFASs can be phased out. *Environmental Science: Processes & Impacts*, 21(11), 1803–1815.

Cousins, I. T., DeWitt, J. C., Glüge, J., Goldenman, G., Herzke, D., Lohmann, R., Miller, M., Ng, C. A., Scheringer, M., Vierke, L., & Wang, Z. (2020). Strategies for grouping per- and polyfluoroalkyl substances (PFAS) to protect human and environmental health. *Environmental Science: Processes & Impacts*, 22(7), 1444–1460. doi: 10.1039/D0EM00147C

Daikin Global. (2020a). *Fluoroelastomers*. Available at: https://www.daikinchemicals.com/solutions/products/dai-el-fluoroelastomers.html (Accessed August 29 2020).

Daikin Global. (2020b). *Fluoropolymer Films*. Available at: https://www.daikinchemicals.com/solutions/products/neoflon-fluoropolymers-films.html (Accessed August 29 2020).
Daikin Global. (2020c). *Fluoropolymers*. Available at: https://www.daikinchemicals.com/solutions/products/fluoropolymers.html (Accessed August 29 2020).
Danish Ministry of Environment and Food. (2019, September 2). *The Minister of Food is ready to ban fluoride substances*. Available at: https://mfvm.dk/nyheder/nyhed/nyhed/foedevareministeren-er-klar-til-at-forbyde-fluorstoffer/ (Accessed August 29 2020).
Dauchy, X. (2019). Per- and polyfluoroalkyl substances (PFASs) in drinking water: Current state of the science. *Current Opinion in Environmental Science & Health*, 7, 8–12. doi: 10.1016/j.coesh.2018.07.004
de la Torre, A., Navarro, I., Sanz, P., & Mártinez, M.. (2019). Occurrence and human exposure assessment of perfluorinated substances in house dust from three European countries. *Science of The Total Environment*, 685, 308–314. doi: 10.1016/j.scitotenv.2019.05.463
Denys, S., Fraize-Frontier, S., Moussa, O., Bizec, B. L., Veyrand, B., & Volatier, J.-L. (2014). Is the fresh water fish consumption a significant determinant of the internal exposure to perfluoroalkylated substances (PFAS)? *Toxicology Letters*, 231(2), 233–238. doi: 10.1016/j.toxlet.2014.07.028
Domingo, J. L. (2012). Health risks of dietary exposure to perfluorinated compounds. *Environment International*, 40, 187–195. doi: 10.1016/j.envint.2011.08.001
DTSC. (2019, November 8). *Department of Toxic Substances Control (DTSC): Treatments Containing Perfluoroalkyl or Polyfluoroalkyl Substances (PFASs) for Use on Converted Textiles or Leathers*. Available at: https://calsafer.dtsc.ca.gov/cms/commentpackage/?rid=12,746&from=search (Accessed August 29 2020).
DTSC. (2020a, February 28). *Department of Toxic Substances Control (DTSC): Regulatory Proposal: 45-Day Comment Period for PFASs in Carpets and Rugs*. Available at: https://calsafer.dtsc.ca.gov/cms/commentpackage/?rid=12,751 (Accessed August 29 2020).
DTSC. (2020b, July 30). *Department of Toxic Substances Control (DTSC): Food Packaging Containing Perfluoroalkyl of Polyfuoroalkyl Substances (PFASs)*. Available at: https://calsafer.dtsc.ca.gov/cms/commentpackage/?rid=12,752 (Accessed August 29 2020).
Dusaj, S. (1988). *Effect of aftermarket carpet protector treatments on the soiling propensity of nylon 6 carpeting* [Master's Thesis, Kansas State University]. Available at: https://krex.k-state.edu/dspace/handle/2097/22426 (Accessed August 29 2020).
ECHA. (2020a). *European Chemicals Agency (ECHA): Candidate list of substances of very high concern for authorisation*. Available at: https://echa.europa.eu/web/guest/candidate-list-table (Accessed August 29 2020).
ECHA. (2020b). *European Chemicals Agency (ECHA): Public activities coordination tool*. Available at: https://echa.europa.eu/pact (Accessed August 29 2020).
ECHA. (2020c, May 11). *European Chemicals Agency (ECHA): Five European states call for evidence on broad PFAS restriction*. Available at: https://echa.europa.eu/–/five-european-states-call-for-evidence-on-broad-pfas-restriction (Accessed August 29 2020).
Egeghy, P. P., & Lorber, M. (2011). An assessment of the exposure of Americans to perfluorooctane sulfonate: A comparison of estimated intake with values inferred from NHANES data. *Journal of Exposure Science & Environmental Epidemiology*, 21(2), 150–168. doi: 10.1038/jes.2009.73
Engrossed Substitute Senate Bill 6413, 2018 Regular Session, 65th Legislature, (2018). Available at: http://lawfilesext.leg.wa.gov/biennium/2017–18/Pdf/Bills/Session%20Laws/Senate/6413-S.SL.pdf?q = 20,200,708,201,908 (Accessed August 29 2020).
Eriksson, U., & Kärrman, A. (2015). World-Wide Indoor Exposure to Polyfluoroalkyl Phosphate Esters (PAPs) and other PFASs in Household Dust. *Environmental Science & Technology*, 49(24), 14503–14511. doi: 10.1021/acs.est.5b00679
Fair, P. A., Wolf, B., White, N. D., Arnott, S. A., Kannan, K., Karthikraj, R., & Vena, J. E. (2019). Perfluoroalkyl substances (PFASs) in edible fish species from Charleston Harbor

and tributaries, South Carolina, United States: Exposure and risk assessment. *Environmental Research*, 171, 266–277. doi: 10.1016/j.envres.2019.01.021

FDA. (2020a). *Food and Drug Administration (FDA): Inventory of Effective Food Contact Substance (FCS) Notifications*. Available at: https://www.accessdata.fda.gov/scripts/fdcc/?set=FCN (Accessed August 29 2020).

FDA. (2020b, February 20). Food and Drug Administration (FDA): Packaging & Food Contact Substances (FCS). *FDA*. Available at: http://www.fda.gov/food/food-ingredients-packaging/packaging-food-contact-substances-fcs (Accessed August 29 2020).

FDA. (2020c, July 31). Food and Drug Administration (FDA): Per and Polyfluoroalkyl Substances (PFAS). *FDA*. Available at: https://www.fda.gov/food/chemicals/and-polyfluoroalkyl-substances-pfas (Accessed August 29 2020).

FDA (Food & Drug Administration). (2018, February 9). *Update on Perfluorinated Grease-proofing Agents* [WebContent]. Available at: https://www.fda.gov/Food/IngredientsPackagingLabeling/PackagingFCS/Notifications/ucm308462.htm (Accessed August 29 2020).

Feng, M., Qu, R., Wei, Z., Wang, L., Sun, P., & Wang, Z. (2015). Characterization of the thermolysis products of nafion membrane: a potential source of perfluorinated compounds in the environment. *Scientific Reports*, 5(1). doi: 10.1038/srep09859

Fiedler, S., Pfister, G., & Schramm, K.-W. (2010). Poly- and perfluorinated compounds in household consumer products. *Toxicological & Environmental Chemistry*, 92(10), 1801–1811. doi: 10.1080/02772248.2010.491482

Fromme, H., Tittlemier, S. A., Völkel, W., Wilhelm, M., & Twardella, D. (2009). Perfluorinated compounds – Exposure assessment for the general population in western countries. *International Journal of Hygiene and Environmental Health*, 212(3), 239–270. doi: 10.1016/j.ijheh.2008.04.007

General Assembly of North Carolina. (2020). *House Bill 1109*. Available at: https://www.ncleg.gov/Sessions/2019/Bills/House/PDF/H1109v0.pdf (Accessed August 29 2020).

Ghisi, R., Vamerali, T., & Manzetti, S. (2019). Accumulation of perfluorinated alkyl substances (PFAS) in agricultural plants: A review. *Environmental Research*, 169, 326–341. doi: 10.1016/j.envres.2018.10.023

Glynn, A., Kotova, N., Dahlgren, E., Lindh, C., Jakobsson, K., Gyllenhammar, I., Lignell, S., & Nälsén, C. (2020). Determinants of serum concentrations of perfluoroalkyl acids (PFAAs) in school children and the contribution of low-level PFAA-contaminated drinking water. *Environmental Science: Processes & Impacts*, 22(4), 930–944. doi: 10.1039/C9EM00497A

Goosey, E., & Harrad, S. (2012). Perfluoroalkyl substances in UK indoor and outdoor air: Spatial and seasonal variation, and implications for human exposure. *Environment International*, 45, 86–90. doi: 10.1016/j.envint.2012.04.007

Gredelj, A., Nicoletto, C., Valsecchi, S., Ferrario, C., Polesello, S., Lava, R., Zanon, F., Barausse, A., Palmeri, L., Guidolin, L., & Bonato, M. (2020). Uptake and translocation of perfluoroalkyl acids (PFAA) in red chicory (*Cichorium intybus* L.) under various treatments with pre-contaminated soil and irrigation water. *Science of The Total Environment*, 708, 134,766. doi: 10.1016/j.scitotenv.2019.134766

Hahn, S. M. (2020, July 31). FDA Announces Voluntary Agreement with Manufacturers to Phase Out Certain Short-Chain PFAS Used in Food Packaging. *FDA*. Available at: https://www.fda.gov/news-events/press-announcements/fda-announces-voluntary-agreement-manufacturers-phase-out-certain-short-chain-pfas-used-food (Accessed August 29 2020).

Harris, M. H., Rifas-Shiman, S. L., Calafat, A. M., Ye, X., Mora, A. M., Webster, T. F., Oken, E., & Sagiv, S. K. (2017). Predictors of per- and polyfluoroalkyl substance (PFAS) plasma concentrations in 6–10 year old American children. *Environmental Science & Technology*, 51(9), 5193–5204. doi: 10.1021/acs.est.6b05811

Herzke, D., Huber, S., Bervoets, L., D'Hollander, W., Hajslova, J., Pulkrabova, J., Brambilla, G., de Filippis, S. P., Klenow, S., Heinemeyer, G., & de Voogt, P. (2013). Perfluorinated alkylated substances in vegetables collected in four European countries; occurrence and human exposure estimations. *Environmental Science and Pollution Research, 20*(11), 7930–7939. doi: 10.1007/s11356-013-1777-8

Herzke, D., Olsson, E., & Posner, S. (2012). Perfluoroalkyl and polyfluoroalkyl substances (PFASs) in consumer products in Norway – A pilot study. *Chemosphere*, 88(8), 980–987. doi: 10.1016/j.chemosphere.2012.03.035

Hopkins, Z. R., Sun, M., DeWitt, J. C., & Knappe, D. R. U. (2018). Recently Detected Drinking Water Contaminants: GenX and Other Per- and Polyfluoroalkyl Ether Acids: JOURNAL AWWA. *Journal - American Water Works Association*, 110(7), 13–28. doi: 10.1002/awwa.1073

House Bill 2658, 2018 *Regular Session, 65th Legislature*, (2018). Available at: http://lawfilesext.leg.wa.gov/biennium/2017-18/Pdf/Bills/House%20Bills/2658.pdf (Accessed August 29 2020).

Hu, X. C., Andrews, D. Q., Lindstrom, A. B., Bruton, T. A., Schaider, L. A., Grandjean, P., Lohmann, R., Carignan, C. C., Blum, A., Balan, S. A., Higgins, C. P., & Sunderland, E. M. (2016). Detection of poly- and perfluoroalkyl substances (PFASs) in U.S. drinking water linked to industrial sites, military fire training areas, and wastewater treatment plants. *Environmental Science & Technology Letters*, 3(10), 344–350. doi: 10.1021/acs.estlett.6b00260

Hu, X. C., Dassuncao, C., Zhang, X., Grandjean, P., Weihe, P., Webster, G. M., Nielsen, F., & Sunderland, E. M. (2018). Can profiles of poly- and perfluoroalkyl substances (PFASs) in human serum provide information on major exposure sources? *Environmental Health: A Global Access Science Source*, 17(1), 11. doi: 10.1186/s12940-018-0355-4

Hung, M. D., Lam, N. H., Jeong, H. H., Jeong, H. J., Jeong, D. J., Park, G. H., Kim, P. J., Oh, J. E., & Cho, H. S. (2018). Perfluoroalkyl Substances (PFASs) in Ten Edible Freshwater Fish Species from Major Rivers and Lakes in Korea: Distribution and Human Exposure by Consumption. *Toxicology and Environmental Health Sciences*, 10(5), 307–320. doi: 10.1007/s13530-018-0379-8

ITRC. (2020). *Interstate Technology and Regulatory Council (ITRC): Naming Conventions and Physical and Chemical Properties of Per- and Polyfluoroalkyl Substances (PFAS).* Available at: https://pfas-1.itrcweb.org/fact_sheets_page/PFAS_Fact_Sheet_Naming_Conventions_April2020.pdf (Accessed August 29 2020).

Jian, J.-M., Chen, D., Han, F.-J., Guo, Y., Zeng, L., Lu, X., & Wang, F. (2018). A short review on human exposure to and tissue distribution of per- and polyfluoroalkyl substances (PFASs). *Science of The Total Environment*, 636, 1058–1069. doi: 10.1016/j.scitotenv.2018.04.380

Jian, J.-M., Guo, Y., Zeng, L., Liang-Ying, L., Lu, X., Wang, F., & Zeng, E. Y. (2017). Global distribution of perfluorochemicals (PFCs) in potential human exposure source–A review. *Environment International*, 108, 51–62. doi: 10.1016/j.envint.2017.07.024

Jusko, T. A., Oktapodas, M., Palkovičová Murinová, L., Babinská, K., Babjaková, J., Verner, M.-A., DeWitt, J. C., Thevenet-Morrison, K., Čonka, K., Drobná, B., Chovancová, J., Thurston, S. W., Lawrence, B. P., Dozier, A. M., Järvinen, K. M., Patayová, H., Trnovec, T., Legler, J., Hertz-Picciotto, I., & Lamoree, M. H. (2016). Demographic, Reproductive, and Dietary Determinants of Perfluorooctane Sulfonic (PFOS) and Perfluorooctanoic Acid (PFOA) Concentrations in Human Colostrum. *Environmental Science & Technology*, 50(13), 7152–7162. doi: 10.1021/acs.est.6b00195

Kabadi, S. V., Fisher, J., Aungst, J., & Rice, P. (2018). Internal exposure-based pharmacokinetic evaluation of potential for biopersistence of 6:2 fluorotelomer alcohol (FTOH) and its metabolites. *Food and Chemical Toxicology*, 112, 375–382. doi: 10.1016/j.fct.2018.01.012

Kabadi, S. V., Fisher, J. W., Doerge, D. R., Mehta, D., Aungst, J., & Rice, P. (2020). Characterizing biopersistence potential of the metabolite 5:3 fluorotelomer carboxylic acid after repeated oral exposure to the 6:2 fluorotelomer alcohol. *Toxicology and Applied Pharmacology*, 388, 114878. doi: 10.1016/j.taap.2020.114878

Karásková, P., Venier, M., Melymuk, L., Bečanová, J., Vojta, Š., Prokeš, R., Diamond, M. L., & Klánová, J. (2016). Perfluorinated alkyl substances (PFASs) in household dust in Central Europe and North America. *Environment International*, 94, 315–324. doi: 10.1016/j.envint.2016.05.031

Keller and Heckman LLP. (2020, June 16). Denmark's PFAS Ban in Paper and Cardboard Effective in July 2020. *The National Law Review*. Available at: https://www.natlawreview.com/article/denmark-s-pfas-ban-paper-and-cardboard-effective-july-2020 (Accessed August 29 2020).

KEMI. (2015). Swedish Chemicals Agency (KEMI): Occurrence and use of highly fluorinated substances and alternatives. *Report from a government assignment. Report 7/15. Stockholm, Sweden* (p. 112). Available at: https://www.kemi.se/global/rapporter/2015/report-7-15-occurrence-and-use-of-highly-fluorinated-substances-and-alternatives.pdf (Accessed August 29 2020).

Kingsley, S. L., Eliot, M. N., Kelsey, K. T., Calafat, A. M., Ehrlich, S., Lanphear, B. P., Chen, A., & Braun, J. M. (2018). Variability and predictors of serum perfluoroalkyl substance concentrations during pregnancy and early childhood. *Environmental Research*, 165, 247–257. doi: 10.1016/j.envres.2018.04.033

Kotthoff, M., & Bücking, M. (2018). Four Chemical Trends Will Shape the Next Decade's Directions in Perfluoroalkyl and Polyfluoroalkyl Substances Research. *Frontiers in Chemistry*, 6, 103. doi: 10.3389/fchem.2018.00103

Kotthoff, M., Müller, J., Jürling, H., Schlummer, M., & Fiedler, D. (2015). Perfluoroalkyl and polyfluoroalkyl substances in consumer products. *Environmental Science and Pollution Research*, 22(19), 14546–14559. doi: 10.1007/s11356-015-4202-7

Kwiatkowski, C. F., Andrews, D. Q., Birnbaum, L. S., Bruton, T. A., DeWitt, J. C., Knappe, D. R. U., Maffini, M. V., Miller, M. F., Pelch, K. E., Reade, A., Soehl, A., Trier, X., Venier, M., Wagner, C. C., Wang, Z., & Blum, A. (2020). Scientific basis for managing PFAS as a chemical class. *Environmental Science & Technology Letters*. doi: 10.1021/acs.estlett.0c00255

Lee, L. S., & Trim, H. (2018). Evaluating perfluoroalkyl acids in composts with compostable food serviceware products in their feedstocks. *Summary sheet*. January 2018 (revised March 9, 2018). Available at: https://zerowastewashington.org/wp-content/uploads/2018/11/one-pager_revised-march-9-2018_lslee_trim.pdf (Accessed August 29 2020).

Liu, X., Guo, Z., Folk, E. E., & Roache, N. F. (2015). Determination of fluorotelomer alcohols in selected consumer products and preliminary investigation of their fate in the indoor environment. *Chemosphere*, 129, 81–86. doi: 10.1016/j.chemosphere.2014.06.012

Lorber, M., & Egeghy, P. P. (2011). Simple Intake and Pharmacokinetic Modeling to Characterize Exposure of Americans to Perfluoroctanoic Acid, PFOA. *Environmental Science & Technology*, 45(19), 8006–8014. doi: 10.1021/es103718h

Maitland, L. (1982, July 1). Finishes to help fabric stay clean. *The New York Times*, 4. Available at: https://www.nytimes.com/1982/07/01/garden/finishes-to-help-fabric-stay-clean.html (Accessed August 29 2020).

Makey, C. M., Webster, T. F., Martin, J. W., Shoeib, M., Harner, T., Dix-Cooper, L., & Webster, G. M. (2017). Airborne precursors predict maternal serum perfluoroalkyl acid concentrations. *Environmental Science & Technology*, 51(13), 7667–7675. doi: 10.1021/acs.est.7b00615

Manzano-Salgado, C. B., Casas, M., Lopez-Espinosa, M.-J., Ballester, F., Martinez, D., Ibarluzea, J., Santa-Marina, L., Schettgen, T., Vioque, J., Sunyer, J., & Vrijheid, M. (2016). Variability of perfluoroalkyl substance concentrations in pregnant women by socio-demographic and dietary factors in a Spanish birth cohort. *Environment International*, 92–93, 357–365. doi: 10.1016/j.envint.2016.04.004

Miralles-Marco, A., & Harrad, S. (2015). Perfluorooctane sulfonate: A review of human exposure, biomonitoring and the environmental forensics utility of its chirality and isomer distribution. *Environment International*, 77, 148–159. doi: 10.1016/j.envint.2015.02.002

Nadal, M., & Domingo, J. (2014). Indoor dust levels of perfluoroalkyl substances (PFASs) and the role of ingestion as an exposure pathway: A review. *Current Organic Chemistry*, 18(17), 2200–2208. doi: 10.2174/1385272819666140804230713

Nelter, T. (2018, June 29). *FDA-approved PFAS and drinking water – Q & A on analytical measurements*. Available at: http://blogs.edf.org/health/2018/06/29/pfas-and-drinking-water-analytical-measurements/ (Accessed August 29 2020).

Nordic Council of Ministers. (2017). *PFAS in paper and board for food contact: Options for risk management of poly- and perfluorinated substances*. Available at: https://backend.orbit.dtu.dk/ws/portalfiles/portal/149769110/Rapport_PFAS_in_paper_and_board_for_food_contact_Options_for_risk_management_of_poly_and_perfluorina.pdf (Accessed August 29 2020).

Nordic Council of Ministers. (2020). *Per- and polyfluoroalkylether substances: Identity, production and use*. Available at: http://norden.diva-portal.org/smash/record.jsf?pid=diva2%3A1392167&dswid=adv_scores (Accessed August 29 2020).

Nøst, T. H., Vestergren, R., Berg, V., Nieboer, E., Odland, J. Ø., & Sandanger, T. M. (2014). Repeated measurements of per- and polyfluoroalkyl substances (PFASs) from 1979 to 2007 in males from Northern Norway: Assessing time trends, compound correlations and relations to age/birth cohort. *Environment International*, 67, 43–53. doi: 10.1016/j.envint.2014.02.011

OECD. (2013). Synthesis paper on per- and polyfluorinated chemicals (PFCs). *Organisation for Economic Cooperation and Development (OECD)*. Available at: https://www.oecd.org/env/ehs/risk-management/PFC_FINAL-Web.pdf (Accessed August 29 2020).

OECD. (2018). Toward a new comprehensive global database of per- and polyfluoroalkyl substances (PFASs): Summary report on updating the OECD 2007 list of per- and polyfluoroalkyl substances (PFASs). *Organisation for Economic Cooperation and Development (OECD)*. Available at: https://www.oecd.org/officialdocuments/publicdisplaydocumentpdf/?cote=ENV-JM-MONO(2018)7&doclanguage=en (Accessed August 29 2020).

Olsen, G. W., Mair, D. C., Lange, C. C., Harrington, L. M., Church, T. R., Goldberg, C. L., Herron, R. M., Hanna, H., Nobiletti, J. B., Rios, J. A., Reagen, W. K., & Ley, C. A. (2017). Per- and polyfluoroalkyl substances (PFAS) in American Red Cross adult blood donors, 2000–2015. *Environmental Research*, 157, 87–95. doi: 10.1016/j.envres.2017.05.013

Padilla-Sánchez, J. A., Papadopoulou, E., Poothong, S., & Haug, L. S. (2017). Investigation of the Best Approach for Assessing Human Exposure to Poly- and Perfluoroalkyl Substances through Indoor Air. *Environmental Science & Technology*, 51(21), 12836–12843. doi: 10.1021/acs.est.7b03516

Papadopoulou, E., Poothong, S., Koekkoek, J., Lucattini, L., Padilla-Sánchez, J. A., Haugen, M., Herzke, D., Valdersnes, S., Maage, A., Cousins, I. T., Leonards, P. E. G., & Småstuen Haug, L. (2017). Estimating human exposure to perfluoroalkyl acids via solid food and drinks: Implementation and comparison of different dietary assessment methods. *Environmental Research*, 158, 269–276. doi: 10.1016/j.envres.2017.06.011

Park, S. K., Peng, Q., Ding, N., Mukherjee, B., & Harlow, S. D. (2019). Determinants of per- and polyfluoroalkyl substances (PFAS) in midlife women: Evidence of racial/ethnic and geographic differences in PFAS exposure. *Environmental Research*, 175, 186–199. doi: 10.1016/j.envres.2019.05.028

Poothong, S., Padilla-Sánchez, J. A., Papadopoulou, E., Giovanoulis, G., Thomsen, C., & Haug, L. S. (2019). Hand wipes: A useful tool for assessing human exposure to poly- and perfluoroalkyl substances (PFASs) through hand-to-mouth and dermal contacts. *Environmental Science & Technology*, 53(4), 1985–1993. doi: 10.1021/acs.est.8b05303

Poothong, S., Papadopoulou, E., Padilla-Sánchez, J. A., Thomsen, C., & Haug, L. S. (2020). Multiple pathways of human exposure to poly- and perfluoroalkyl substances (PFASs): From external exposure to human blood. *Environment International*, 134, 105244. doi: 10.1016/j.envint.2019.105244

POPRC. (2020). *Persistent Organic Pollutants Review Committee (POPRC): POPRC recommendations for listing chemicals*. Available at: http://chm.pops.int/tabid/243/Default.aspx (Accessed August 29 2020).

Posner, S., Roos, S., Poulsen, P. B., Jörundsdottir, H. Ó., Gunnlaugsdóttir, H., Trier, X., Jensen, A. A., Katsogiannis, A. A., Herzke, D., Bonefeld-Jörgensen, E. C., Jönsson, C., Pedersen, G. A., Ghisari, M., & Jensen, S. (2013). Per- and polyfluorinated substances in the Nordic Countries: Use, occurence and toxicology. *Nordic Council of Ministers*. Available at: https://orbit.dtu.dk/files/56198991/Per%20and%20polyfluorinated%20substances%20in%20the%20Nordic%20Countries.pdf (Accessed August 29 2020).

Ramli, M. R., Yoneda, M., Ali Mohd, M., Mohamad Haron, D. E., & Ahmad, E. D. (2020). Level and determinants of serum perfluoroalkyl acids (PFAAs) in a population in Klang Valley, Malaysia. *International Journal of Hygiene and Environmental Health*, 223(1), 179–186. doi: 10.1016/j.ijheh.2019.09.005

Renner, R. (2006). The long and the short of perfluorinated replacements. *Environmental Science & Technology*. 40(1), 12–13.

Rice, P. A., Aungst, J., Cooper, J., Bandele, O., & Kabadi, S. V. (2020). Comparative analysis of the toxicological databases for 6:2 fluorotelomer alcohol (6:2 FTOH) and perfluorohexanoic acid (PFHxA). *Food and Chemical Toxicology*, 138, 111210. doi: 10.1016/j.fct.2020.111210

RIVM. (2018). *Dutch National Institute for Public rHealth and the Environment (RIVM): Mixture exposure to PFAS: A Relative Potency Factor approach*. Available at: http://rivm.openrepository.com/rivm/handle/10029/622164 (Accessed August 29 2020).

RIVM. (2019). Dutch National Institute for Public Health and the Environment (RIVM): Per- and polyfluoroalkyl substances (PFASs) in food contact materials (RIVM Letter report 2018–0181; p. 112). *National Institute for Public Health and the Environment*. Available at: https://rivm.openrepository.com/bitstream/handle/10029/623023/2018–0181.pdf?sequence=1&isAllowed=y (Accessed August 29 2020).

Russell, M. H., Berti, W. R., Szostek, B., & Buck, R. C. (2008). Investigation of the biodegradation potential of a fluoroacrylate polymer product in aerobic soils. *Environmental Science & Technology*, 42(3), 800–807. doi: 10.1021/es0710499

Russell, M. H., Berti, W. R., Szostek, B., Wang, N., & Buck, R. C. (2010). Evaluation of PFO formation from the biodegradation of a fluorotelomer-based urethane polymer product in aerobic soils. *Polymer Degradation and Stability*, 95(1), 79–85. doi: 10.1016/j.polymdegradstab.2009.10.004

Sagiv, S. K., Rifas-Shiman, S. L., Webster, T. F., Mora, A. M., Harris, M. H., Calafat, A. M., Ye, X., Gillman, M. W., & Oken, E. (2015). Sociodemographic and Perinatal Predictors of Early Pregnancy Per- and Polyfluoroalkyl Substance (PFAS) Concentrations. *Environmental Science & Technology*, 49(19), 11849–11858. doi: 10.1021/acs.est.5b02489

Schaider, L. A., Balan, S. A., Blum, A., Andrews, D. Q., Strynar, M. J., Dickinson, M. E., Lunderberg, D. M., Lang, J. R., & Peaslee, G. F. (2017). Fluorinated compounds in U.S. fast food packaging. *Environmental Science & Technology Letters*, 4(3), 105–111. doi: 10.1021/acs.estlett.6b00435

Scher, D. P., Kelly, J. E., Huset, C. A., Barry, K. M., Hoffbeck, R. W., Yingling, V. L., & Messing, R. B. (2018). Occurrence of perfluoroalkyl substances (PFAS) in garden produce at homes with a history of PFAS-contaminated drinking water. *Chemosphere*, 196, 548–555. doi: 10.1016/j.chemosphere.2017.12.179

Schlummer, M., Sölch, C., Meisel, T., Still, M., Gruber, L., & Wolz, G. (2015). Emission of perfluoroalkyl carboxylic acids (PFCA) from heated surfaces made of polytetrafluoroethylene (PTFE) applied in food contact materials and consumer products. *Chemosphere*, 129, 46–53. doi: 10.1016/j.chemosphere.2014.11.036

Schultes, L., Peaslee, G. F., Brockman, J. D., Majumdar, A., McGuinness, S. R., Wilkinson, J. T., Sandblom, O., Ngwenyama, R. A., & Benskin, J. P. (2019). Total Fluorine Measurements in Food Packaging: How Do Current Methods Perform? *Environmental Science & Technology Letters*. doi: 10.1021/acs.estlett.8b00700

Shoeib, M., Harner, T., Webster, M., & Lee, S. C. (2011). Indoor sources of poly- and perfluorinated compounds (PFCs) in Vancouver, Canada: Implications for human exposure. *Environmental Science & Technology*, 45(19), 7999–8005. doi: 10.1021/es103562v

Shu, H., Lindh, C. H., Wikström, S., & Bornehag, C.-G. (2018). Temporal trends and predictors of perfluoroalkyl substances serum levels in Swedish pregnant women in the SELMA study. *PLOS ONE*, 13(12), e0209255. doi: 10.1371/journal.pone.0209255

Singh, J., Jayanti, R. K., Kilgore, J. E., Agarwal, K., & Gandarvakottai, R. R. (2005). What goes around comes around: Understanding trust–value dilemmas of market relationships. *Journal of Public Policy & Marketing, 24*(1), 38–62. doi: 10.1509/jppm.24.1.38.63882

Solvay. (2020). *Solvera® PFPE*. Available at: https://www.solvay.com/en/brands/solvera-pfpe (Accessed August 29 2020).

Still, M., Schlummer, M., Gruber, L., Fiedler, D., & Wolz, G. (2013). Impact of Industrial Production and Packaging Processes on the Concentration of Per- and Polyfluorinated Compounds in Milk and Dairy Products. *Journal of Agricultural and Food Chemistry*, 61(38), 9052–9062. doi: 10.1021/jf4020137

Sunderland, E. M., Hu, X. C., Dassuncao, C., Tokranov, A. K., Wagner, C. C., & Allen, J. G. (2019). A review of the pathways of human exposure to poly- and perfluoroalkyl substances (PFASs) and present understanding of health effects. *Journal of Exposure Science & Environmental Epidemiology*, 29(2), 131–147. doi: 10.1038/s41370-018-0094-1

Tian, Y., Zhou, Y., Miao, M., Wang, Z., Yuan, W., Liu, X., Wang, X., Wang, Z., Wen, S., & Liang, H. (2018). Determinants of plasma concentrations of perfluoroalkyl and polyfluoroalkyl substances in pregnant women from a birth cohort in Shanghai, China. *Environment International*, 119, 165–173. doi: 10.1016/j.envint.2018.06.015

Tian, Z., Kim, S. K., Shoeib, M., & Oh, J. E. (2016). Human exposure to per- and polyfluoroalkyl substances (PFASs) via house dust in Korea: Implication to exposure pathway. *Science of The Total Environment*, 553, 266–275. doi: 10.1016/j.scitotenv.2016.02.087

Tittlemier, S. A., Pepper, K., Seymour, C., Moisey, J., Bronson, R., Cao, X.-L., & Dabeka, R. W. (2007). Dietary Exposure of Canadians to Perfluorinated Carboxylates and Perfluorooctane Sulfonate via Consumption of Meat, Fish, Fast Foods, and Food Items Prepared in Their Packaging. *Journal of Agricultural and Food Chemistry*, 55(8), 3203–3210. doi: 10.1021/jf0634045

Tokranov, A. K., Nishizawa, N., Amadei, C. A., Zenobio, J. E., Pickard, H. M., Allen, J. G., Vecitis, C. D., & Sunderland, E. M. (2019). How do we measure poly- and perfluoroalkyl substances (PFASs) at the surface of consumer products? *Environmental Science & Technology Letters*, 6(1), 38–43. doi: 10.1021/acs.estlett.8b00600

Trier, X., Nielsen, N. J., & Christensen, J. H. (2011a). Structural isomers of polyfluorinated di- and tri-alkylated phosphate ester surfactants present in industrial blends and in microwave popcorn bags. *Environmental Science and Pollution Research*, 18(8), 1422–1432. doi: 10.1007/s11356-011-0488-2

Trier, X., Granby, K., & Christensen, J. H. (2011b). Polyfluorinated surfactants (PFS) in paper and board coatings for food packaging. *Environmental Science and Pollution Research*, 18(7), 1108–1120. doi: 10.1007/s11356-010-0439-3

Trier, X., Granby, K., & Christensen, J. H. (2011c). Polyfluorinated surfactants (PFS) in paper and board coatings for food packaging. *Environmental Science and Pollution Research*, 18(7), 1108–1120. doi: 10.1007/s11356-010-0439-3

UNEP. (2020). *United Nations Environment Programme (UNEP): All POPs listed in the Stockholm Convention*. Available at: http://chm.pops.int/TheConvention/ThePOPs/AllPOPs/tabid/2509/Default.aspx (Accessed August 29 2020).

U.S. EPA. (2009a). *U.S. Environmental Protection Agency (U.S. EPA): Long-chain perfluorinated chemicals (PFCs) action plan*. December 30, 2009. Available at: https://www.epa.gov/sites/production/files/2016-01/documents/pfcs_action_plan1230_09.pdf (Accessed August 29 2020).

U.S. EPA. (2009b). *U.S. Environmental Protection Agency (U.S. EPA): Perfluorocarboxylic Acid Content in 116 Articles of Commerce. EPA/600/R-09/33, 51*. Available at: https://www.oecd.org/env/48125746.pdf (Accessed August 29 2020).

U.S. EPA. (2012, July 27). *U.S. Environmental Protection Agency (U.S. EPA): Long-chain perfluorinated chemicals (LCPFCs) used in carpets*. Available at: http://www.fluoride-alert.org/wp-content/uploads/pfcs.carpet.report.epa_.2012.pdf (Accessed August 29 2020).

U.S. EPA. (2019, February). *U.S. Environmental Protection Agency (U.S. EPA): EPA's Per- and Polyfluoroalkyl Substances (PFAS) Action Plan*. Available at: https://www.epa.gov/sites/production/files/2019-02/documents/pfas_action_plan_021319_508compliant_1.pdf (Accessed August 29 2020).

U.S. EPA. (2020a). U.S. Environmental Protection Agency (EPA). *2010/2015 PFOA Stewardship Program*. Available at: https://www.epa.gov/assessing-and-managing-chemicals-under-tsca/and-polyfluoroalkyl-substances-pfass-under-tsca#tab-3 (Accessed August 29 2020).

U.S. EPA. (2020b). *U.S. Environmental Protection Agency (U.S. EPA): Chemistry Dashboard, PFAS Master List of PFAS Substances*. Available at: https://comptox.epa.gov/dashboard/chemical_lists/pfasmaster (Accessed August 29 2020).

U.S. EPA. (2020c). *U.S. Environmental Protection Agency (U.S. EPA): Risk Management for Per- and Polyfluoroalkyl Substances (PFASs) under TSCA* [Overviews and Factsheets]. Available at: https://www.epa.gov/assessing-and-managing-chemicals-under-tsca/risk-management-and-polyfluoroalkyl-substances-pfas (Accessed August 29 2020).

VanNoy, B. N., Lam, J., & Zota, A. R. (2018). Breastfeeding as a Predictor of Serum Concentrations of Per- and Polyfluorinated Alkyl Substances in Reproductive-Aged Women and Young Children: A Rapid Systematic Review. *Current Environmental Health Reports*, 5(2), 213–224. doi: 10.1007/s40572-018-0194-z

Vermont Legislature. (2020). *S-0295 Bill As Introduced*. Available at: https://legislature.vermont.gov/Documents/2020/Docs/BILLS/S-0295/S-0295%20As%20Introduced.pdf (Accessed August 29 2020).

Vestergren, R., Herzke, D., Wang, T., & Cousins, I. T. (2015). Are imported consumer products an important diffuse source of PFASs to the Norwegian environment? *Environmental Pollution*, 198, 223–230. doi: 10.1016/j.envpol.2014.12.034

Wang, Z., Cousins, I. T., Scheringer, M., & Hungerbuehler, K. (2015). Hazard assessment of fluorinated alternatives to long-chain perfluoroalkyl acids (PFAAs) and their precursors: Status quo, ongoing challenges and possible solutions. *Environment International*, 75, 172–179. doi: 10.1016/j.envint.2014.11.013

Wang, Z., DeWitt, J. C., Higgins, C. P., & Cousins, I. T. (2017). A never-ending story of per- and polyfluoroalkyl substances (PFASs)? *Environmental Science & Technology*, 51(5), 2508–2518. doi: 10.1021/acs.est.6b04806

Washington, J. W., Ellington, J. J., Jenkins, T. M., Evans, J. J., Yoo, H., & Hafner, S. C. (2009). Degradability of an acrylate-linked, fluorotelomer polymer in soil. *Environmental Science & Technology*, 43(17), 6617–6623. doi: 10.1021/es9002668

Washington, J. W., Jenkins, T. M., Rankin, K., & Naile, J. E. (2015). Decades-scale degradation of commercial, side-chain, fluorotelomer-based polymers in soils and water. *Environmental Science & Technology*, 49(2), 915–923. doi: 10.1021/es504347u

Washington, J. W., Rankin, K., Libelo, E. L., Lynch, D. G., & Cyterski, M. (2019). Determining global background soil PFAS loads and the fluorotelomer-based polymer degradation rates that can account for these loads. *Science of The Total Environment*, 651, 2444–2449. doi: 10.1016/j.scitotenv.2018.10.071

Washington State Legislature. (2019, July 28). *Substitute Senate Bill 5135*. Available at: http://lawfilesext.leg.wa.gov/biennium/2019–20/Pdf/Bills/Session%20Laws/Senate/5135-S.SL.pdf (Accessed August 29 2020).

Winkens, K., Giovanoulis, G., Koponen, J., Vestergren, R., Berger, U., Karvonen, A. M., Pekkanen, J., Kiviranta, H., & Cousins, I. T. (2018). Perfluoroalkyl acids and their precursors in floor dust of children's bedrooms – Implications for indoor exposure. *Environment International*, 119, 493–502. doi: 10.1016/j.envint.2018.06.009

Winkens, K., Koponen, J., Schuster, J., Shoeib, M., Vestergren, R., Berger, U., Karvonen, A. M., Pekkanen, J., Kiviranta, H., & Cousins, I. T. (2017). Perfluoroalkyl acids and their precursors in indoor air sampled in children's bedrooms. *Environmental Pollution*, 222, 423–432. doi: 10.1016/j.envpol.2016.12.010

Wood Environment & Infrastructure Solutions UK Limited. (2020, February). *Socio-Economic Assessment of the US Fluoropolymer Industry - Executive Summary*. Available at: https://fluorocouncil.com/wp-content/uploads/2020/03/Socio-Economic-Assessment-of-the-US-Fluoropolymer-Industry-Executive-Summary.pdf (Accessed August 29 2020).

Yao, Y., Zhao, Y., Sun, H., Chang, S., Zhu, L., Alder, A. C., & Kannan, K. (2018). Per- and polyfluoroalkyl substances (PFASs) in indoor air and dust from homes and various microenvironments in China: Implications for human exposure. *Environmental Science & Technology*, 52(5), 3156–3166. doi: 10.1021/acs.est.7b04971

Ye, F., Zushi, Y., & Masunaga, S. (2015). Survey of perfluoroalkyl acids (PFAAs) and their precursors present in Japanese consumer products. *Chemosphere*, 127, 262–268. doi: 10.1016/j.chemosphere.2015.02.026

Zhang, M., Wang, P., Lu, Y., Lu, X., Zhang, A., Liu, Z., Zhang, Y., Khan, K., & Sarvajayakesavalu, S. (2020). Bioaccumulation and human exposure of perfluoroalkyl acids (PFAAs) in vegetables from the largest vegetable production base of China. *Environment International*, 135, 105347. doi: 10.1016/j.envint.2019.105347

Zhang, T., Sun, H. W., Wu, Q., Zhang, X. Z., Yun, S. H., & Kannan, K. (2010). Perfluorochemicals in Meat, Eggs and Indoor Dust in China: Assessment of Sources and Pathways of Human Exposure to Perfluorochemicals. *Environmental Science & Technology*, 44(9), 3572–3579. doi: 10.1021/es1000159

Zheng, G., Boor, B. E., Schreder, E., & Salamova, A. (2020). Indoor exposure to per- and polyfluoroalkyl substances (PFAS) in the childcare environment. *Environmental Pollution*, 258, 113714. doi: 10.1016/j.envpol.2019.113714

Zhou, J., Yang, Z., Liu, Q., Liu, Y., Liu, M., Wang, T., & Zhu, L. (2019). Insights into uptake, translocation, and transformation mechanisms of perfluorophosphinates and perfluorophosphonates in wheat (*Triticum aestivum L.*). *Environmental Science & Technology*. doi: 10.1021/acs.est.9b05656

4 Regulatory Implications of PFAS

J. Alan Roberson

Association of State Drinking Water Administrators, Virginia

CONTENTS

4.1 Introduction to the Safe Drinking Water Act (SDWA) 87
4.2 The Federal Perspective 90
4.3 The States' Perspective 94
4.4 What the Future Holds 95
References 96

4.1 INTRODUCTION TO THE SAFE DRINKING WATER ACT (SDWA)

The 1996 Amendments to Safe Drinking Water Act (SDWA) fundamentally changed Environmental Protection Agency's (EPA's) regulatory development process. The 1996 SDWA Amendments (P.L. 104–182) provided a step-by-step framework for both the development of new regulations (on five-year cycles) and for the review of all existing regulations (on six-year cycles). The development of a new regulation follows a multi-step process, as shown below in Figure 4.1. The process starts with a list of contaminants to be considered for potential regulation, moves to data collection and decision-making on whether to regulate or not, and then moves to a national regulation if that is the Agency's decision. Since 1996, the timeframe between each step in the process has varied. The public can review and comment throughout this process on each step shown with stars in Figure 4.1.

The Contaminant Candidate List (CCL) is the starting point for the EPA's regulatory development process. If there is a need for national occurrence data for CCL contaminants (and, in most cases, there is such a need), EPA will require monitoring, through the Unregulated Contaminant Monitoring Rule (UCMR), for systems serving >10,000 people (approximately 4200 systems) for up to 30 contaminants to develop national occurrence data for its decision-making process. Water systems collect samples and pay for the UCMR monitoring. Water systems have a three-year window to conduct the required UCMR monitoring. It should be noted that the America's Water Infrastructure Act of 2018 (AWIA, P.L. 115–270) extended UCMR monitoring down to systems serving 3300–10,000 people, but Congress must

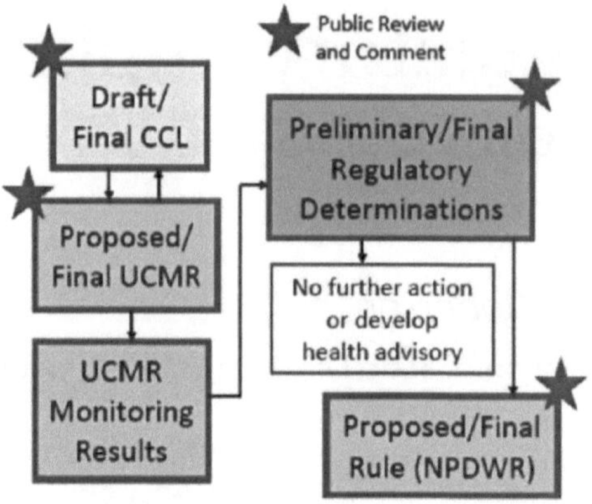

FIGURE 4.1 SDWA Regulatory Development Process.

appropriate the funding (which has not been yet appropriated) for that additional future monitoring to occur.

Regulatory determinations are the next steps in the EPA's regulatory development process. The EPA must consolidate and consider the best available data regarding health effects, occurrence, and treatment information and decide if a national regulation is warranted (or not), or another action such as developing guidance or conducting more research is more appropriate. The SDWA mandates that EPA consider three criteria when making regulatory determinations:

1. The contaminant may have an adverse effect on the health of persons;
2. The contaminant is known to occur or there is a substantial likelihood the contaminant will occur in public water systems with a frequency and at levels of public health concern; and
3. In the sole judgment of the Administrator, regulation of the contaminant presents a meaningful opportunity for health risk reductions for persons served by public water systems.

Another SDWA requirement for EPA's regulatory determinations is that the Agency must make decisions on at least five contaminants every five years. It should also be noted that each Regulatory Determination is based on the previous UCMR data; i.e., the second regulatory determination in 2008 was based on monitoring data from the First UCMR (UCMR1), and so on. The first regulatory determination in 2003 was based on pre-1996 monitoring data. This lag time allows EPA to appropriately address the above three criteria in its regulatory determinations.

While regulatory determinations are a critical step within the SDWA, these decisions do not establish any numbers that are legally enforceable. These decisions are simply another step in the regulatory development process (in the case of a positive regulatory determination), or the end of that process (in the case of a negative regulatory determination). If EPA moves forward with a National Primary Drinking Water Regulation (NPDWR), then the SDWA requires EPA to propose the regulation within

24 months of the regulatory determination. The SDWA requires EPA to finalize the regulation 18 months after the proposal, but also provides a provision for the EPA Administrator to ask for additional time to finalize the regulation. A NPDWR has a numerical standard, known as the Maximum Contaminant Level (MCL), or a Treatment Technique (TT) for those contaminants without a robust and reliable analytical method.

Additionally, the 1996 SDWA Amendments expanded the issues to be analyzed for EPA's cost–benefit analysis, known as the Health Risk Reduction and Cost Analysis (HRRCA). EPA must analyze the quantifiable and non-quantifiable benefits that are likely to occur as the result of compliance with the proposed standard. EPA must analyze certain increased costs that will result from the proposed drinking water standard. As part of the HRRCA, EPA must also consider:

- Incremental costs and benefits associated with the proposed and alternative MCL values;
- The contaminant's adverse health effects on the general population and sensitive subpopulations;
- Any increased health risk to the general population that may occur as a result of the new MCL; and
- Other relevant factors such as data quality and the nature of the risks.

Given the SDWA requirements, development of the HRRCA takes significant Agency time and resources as part of EPA's regulatory development process. The HRRCA, also known as the Economic Analysis, developed for the 2019 proposed Lead and Copper Rule Revisions was over 1100 pages long with appendices.

While the 1996 SDWA provided for a robust regulatory development process that uses the best available, peer-reviewed science, the sequential steps lead to a significant number of years between a contaminant first being identified on the draft CCL and a final regulation. If everything goes "smoothly," then this process can take a decade or longer. Table 4.1 provides history of the post-1996 regulatory actions (except for the separate regulatory determination for perchlorate, which is discussed below), and one can infer two conclusions from Table 4.1. First, EPA has completed

TABLE 4.1
Post-1996 SDWA Regulatory History.

	FIRST	SECOND	THIRD	FOURTH
CCL	1998 60 contaminants	2005 51 contaminants	2009 116 contaminants	2016 109 contaminants
UCMR	1999 26 contaminants	2007 25 contaminants	2012 30 contaminants	2016 30 contaminants
RegDet	2003 9-Not regulated	2008 11-Not regulated	2016 4-Not regulated 1-Needs more research	2021 2-Regulate 6-Not regulate

12 final regulatory actions while following the 1996 SDWA regulatory development process, and that is quite a body of work. Second, the timeframe varies between listing contaminants and the Agency making its decisions based on that list. While each CCL is approximately five years apart, the timeframe between the CCL and the resultant regulatory determination has varied from 10 years (CCL1 in 1998 to the final second regulatory determination in 2008) to 12 years (CCL3 in 2009 to 2021 for the final fourth regulatory determination).

Perchlorate is the "poster child" of the twists and turns of the SDWA regulatory development process. Perchlorate was listed on the final CCL1 in 1998 and included for monitoring in UCMR1 in 1999 (EPA 1998, 1999). EPA developed an Interim Health Advisory of 15 μg/L in 2008 and made a final positive regulatory determination in 2011 (EPA 2008a, 2011). EPA then conducted additional health research and in June 2020, and the Agency reversed the 2011 positive regulatory determination as the EPA Administrator made the judgment call that a national perchlorate regulation would not provide "a meaningful opportunity for health risk reductions for persons served by public water systems" as required by the SDWA (EPA 2020). This was 22 years after the listing of perchlorate on CCL1.

No new contaminants have yet to be regulated under this regulatory development process as all regulations that have been finalized since 1996 were either in progress prior to 1996 or had deadlines in the 1996 SDWA Amendments. However, it should be noted that EPA has met the SDWA requirements for five regulatory determinations every five years with 24 negative regulatory determinations since 2003 in the first three rounds of regulatory determinations in 2003, 2008, and 2016 (EPA 2003, 2008b, 2016a). EPA also decided to conduct additional research on strontium as part of the third regulatory determination, and that decision counted as one of its five decisions. In 2021, in the final Fourth Regulatory Determinations, EPA decided a national regulation was warranted for perfluorooctanoic acid (PFOA) and perfluorooctane sulfonate (PFOS), along with making six negative regulatory determinations (EPA, 2021a). EPA is meeting the statutory requirements, but until 2021, the Agency had been unable to find a new contaminant where a national regulation provides "a meaningful opportunity for health risk reductions for persons served by public water systems."

The lack of new contaminants being regulated is becoming a more significant public perception problem in the face of several emerging contaminants such as 1,4-dioxane and per- and polyfluoroalkyl substances (PFAS). The ten- to twenty-year EPA regulatory development process is leading States to take actions on their own, which may not be the optimal solution to national public health protection.

4.2 THE FEDERAL PERSPECTIVE

Some might argue that the Federal government has been slow to react to PFAS problems. From the drinking water perspective, the Federal government began taking steps to address two PFAS in 2009 with the release of Provisional Health Advisories for perfluorooctanoic acid (PFOA) at 400 ppt and perfluorooctane sulfonate (PFOS) at 200 ppt (EPA 2009a). Additionally, in 2009, EPA listed PFOA and PFOS on the final Third Contaminant Candidate List (CCL3, EPA 2009b). As the next step, in

2012, to develop national occurrence data as previously discussed, EPA included six PFAS on the final Third Unregulated Contaminant Monitoring Rule (UCMR3, EPA 2012). The six PFAS were PFOA, PFOS, PFNA, PFHxS, PFHpA, and PFBS. The UMCR3 monitoring started in January 2013 and was completed in December 2015.

EPA publishes the results of UCMR monitoring quarterly, and this publication typically lags eight to nine months behind monitoring. This lag time became important in 2016, as while EPA was wrapping up the publication of the final UCMR3 monitoring data, on May 25, 2016, the Agency published a *Federal Register* notice of Lifetime Health Advisories (HAs) and Health Effects Support Documents for PFOA and PFOS (EPA 2016b). The Lifetime HAs are 70 ppt for PFOA and for PFOS, as well as 70 ppt for the sum of both. These levels were substantial reductions from the Provisional HAs, and public concern over PFAS exposure continued to increase due to these lower numbers which triggered additional sampling in non-UCMR systems so that PFAS were found in more water systems. Additionally, water systems and state primacy agencies were not sure how to react to Lifetime HAs, as they are not legally enforceable standards. Is public notification enough, or should primacy agencies require the water systems to install treatment? How can a primacy agency require treatment without a legally enforceable standard? If EPA is not going to set a national standard in a timely manner, then should States then set their own standard, even if they have never set their own standard before?

In May 2018, EPA hosted a PFAS Leadership Summit in Washington, D.C. The leadership summit included representatives from over 40 States, tribes, and territories; 13 federal agencies; Congressional staff; associations; industry groups; and nongovernmental organizations. During the summit, through a combination of panel presentations, digital brainstorming, plenary sessions, and small table discussions, the participants:

- Shared information on ongoing efforts to monitor and characterize risks from PFAS;
- Discussed specific near-term actions, beyond those already underway, that are needed to address challenges currently facing States and local communities; and
- Discussed risk communication strategies to address public concerns with PFAS.

Coming on the heels of this Summit in June 2018, the debate over the appropriate levels for "safe" drinking water containing PFAS increased with the release of draft PFAS toxicological profiles by the Agency for Toxic Substances and Disease Registry (ATSDR) (ATSDR 2018). ATSDR's profiles are reference guides and are not legally enforceable standards. These profiles provide information about a toxic substance, such as its chemical and physical properties, sources of exposure, routes of exposure, minimal risk levels, children's health, and general health effects, as well as how the substance might interact in the environment. These profiles led to lower numbers in drinking water than EPA's HAs of 70 ppt for PFOA and PFOS (and the sum of the two).

The EPA PFAS Leadership Summit led to an EPA PFAS Action Plan that was released in February 2019 (EPA 2019). This Action Plan was a multi-pronged approach with EPA committing to research and actions in source control, drinking water, treatment, cleanup, enforcement, and risk communications.

Throughout 2019, EPA had been working on the preliminary regulatory determinations for PFOA and PFOS and published the preliminary regulatory determinations in the March 10, 2020 *Federal Register* (EPA 2020). EPA received over 11,000 public comments on these regulatory determinations which shows the continued level of public concern about PFAS – noting that the preliminary Third Regulatory Determinations received only 15 public comments and the preliminary Second Regulatory Determinations received only 11 public comments.

In this *Federal Register* notice, EPA asked for how to make the best use of the available information when developing potential PFAS regulatory approaches and asked for input on a few critical regulatory policy issues over and above the preliminary positive regulatory determinations for PFOA and PFOS. Three potential regulatory approaches were discussed:

1. Evaluate each additional PFAS on an individual basis, i.e., individual MCLs for each PFAS;
2. Evaluate additional PFAS by different grouping approaches; and
3. Evaluate PFAS based on drinking water treatment techniques.

As previously mentioned, in 2021, in the final Fourth Regulatory Determinations, EPA decided a national regulation was warranted for PFOA and PFOS. EPA has left its options open on potential regulatory approaches for these two PFAS.

EPA's experience with regulating by class has been a mixed bag in the past. For disinfection by-products (DBPs), regulating by class has worked successfully for total trihalomethanes (TTHMs) and five haloacetic acids (HAA5) with MCLs of 80 µg/l and 60 µg/L, respectively. A similar effort for carcinogenic Volatile Organic Chemicals (cVOCs) simply did not work out. Former EPA Administrator Lisa Jackson made regulating contaminants as a group (as opposed to one compound at a time) a part of the Agency's new drinking water strategy and identified carcinogenic volatile organic chemicals (cVOCs) as the first group to be regulated. The strategy did not work out as intended as the cVOC with the most adverse health effects (i.e., 1,2,3- trichloropropane (TCP) with the lowest potential MCL at 0.02 µg/L) was the driver for the regulation. Essentially, it resulted in a situation where if you had TCP, you had a problem and additional treatment would have to be installed, and if you did not, then the balance of cVOCs did not need to be regulated.

EPA has one additional future PFAS regulatory action with the proposed Fifth Unregulated Contaminant Monitoring Rule (UCMR5, EPA, 2021b). EPA proposed UCMR5 in 2021, will finalize UCMR5 in 2022, and then the water systems will conduct UCMR5 monitoring in 2023–2025. Another round of PFAS monitoring in the proposed UCMR5 includes additional PFAS beyond the six compounds in UCMR3 and has lower minimum reporting levels. EPA's recently validated Method 533 focuses on "short chain" per- and polyfluoroalkyl substances (PFAS) (i.e., those with carbon chain lengths of five or six depending on the functional group (sulfonate or carboxylate, respectively)). EPA Method 533 complements the earlier EPA Method 537.1 and can be used to test for 11 additional PFAS. Using both methods, a total of 29 unique PFAS can be measured in drinking water and these 29 PFAS are included in the proposed UCMR5. Additionally, as previously mentioned, the America's Water Infrastructure Act of 2018 (AWIA, P.L. 115–270) allows EPA to extend the UCMR monitoring requirements down to systems serving 3300–10,000 people if additional funding is appropriated by Congress, so that EPA can pay for these additional systems to conduct UCMR monitoring.

Congress is still very much interested in addressing PFAS and is not letting EPA "off the hook" on PFAS. In late 2019, Congress passed the 2020 National Defense Authorization Act (NDAA, P.L. 116–92). The final bill included language to deal with PFAS under the SDWA, the Clean Water Act (CWA), and other environmental laws. Several important new actions for EPA to address PFAS were included in the final NDAA:

- Required the inclusion of all PFAS for which EPA has validated a method to measure the level in drinking water in UCMR5. The PFAS included in UCMR 5 will not count towards the limit of 30 contaminants to be monitored.
- Additional funds of $100,000,000 for each of fiscal years 2020 through 2024 were authorized (noting that these funds are not yet appropriated and nothing is guaranteed for appropriations) for the Drinking Water State Revolving Fund (DWSRF) to provide grants to water systems for addressing emerging contaminants, with a focus on PFAS. The NDAA mandated that 25% of this funding be directed to disadvantaged communities and water serving less than 25,000 people.
- Added several PFAS chemicals to the Toxic Release Inventory (TRI) with a reporting threshold of 100 pounds (noting that this is a relatively large amount).
- Within one year, EPA must publish interim guidance on the destruction and disposal of PFAS and materials containing PFAS.
- PFAS manufacturers will be required to submit information under the Toxic Substances Control Act (TSCA) about where PFAS chemicals were previously manufactured and sold.
- Directed EPA to review Federal efforts to identify, monitor, and assist in the development of treatment methods for emerging contaminants and to assist States in responding to the human health risks posed by contaminants of emerging concern; EPA, in collaboration with stakeholders, must then establish a strategic plan for improving such Federal efforts.
- Directed EPA to develop a National Emerging Contaminant Research Initiative in coordination with other Federal agencies.
- Directed EPA to conduct a study on actions they can take to increase technical assistance and support for States with respect to emerging contaminants in drinking water. EPA will then implement a program for States to apply for technical assistance on emerging contaminants. Part of this program will include the development of an EPA database of tools and resources to assist States with emerging contaminants.

The 2020 NDAA also included additional mandates for the US Geological Survey (USGS) and the Department of Defense (DoD). The combined Federal Agency efforts have the potential to significantly change the future path for PFAS regulation and control.

4.3 THE STATES' PERSPECTIVE

Since Federal actions to address PFAS contamination have been minimal, States are taking action on their own. Many of these actions have different names to address different environmental issues such as drinking water, ambient water, groundwater protection, industrial discharges, etc. The Interstate Technology and Regulatory Council (ITRC) regularly compiles the States' actions, and has found that 22 States have taken their own actions to reduce PFAS contamination in drinking water, groundwater, and surface water/effluent (wastewater), using a wide variety of both numbers and nomenclature (ITRC 2020):

- AGQS = ambient groundwater quality standard
- AL = private well action level
- BCL = basic comparison level
- CL = groundwater cleanup level
- GCC = Generic Cleanup Criteria
- GWQS = Groundwater Water Quality Standard
- HA = lifetime health advisory
- HNV = human noncancer value for drinking water
- HBV = health-based value
- HRL = health risk limit
- ISGWQS = Interim Specific Ground Water Quality Standard
- MCL = maximum contaminant level
- NL = notification level
- PAL = preventive action level
- PCL = protective concentration level
- RAG = remedial action guideline
- RL (CA) = Response Level (California only)
- SL = Screening Level

Additional terms used by States include Drinking Water Values and Water Quality Standards. Clearly, the public can easily become confused when confronted with these different names, acronyms, and numbers. The 22 States have developed 33 different numbers or actions to address PFAS contamination in drinking water, groundwater, and surface water/effluent (wastewater). Many of these actions address different PFAS. Additionally, in contrast to a single national standard, the different PFAS as well as different numbers can often create risk communication challenges for adjoining States.

Under the SDWA, States are required to set their own standards at least as strict as EPA's to maintain primary enforcement authority, known as primacy. Some States, either by statute or by policy, are never stricter than the EPA standard, nor can they set their own standards for a contaminant that EPA does not regulate. In the past, only a small number of States, such as California, New York, New Jersey, and

Massachusetts, set their own drinking water standards. Now, many States, such as New Hampshire, Michigan and others, are facing public pressure and/or mandates from their state legislatures to set their own standards in the absence of federal action. For example, Michigan established drinking water MCLs for seven PFAS chemicals in August 2020.

States are facing significant challenges in setting their own standards for the first time. Many States do not have the technical staff and the resources to analyze health effects data, assess the robustness of analytical methods, assess the efficacy of potential treatment technologies, and conduct a state-level cost–benefit analysis. In 2020, the Association of State Drinking Water Administrators (ASDWA) released a toolkit to assist States with the state-level standard-setting or other risk management actions (ASDWA 2020). This toolkit generally follows EPA's regulatory development process and provides tools and resources for eight categories of potential actions that might be considered by an individual State:

1. Self-assessment of state resources
2. Health effects
3. Occurrence
4. Analytical methods
5. Treatment and compliance options
6. Benefits, costs, and economic considerations
7. Intermediate management strategies
8. Rule options – Maximum Contaminant Level (MCL) or Treatment Technique (TT)

Each one of these categories requires careful analysis and decision-making by State staff, that may or may not have experience with setting their own standards. It is clear to see that setting their own standard due to the lack of a national standard is a significant effort for States.

4.4 WHAT THE FUTURE HOLDS

It appears almost certain that EPA will finalize the preliminary positive regulatory determinations for PFOA and PFOS and move forward with the development of proposed/final regulations for these two PFAS. The final regulatory determinations was published in 2021. Then, following its regulatory development process, the Agency has two years to propose the regulations (2023) and 18 months after the proposal to finalize the regulation (2025). The States then have three years to adopt the Federal standard (2028). In the meantime, a few more States are likely to move forward with their own State-level standard. Beyond that, predictions are challenging to make.

In its comments on the preliminary regulatory determinations, ASDWA recommended that in its final regulatory determinations, EPA also include positive determinations for four additional PFAS compounds with PFOA and PFOS: perfluoroonanoic acid (PFNA), perfluorohexanesulfonic acid (PFHxS), perfluoroheptanoic acid (PFHpA), and perfluorodecanoic acid (PFDA). Including all six PFAS would be similar to current state regulatory approaches by Massachusetts, Vermont, Connecticut, New Hampshire, Minnesota, and Michigan. EPA elected to not include these four PFAS in its final regulatory determinations. ASDWA supports EPA in using its flexibility as detailed in the *Federal Register* notice to expedite the regulatory development process based on a positive final regulatory determination for the six PFAS so that the proposed and final regulation is developed as soon as possible. The States need EPA to take a leadership role in regulating these six PFAS in a timely manner versus the current conundrum of different state standards.

Beyond that, the future depends on what the public wants for PFAS. Some communities want "zero PFAS" without an understanding of the analytical and treatment challenges, as well as the resultant costs – both capital costs and operations and maintenance (O&M) costs, which are in perpetuity. "Zero PFAS" could lead to the installation of two types of advanced treatment – Granular Activated Carbon (GAC) and ion-exchange. Both of these have media that would need to be replaced on a regular basis, and it's not clear what the replacement frequency would be without knowledge of what PFAS are present in the water being treated, and at what levels. Installation of both treatments would create waste streams of the spent media that need disposal, and some of the disposal options, such as high-temperature incineration, have their own associated risks and costs.

Increased concern from constituents can force State legislatures to pass legislation mandating that State-level standards be set in a tight timeline. The same pressure could force Congress to take additional action at the Federal level. New PFAS formulations are continuing to be developed and additional actions are needed through the Toxic Substances and Control Act (TSCA) to keep these substances out of the environment. Appropriately addressing PFAS in drinking water and in the environment is going to be a marathon.

REFERENCES

Agency for Toxic Substances and Disease Registry, 2018. *Toxicological Profiles for Perfluoroalkyls*. https://www.atsdr.cdc.gov/toxprofiles/tp.asp?id=1117&tid=237 Accessed August 23, 2020.

Association of State Drinking Water Administrators, 2020. *State CEC Rule Development and Management Strategies*. https://www.asdwa.org/wp-content/uploads/2020/03/State-CEC-Rule-Development-and-Management-Strategies-Toolkit.pdf Accessed August 23, 2020.

Environmental Protection Agency, 1998. Announcement of the Drinking Water Contaminant Candidate List; Notice. 63 FR 10274.

Environmental Protection Agency, 1999. Revisions to the Unregulated Contaminant Monitoring Rule for Public Water Systems; Final Rule. 64 FR 50556.

Environmental Protection Agency, 2003. Announcement of Regulatory Determinations for Priority Contaminants on the First Contaminant Candidate List; Notice. 68 FR 42897.

Environmental Protection Agency, 2008a. Interim Drinking Water Health Advisory for Perchlorate. EPA-822-R-08-025, December.

Environmental Protection Agency, 2008b. Drinking Water: Regulatory Determinations Regarding Contaminants on the Second Contaminant Candidate List; Notice. 73 FR 44251.

Environmental Protection Agency, 2009a. Provisional Health Advisories for Perfluorooctanoic Acid and Perfluorooctane Sulfonate. January 8, 2009.
Environmental Protection Agency, 2009b. Drinking Water Contaminant Candidate List 3-Final; Notice. 74 FR 51850.
Environmental Protection Agency, 2011. Drinking Water: Regulatory Determinations for Perchlorate; Regulation Determinations. 76 FR 7762.
Environmental Protection Agency, 2012. Revisions to the Unregulated Contaminant Monitoring Rule for Public Water Systems; Final Rule. 77 FR 43523.
Environmental Protection Agency, 2016a. Announcement of Final Regulatory Determinations for the Third Contaminant Candidate List; Final Regulatory Determinations. 81 FR 13.
Environmental Protection Agency, 2016b. Lifetime Health Advisory Health Advisories and Health Effect Support Documents for Perfluorooctanoic Acid and Perfluorooctane Sulfonate. 81 FR 33250.
Environmental Protection Agency, 2019. *EPA's PFAS Action Plan.* https://www.epa.gov/pfas/epas-pfas-action-plan accessed August 23, 2020.
Environmental Protection Agency, 2020. Drinking Water: Final Action on Perchlorate; Final Action. 85 FR 43990.
Environmental Protection Agency, 2021a. Final Regulatory Determinations – Announcement of Final Regulatory Determinations for Contaminants on the Fourth Contaminant Candidate List. 86 FR 12272.
Environmental Protection Agency, 2021b. Revisions to the Unregulated Contaminant Monitoring Rule (UCMR5) for Public Water Systems and Announcement of Public Meeting. 86 FR 13846.
Interstate Technology and Regulatory Council, 2020. *PFAS: Per- and Polyfluoroalkyl Substances.* https://pfas-1.itrcweb.org/Accessed August 23, 2020.

5 The Analytical Conundrum

Taryn McKnight
Eurofins Environment Testing America, California

CONTENTS

5.1 Introduction to the Chemistry ... 99
5.2 Method Development Timeline ... 100
5.3 Where the Industry is at Today ... 104
5.4 A Look at Specific Matrices ... 113
5.4.1 Landfill Leachate ... 113
5.4.2 Biosolids ... 113
5.4.3 Air ... 114
5.5 Industry Trends ... 115
5.6 Conclusions ... 117
References ... 117

5.1 INTRODUCTION TO THE CHEMISTRY

Laboratories require standard reference material to quantify chemicals in environmental samples. Currently, there are only a handful of branched standards available as opposed to their linear isomers. This means we are potentially underestimating the true mass of PFAS in the environment. Although we can detect what may be additional branched isomers present in the sample, they cannot be confirmed or quantified. Only the isomers with standards can be accurately identified and concentrations measured. Laboratories are instructed by the US Environmental Protection Agency (EPA) to add branched standards to their calibration as they become commercially available, so over time concentrations of target analytes could increase (US EPA 2019).

Production of branched versus linear isomers is unique to different manufacturing processes. It is believed that measuring the ratios of these isomers could be useful in a forensics investigation of sources.

Formation of these chemicals began with the electrochemical fluorination (ECF) process back in the 1940s. Industry had been producing approximately 70% linear isomers with the remaining production resulting in branched isomers. Unintended by-products were also being produced in percent levels, which were PFAS compounds of varying chain lengths.

In the early 2000s, telomerization became the dominant process. This process was an improvement over ECF in one regard, yielding nearly 100% linear isomers, but the intended products produced are precursor compounds and also included the production of unintended by-products. Precursors are polyfluorinated chemicals that have the potential to transform in the environment.

There is a myriad of reasons why PFAS are challenging to analyze. These chemicals are believed to be ubiquitous given that they are detected in the blood of humans and wildlife worldwide. They are typically found at ultra-trace levels in drinking water, but they can also be found at gross levels in wastewater and where AFFF (aqueous film forming foam) was applied. The ubiquitous and surface-active nature of these chemicals makes them very challenging to prevent contamination in the field and in the laboratory.

While there are many challenges associated with this family of chemicals, one of the most immediate and largest challenges facing us today is the lack of consensus regarding a "best" method for non-potable water and solid matrices.

5.2 METHOD DEVELOPMENT TIMELINE

For 10 years we had a single drinking water method for the analysis for PFAS, EPA Method 537 (Shoemaker et al. 2009). This method was updated to version 537.1 at the end of 2018, primarily to add four PFAS replacement chemicals the EPA had selected to address (Shoemaker and Tettenhorst 2018). "Replacement chemicals" is the term given to the short-chain PFAS chemicals that were designed to replace their long-chain counterparts.

At the end of 2019, the EPA published a second drinking water method, EPA 533 (Rosenblum and Wendelken 2019). This method was specifically generated to target the short-chain PFAS compounds. This "Key Feature Comparison" in Table 5.1 highlights the differences between the two drinking water methods. A few items are the same in both methods, but many are divergent. Most notable is the use of isotope dilution in method 533. It is anticipated that the two methods would be utilized in concert as this is the only way to capture the full range of target analytes proposed by the EPA. This requires awareness of the different requirements for each sample

TABLE 5.1
Comparison between key features of EPA drinking water methods 537.1 and the newer 533 method.

533	537.1
Drinking Water	Drinking Water
Branched/Linear Isomers -YES	Branched/Linear Isomers -YES
14 of the same and 11 unique compounds	14 of the same and 4 unique compounds
SPE WAX	SPE SDVB
Hold Time: 28/28 days	Hold Time: 14/28 days
LCMSMS with confirmation ion	LCMSMS - no confirmation ion
Isotope Dilution	Internal standard
Recovery Correction - YES	Recovery Correction – NO
RLs: Not defined	RLs: 2 ppt - 40 ppt

collected, one for 537.1 and the other for 533. This also means the cost of analysis increases significantly. Note that these methods are for potable water only and applicable to the limited number of PFAS chemicals addressed by the methods.

Figure 5.1 summarizes the timeline of analytical method development. When we first started testing for these chemicals in environmental matrices, the methods were based off the manufacturers' methods, and they were classified as "user-defined methods." However, they were certifiable under EPA method reference 8321 (US EPA 1994). EPA method 8321 is simply a catch-all method for anything utilizing liquid chromatography/mass spectrometry (LC/MS) technology. Method 8321 can be employed for herbicides, explosives, and PFAS to name a few.

In 2008 the EPA published a method for PFAS in potable water only. To date, this and the latest drinking water method, 533, are the only EPA published methods for PFAS. We still do not have a published EPA method for any other matrix. By early 2016, PFAS started to gain much attention. The UCMR3 rule had been implemented, the EPA published lifetime health advisory limits, and the Department of Defense (DoD) took action across the nation to address PFAS in drinking water. By late 2016, stakeholders indicated that user-defined methods were no longer acceptable, and an EPA reference method was desired. It was around that time that a method referenced as "537 Modified" for all the non-potable water and solid matrices emerged. In 2017, the DoD addressed PFAS data quality assurance criteria for the first time in Table B-15 of the Quality Systems Manual (QSM), giving the industry something to standardize PFAS methodology against (DoD 2019). Around the same time, EPA Region 5's laboratory in Chicago, IL developed and published two methods for other matrices, American Society for Testing and Materials (ASTM) methods ASTM D7979 and ASTM D7968 (ASTM 2017, 2020). In 2018 and 2019, the EPA published an update to method 537 (Shoemaker and Tettenhorst 2018), adding four new target analytes, and the added method 533 (Rosenblum and Wendelken 2019). However, again these methods were only applicable to drinking water.

From the time period between 2009 and 2019 a number of methods were generated to address PFAS in matrices other than drinking water, but each of these methods had limitations and were not widely adopted by the commercial laboratory community. The first method was published in 2009 by the International Organization for Standardization (ISO) which works to develop consensus-based standards. The ISO 25101 method was validated for PFOA and PFOS only in drinking water, groundwater and surface water (ISO 2009). Due to the limited application and lack of prescriptive measures, this method has only been recognized in the US by the state of New York.

In 2014, a method for PFAS in soil was developed by EPA's Chicago laboratory and was published by ASTM, a global standards development organization, as ASTM D7968. They followed up this effort with a method for PFAS in water, sludge, influent, effluent and wastewater under method ASTM D7979. Although both methods utilize liquid chromatography with tandem mass spectrometry (LC/MS/MS) technology, they also utilize antiquated techniques, such as external standard calibration, as opposed to today's gold standard technique of isotope dilution. In order to minimize the introduction of background artifacts during the preparation of samples, the authors of the ASTM methods opted to bypass the solid phase extraction (SPE) procedures prescribed in method 537. Instead they directly injected the sample

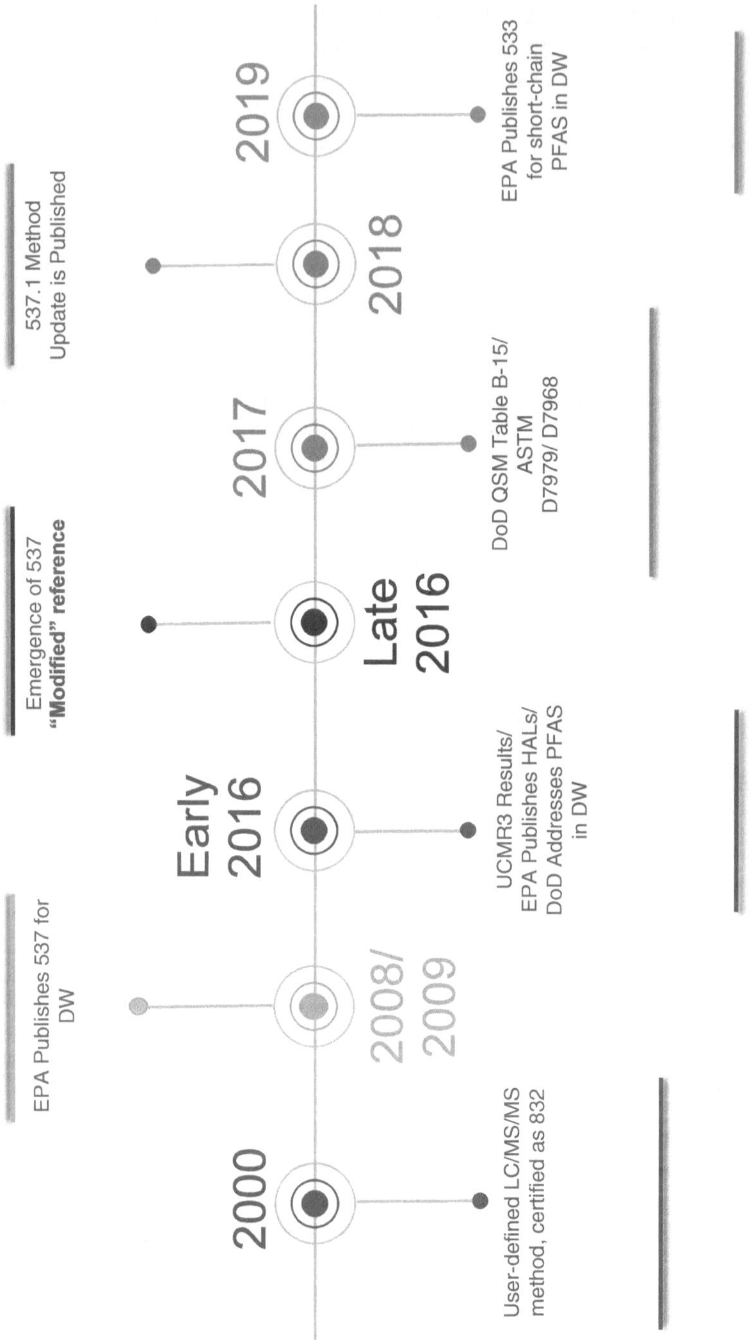

FIGURE 5.1 Historical timeline showing important steps in PFAS analytical method development.

"as is" onto the instrument. This approach required the use of smaller sample volumes leading to higher detection limits. Although these methods were developed by the EPA, they did not go through the regulated process to be published as EPA standard methods, which means they did not go through a multi-laboratory third-party validation process. For all of these reasons, the ASTM methods were never widely adopted outside of state-run laboratories.

Over the course of 2018 and 2019, the EPA developed a SW-846 method, 8327, that was based off of ASTM Method 7979 (US EPA 2019). This effort was essentially the regulated process for validating and publishing the ASTM method. EPA 8327 was released in the summer of 2019 for public comment, and that period has since closed. As of March 2020, the EPA indicated they were still in the process of responding to the public comments received and making appropriate adjustments to the method, so we await the final method to be published in the SW-846 Compendium. This method received a number of criticisms through the public comment process, most notably from the DoD, indicating this method falls short of the data quality objectives established by the DoD QSM.

We also have the somewhat controversial "modified" method 537. "Modified" is somewhat of a misnomer because strict drinking water methods do not lend themselves to modifications. It is helpful to recognize that the modified method 537 was not developed or intended for compliance drinking water samples, but rather for other matrices that lack a standard method. This is a laboratory specific, user-defined method, based on the 537 methodology but adapted to support other matrices. Industry adopted this unconventional nomenclature under the unusual circumstances where there was no EPA source method to reference yet emerging regulations, and demand for testing was increasing.

Next, the EPA began the process of generating a method in coordination with the DoD to support the data quality objectives for the DoD. This method is expected to adopt many of the techniques the commercial laboratory industry has incorporated into their user-defined methods. This method is being developed by EPA's Office of Water and is expected to be published as a 1600 series method. The EPA refers to this future method as the "SPE-ID" method, signifying that it will be a SPE method using isotope dilution quantitation (ID). According to the EPA's latest Technical Brief dated January 2020, the estimated timeframe for releasing this method is no sooner than 2021. See Table 5.2 for a listing of available methods and techniques.

TABLE 5.2

A compilation of available and pending methods with associated analytical techniques for non-potable water and solid matrices as of April 2021.

Method	Technique
ISO 25101	LCMSMS SPE Isotope Dilution
ASTM D7979–18	LCMSMS Direct Injection External Standard
ASTM D7968–17	LCMSMS Direct Injection External Standard
EPA 8327	LCMSMS Direct Injection External Standard
User-Defined Method "537 Modified"	LCMSMS SPE Isotope Dilution
EPA 1600 series method	LCMSMS SPE Isotope Dilution

5.3 WHERE THE INDUSTRY IS AT TODAY

Today, the question about which analytical method to use for PFAS in non-potable water or solid matrices is more perplexing than is typical for method selection. The illustration in Figure 5.2 shows the methods that are available today and the ones that have not been published yet and therefore are not available.

We saw the emergence of an additional methods issue at of the beginning of 2020. With EPA publishing method 533 for short-chain PFAS compounds that were not historically part of any laboratory's target analyte list, we have already seen states and consultants looking for laboratories to support the 533 compounds in non-potable water and solid matrices. What further variability this will lead to in the industry is still unknown. It remains to be seen whether laboratories will develop yet another user-defined method to support these compounds in non-potable water and solid matrices. We also await possible propagation of the "modified" reference for method 533 and further modification of laboratories' 537 method to capture these analytes. As of July 2020, the DoD has confirmed that their joint method validation efforts with the EPA have been expanded to include a target list of 40 PFAS compounds under the future 1600 series method (US EPA 2020). This list of 40 compounds includes all five of the unique short-chain PFAS that were included in method 533. Unfortunately, that method is years away from being published, and the demand for assessing these chemicals in environmental samples exists today.

FIGURE 5.2 Available methods versus methods under development as of April 2021.

The DoD QSM has become an attractive option for practitioners and regulators who want defensible data, but lack a source method for comparison. It is important to understand that the DoD does not publish methods. They typically reference a source method and then build upon it with additional criteria. The DoD did not have this option with PFAS in non-potable water or solid matrices, so they ended up developing such robust and comprehensive parameters that it reads somewhat like a method, and it ensured defensible and consistent data were being generated across program labs.

Over the course of 2019, the DoD revised their QSM from version 5.1, to 5.2 and finally 5.3 where we saw additional revisions to Table B-15 which addresses PFAS in non-potable water and solid matrices (DoD 2019). The applicability of the DoD QSM is to more than just those associated with work performed for or by the DoD. States such as Colorado and California are defaulting to the criteria outlined in the QSM for programs like wastewater monitoring and investigations at landfills and airports to ensure they are acquiring defensible PFAS results.

With the lack of Standard Methods, States have begun taking their own approaches to analytical requirements, as depicted in Figure 5.3. This state-specific approach is a continuation of similar efforts to set their own health-based limits for certain PFAS compounds. For example, Michigan started out with a much softer approach by recommending the use of laboratories with National Environmental Laboratory Accreditation Council (NELAC) or DoD Environmental Laboratory Accreditation Program (ELAP) accreditation. California took it a step further by requiring compliance with QSM Table B-15 at non-DoD sites. There are also States that have been developing their own unique method criteria. Minnesota refers to these as analytical requirements. In Wisconsin, they have established performance-based criteria similar to the DoD approach. New Jersey and Pennsylvania have developed criteria which are largely based on the drinking water method, 537.1, but with the allowance of isotope dilution.

At this point, laboratories must maintain multiple standard operating procedures (SOPs), and 2019 brought a great deal of unrest when it came to certifications as each State or program rolled out their unique requirements. Other challenges, seemingly smaller but still very impactful, are the restrictions around naming conventions for each of these unique method approaches. One State might certify a laboratory's method as "537 Modified" while another State may not accept this nomenclature and require that the SOP references "537 Isotope Dilution". Not only does this challenge laboratories to maintain state certifications, it leaves data users wondering how to compare one laboratory's certification to another's.

The unfortunate part of all this activity is that none of it leads to improving consistency or reducing variability across state lines. This variability is not expected to slow down anytime soon.

As part of some of the method development efforts, the target analyte lists for PFAS continue to grow and change as well. The EPA's draft target analyte list for non-potable water and solid matrices is captured in Table 5.3. For a number of years, it was the 24 compounds listed in the EPA target analyte list. As of 2019, it included the primary replacement chemicals that were added to method 537.1, so that totaled 28 compounds. As previously mentioned, the target analyte list for the method

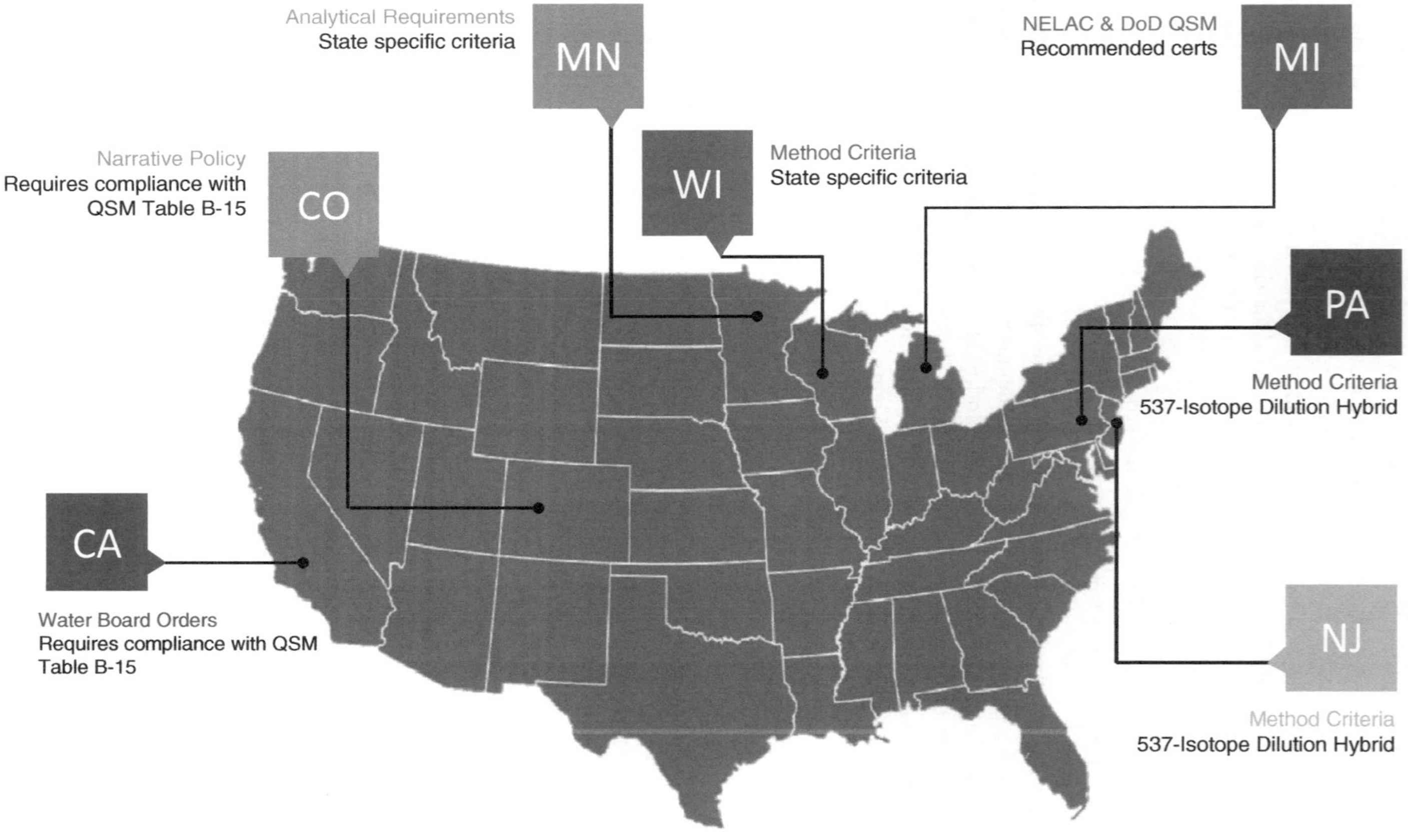

FIGURE 5.3 A varied approach with state specific analytical requirements.

TABLE 5.3
EPA's draft target analyte lists for non-potable water and solid matrices as of October 2020.

Perfluorobutanoic acid (PFBA)	EPA Draft Target Analyte List
Perfluoropentanoic acid (PFPeA)	
Perfluorohexanoic acid (PFHxA)	
Perfluoroheptanoic acid (PFHpA)	
Perfluorooctanoic acid (PFOA)	
Perfluorononanoic acid (PFNA)	
Perfluorodecanoic acid (PFDA)	
Perfluoroundecanoic acid (PFUnA)	
Perfluorododecanoic acid (PFDoA)	
Perfluorotridecanoic Acid (PFTriA)	
Perfluorotetradecanoic acid (PFTeA)	
Perfluorobutanesulfonic acid (PFBS)	
Perfluorohexanesulfonic acid (PFHxS)	
Perfluoroheptanesulfonic Acid (PFHpS)	
Perfluorooctanesulfonic acid (PFOS)	
Perfluorodecanesulfonic acid (PFDS)	
Perfluorooctane Sulfonamide (FOSA)	
N-methyl perfluorooctane sulfonamidoacetic acid (NMeFOSAA)	
N-ethyl perfluorooctane sulfonamidoacetic acid (NEtFOSAA)	
Perfluoro-1-pentanesulfonate (PFPeS)	
Perfluoro-1-nonanesulfonate (PFNS)	
6:2FTS	
8:2FTS	
4:2FTS	
DONA	Replacement Chemicals
HFPO-DA (GenX)	
F-53B Major	
F-53B Minor	

validation efforts for the future 1600 series method continue to expand. A number of States have adopted some version of the EPA's target analyte list, although Wisconsin leads the way with a target list of 33 compounds. This list was established in October 2019, and as of February 2021 it remains the most comprehensive required target analyte list for any of the states.

Until such time that we have a standard EPA method that is widely adopted, we must take care ourselves to ensure that defensible data are being generated and used to make what are significant decisions in terms of cost and implications to regulation, health and liability.

In order to have confidence in the defensibility of data being utilized, one has to understand how the data are generated. When asking PFAS data consumers about their confidence in laboratory PFAS results, often times the response is centered on quality control (QC). If the QC metrics are achieved, the customer is satisfied that the data are defensible quality. There is much that goes on throughout the laboratory process that can impact results that are not evident by reviewing quality control metrics. Ensuring defensible data are being acquired starts with having a reason to have confidence in the laboratory. Once a customer is satisfied with a laboratory's experience with technologies and techniques such as liquid chromatography, high resolution mass spectrometry, or isotope dilution, the next step is understanding the laboratory's SOPs. This is where details regarding sample preparation procedures and complex matrix mitigation measures are found. The SOP includes impactful parameters a customer would not know from a data package. With a user-defined method, it is also important to recognize that the quality control parameters a laboratory meets are the ones they established for themselves.

Some of those impactful parameters can be understood with a comparison of the various method options, considering their key features and significance as delineated in Table 5.4. Consider a comparison of the drinking water method, 537.1, the draft EPA 8327 method, and the commonly employed user-defined method. The primary differences among these methods are sample size, quantitation schemes, and preparation procedures. Method 537 uses an internal standard, the EPA method uses an external standard, and the user-defined versions typically use isotope dilution. Method 8327 is a direct injection method, bypassing the solid phase extraction step in 537.1 and the user-defined method. There is also a difference in sensitivity. Historically. this was of no consequence, but with the downward trend of screening levels and action limits, this is becoming more relevant.

Trace concentrations, such as in the parts per trillion (ppt) range, are possible with LC/MS/MS. This technology employs a soft ionization technique which is very selective. We gain sensitivity through great selectivity. Even though the magnitude of the response is not great, the noise is minimized and the signal to noise ratio is optimized.

There are various approaches to sample preparation in the methods listed in Table 5.1. One is a direct inject method, such as the 8327 method which bypasses the SPE step. The next is an external SPE, and the last is an in-line SPE. In the external method, the aqueous sample is loaded on and eluted off an SPE cartridge like the one depicted in Figure 5.4. In the in-line approach this separation chemistry is automated inside the instrument. During the SPE process, the target analytes are retained and

TABLE 5.4

Comparison of key features of three different analytical methods for detecting PFAS in aqueous matrices.

Features	Method 537.1	EPA DRAFT 8327	User Defined Method
Matrices	Drinking water	Non-potable aqueous	All aqueous matrices
Sample size	250 mL	Any allowed (5 mL used)	250 mL
Analysis	LCMSMS	LCMSMS	LCMSMS
Aqueous Extraction	SPE SDVB	Direct Injection	SPE Waters WAX
Branched/Linear Isomers	Yes, for available standards	Yes, for available standards	Yes, for available standards
Confirmation Ion	No	Yes	Yes
Quantitation	Internal standard	External standard	Isotope dilution
Reporting Limits	(2 ppt - 40 ppt)	(10 ppt - 8000 ppt)	(2 ppt - 10 ppt)
Recovery Correction	No	No	Yes

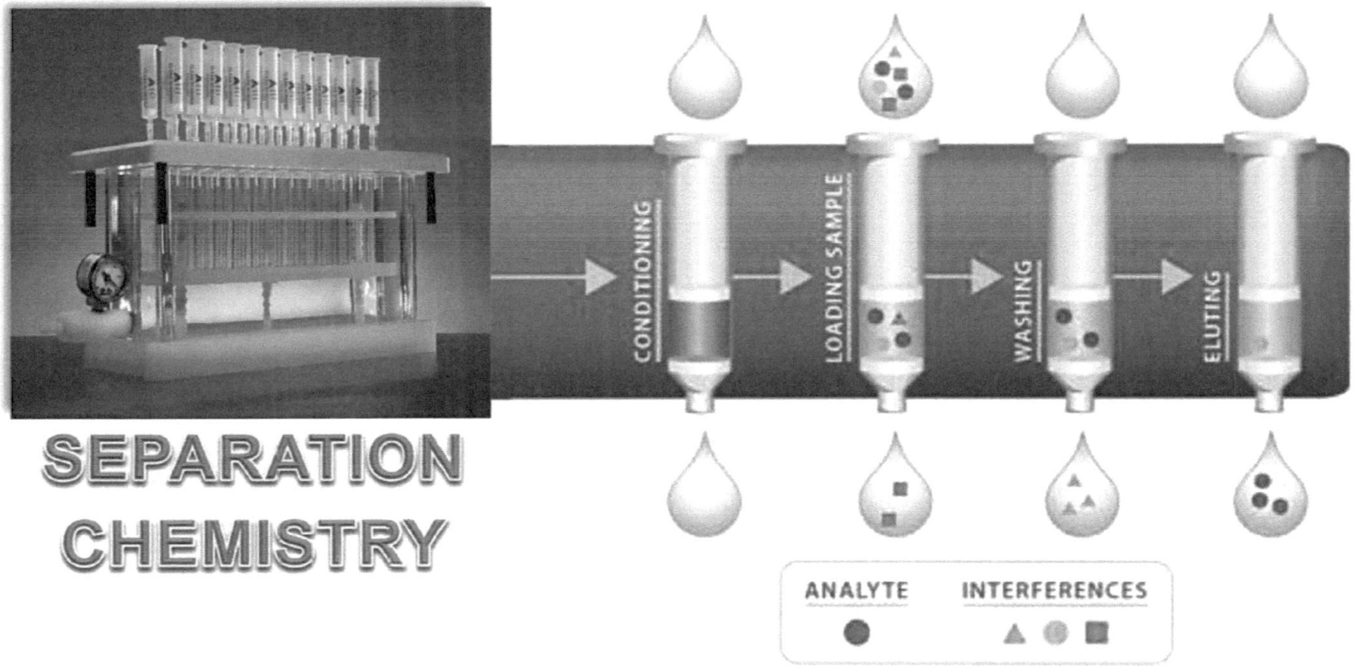

FIGURE 5.4 Process of conducting a manual sample extraction using, for example, a solid phase extraction cartridge.

then eluted with a basic methanol solution. This is the first step in basic separation chemistry applied to the samples. For matrices beyond pristine drinking water, this can become an important element.

Next is an important factor that can impact reported concentrations. In the example shown in Figure 5.5, PFOS is on the left and PFOA is on the right. Method 537.1 states that branched and linear isomers should be reported when quantitative branched and linear standards are available. On the left we have a PFOS standard with branched and linear isomers, so both branched and linear isomers have a response factor and are summed in the sample data. If we look at PFOA, only a quantitative linear standard is commercially available, so some laboratories might only report the linear peak. However, a qualitative branched standard for PFOA is available and can be used to sum the branched and linear peaks. The significance is that one laboratory could report a concentration of 0.99 ng/ml versus 1.20 ng/ml.

If we are looking at a compound such as perflurorbutanesulfonic acid (PFBS), where the literature indicates branched isomers do exist, we do not have any kind of standard that includes the branched isomer. Therefore, no laboratories can report a result for the true mass of this compound.

Consider an even more important factor that can impact reported concentrations. Compounds are identified based on their parent > daughter transition masses and the ratios between the signals of each. Selective ion monitoring (SIM) analysis is a similar concept. Figure 5.6 gives a PFOS analysis. On the top is the reference standard, and the environmental sample is on the bottom. When monitoring the secondary transition, we expect to see the same response as observed in the standard, but looking at the sample we see a signal that does not match the primary transition. If we were only looking at the primary transition in this sample, we would accept it and report a result of 80 ppt for PFOS. If we monitored the secondary transition, we would know that the ion ratio does not meet our acceptance criteria, and we would qualify the 80 ppt result as estimated due to matrix interference. This is an example of where the laboratory might meet their own QC criteria if the SOP does not specify criteria for monitoring the secondary ion ratios.

Isotope dilution is an approach that helps with proper identification and accurate quantification. This scheme has been used for decades to support the analysis of ultra-trace level dioxins and PCB congeners, but is unfamiliar to many. With isotope dilution, we add known amounts of labeled analogues of the native compounds into the sample. We use these analogues for quantitation as well as recovery correction. For example, PFOS amended with carbon-13 can be distinguished from the native PFOS, which is comprised of carbon-12. A known amount of PFOS labeled with carbon-13 is added to the sample, and its behavior is observed through the preparation and analytical process. It may be regarded as a tracer compound. Whatever affects the native analyte in the sample will equally affect the isotope. This means we can accurately measure impacts from matrix interference and make corrections for them when quantifying the sample of interest.

The isotope dilution technique offers many quantitative and qualitative benefits. Compared to external and internal standard applications, isotope dilution is the most accurate and the most precise. Accuracy is improved because the concentration of target analytes is quantitated against structurally very similar standard materials,

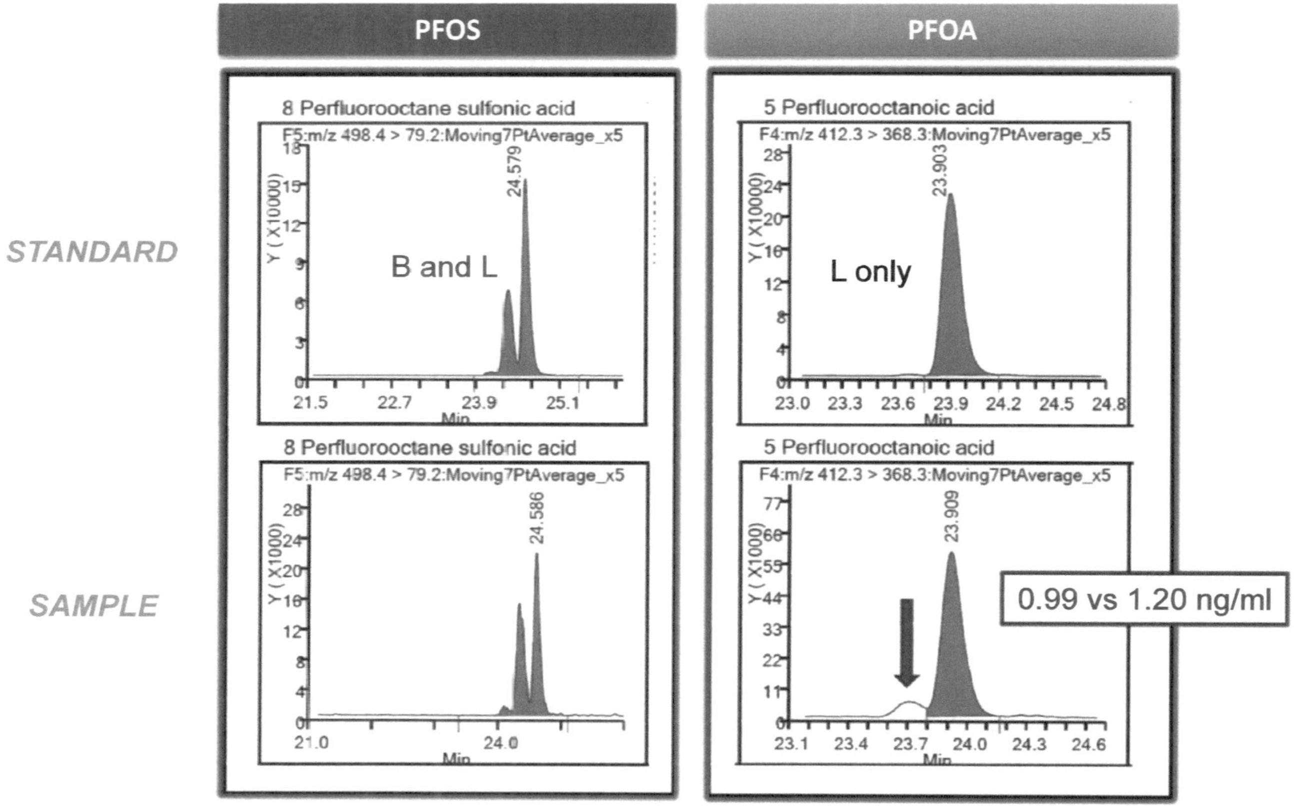

FIGURE 5.5 The importance of considering both branched and linear isomers in quantifying PFAS concentrations. Unlike PFOS which contains a quantitative branched and linear standard. PFOA contains only a qualitative branched standard but both peaks should be summed in the reported concentrations.

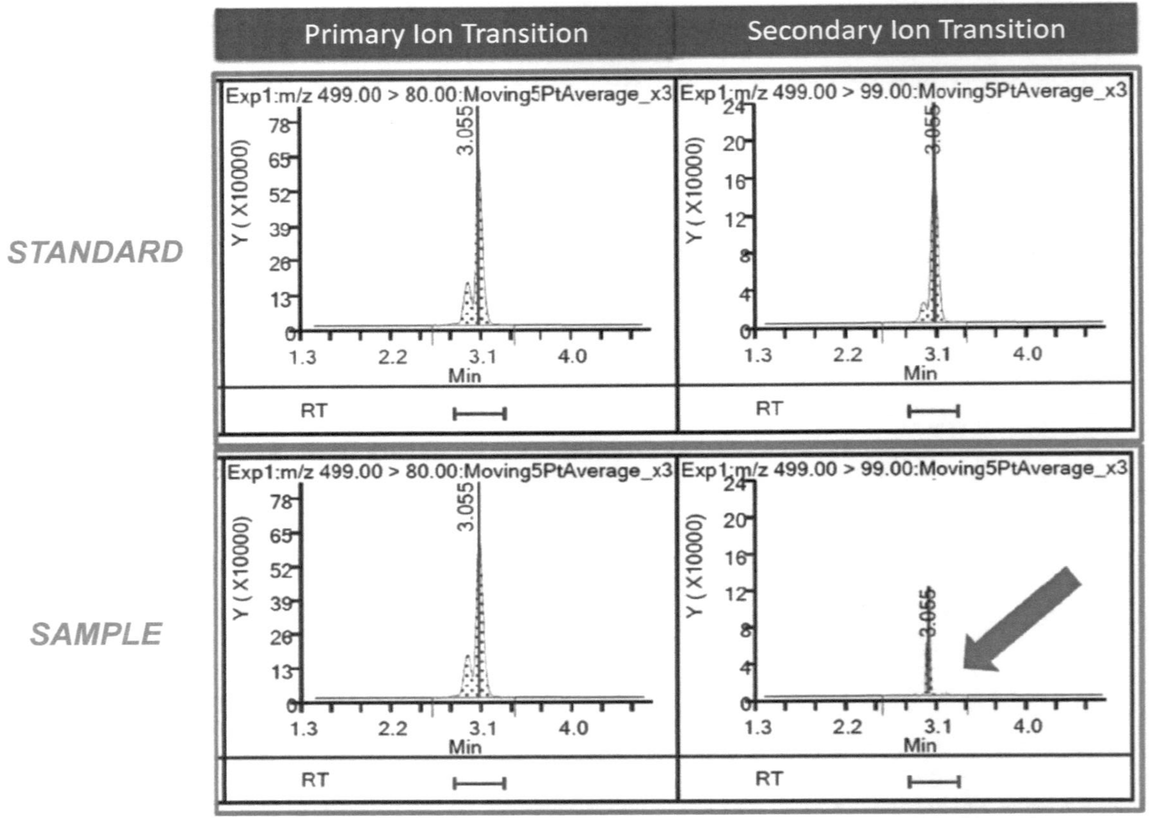

FIGURE 5.6 PFOS ion transitions. The importance of monitoring secondary ion transitions in quantifying PFAS concentrations. If only monitoring primary ion transitions, matrix interference indicated by the secondary ion transition would be missed.

namely the isotopes themselves. It allows for compensation of matrix interference and routine instrument signal drift. Qualitative benefits include increased confidence in peak identifications as matrix-related retention time shifts are easily resolved. This reduces the chances of both false positives and false negatives. All of this becomes increasingly more critical in handling complex matrices.

5.4 A LOOK AT SPECIFIC MATRICES

PFAS in the environment and environmental regulations involve matrices such as surface water, groundwater, and soils. There are fewer regulations and guidance that address the complexities of matrices such as biosolids, landfill leachate, or source air. The following are complex matrices that have become increasingly important from a risk perspective.

5.4.1 Landfill Leachate

The end-of-life cycle for the PFAS chemicals widely used in industrial and consumer products is often at a landfill. Landfill leachate is then discharged to wastewater treatment plants (WWTPs) where, in large part, they may not be captured by traditional treatment systems and are released to the environment. This process garnered growing attention in 2018 as some states worked to understand this life cycle and identify potential sources. This is a complex matrix which presents unique challenges. Michigan has a surface water discharge limit of 11 ppt for PFOS, a limit well below the health advisory level for drinking water. It is problematic to achieve trace level detection limits applicable to pristine drinking water in a concentrated material such as landfill leachate. Concentrations can vary widely depending on what was disposed of at the landfill. Various studies of landfills have concluded that they are predominantly made up of short chain PFAS, like perfluorobutanoic acid (PFBA) and PFBS, rather than the longer chains. In order to understand the potential impact of landfill leachate on the aqueous environment, it is critical to know what percentage of mass the leachate represents of the mass flow from the WWTP. Busch et al. (2010) noted that landfill leachate can represent less than 1% of mass flow from a WWTP, so it can be a minor source of PFAS into the aqueous environment. Regardless of the ultimate impact on the environment, this matrix represents an analytical challenge for laboratories. There is the potential for high-level target analytes as well as bulk interferences from a variety of co-contaminants. Laboratories must have a procedure to manage the potential for high salinity, total organic carbon, and suspended solids. In many cases, the laboratory will mitigate these negative matrix impacts by extracting smaller sample sizes which results in elevated reporting limits. These factors contribute to the problematic nature of achieving low ppt action limits in a complex matrix.

5.4.2 Biosolids

Biosolids are another related matrix that garnered increased attention in 2019. The application of biosolids to agricultural land is a cost-effective method of waste disposal by beneficially recycling organic matter. This is a practice that is employed across the US. The most notable action that transpired in 2019 was Maine's response to a contaminated dairy farm from the land application of contaminated biosolids. This was the first state to set a limit for PFOS and PFOA in biosolids at 2.5 ppb and 5.2 ppb, respectively. There are multiple concerns that stem from the spreading of biosolids on agricultural land which include PFAS contaminants leaching into groundwater, surface water run-off, and plant uptake of the contaminants. The mechanisms that control these outcomes for PFAS, like solubility, leachability, and plant transfer factors, are not well understood or well-studied. When land application is restricted as it is in Maine, WWTPs are generally limited to sending their sludge back to a landfill for disposal or a sewage sludge incinerator. Both of these alternatives have growing limitations. Landfills are understandably more cautious about accepting known PFAS contaminated waste.

Biphasic samples, especially samples with high particulates, pose an analytical challenge. A determination must be made as to whether the data need to represent the PFAS concentration in the whole sample or in the dissolved phase only. High particulate content can clog the SPE cartridge that samples are run through as part of the sample extraction process. When this occurs, if not enough sample volume passes through the cartridge for subsequent analysis, the sample requires phase separation processing. Once the laboratory has generated two separate phases, solid and aqueous, the data user must decide if they want to see the distribution of PFAS across these phases for the total PFAS in the whole sample or PFAS in only the dissolved phase.

5.4.3 Air

PFAS from source air emissions is gaining increased attention. The facilities listed in Figure 5.7 play a critical role in the beginning and end-of-life-cycle for PFAS. The cycle begins with the manufacturing plants and ends with the treatment facilities like thermal oxidizers and incinerators. Source air emissions travel via short- and long-range transport as evident by detections of PFAS in the Arctic Ocean and in soil and

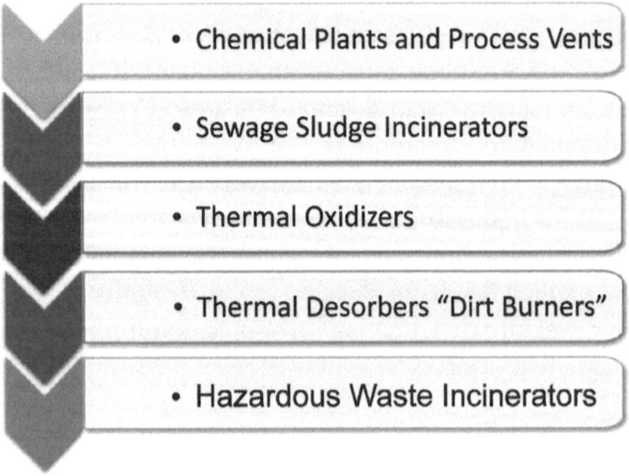

FIGURE 5.7 Critical industrial processes involved in the beginning and the end of life cycle of PFAS.

groundwater surrounding these facilities. Point source emissions can eventually lead to contaminated ambient air with measurable impacts to surface water, soil and eventually groundwater. In the US, the National Defense Authorization Act (NDAA) has become a tool for enacting PFAS legislation. The 2020 NDAA included a provision that addressed the incineration of AFFF. All incineration of AFFF must be conducted at a temperature range believed to be adequate to break down PFAS, while ensuring the maximum degree of reduction in emissions and all incineration must be conducted in accordance with the Clean Air Act (CAA) and at a permitted facility that has been permitted to receive waste regulated under the Solid Waste Disposal Act. The CAA requires the EPA to establish new source performance standards for new incineration facilities and emission guidelines for existing facilities. As is the case with all non-drinking water matrices, we lack an EPA standard reference method specific to PFAS in air, and in order to enforce emissions guidelines EPA will need to publish a method. In the interim, The EPA has published Other Test Method (OTM) 45 for PFAS characterization in stack gas. OTMs are test methods which have not yet been subject to the Federal rulemaking process.

PFAS in ambient air is also gaining more recognition these days. We have limited data sets for what is present at background levels in ambient air. This is important information to have when trying to understand the impacts that are attributable to point sources in a particular area. With the lack of published methods for PFAS in source or ambient air, commercial laboratories have conducted the necessary method development to support existing consent orders, research studies and the government in their method development efforts. These methods are built from EPA standard methods for semivolatile and volatile compounds in air, but modified to support the complexities unique to PFAS chemicals. The method development process has illuminated much about the unique chemical characteristics of these compounds and how they behave under different conditions, but we still have much to learn about what is present in the environment.

5.5 INDUSTRY TRENDS

While standard methods are available for the analysis of a few dozen PFAS compounds, the quantitative analysis of other PFAS is difficult due to the sheer quantity of compounds and the lack of reference materials. Because of this, the full extent and distribution of PFAS precursors, intermediates, and terminal products in the environment have generally not been assessed. There are techniques available for capturing a total number as opposed to speciated results for 4700 chemicals.

A largely academic technique, the Particle-Induced Gamma-ray Emission test (PIGE) measures total fluorine on consumer products or industrial product surfaces. It can also be used to measure total fluorine in water with possible detection limits of 1 part per billion (ppb). Often times when journal and news articles reference results for Total Fluorine from a PIGE analysis in the US, these data stem from a single laboratory run by Dr. Graham Peaslee at the University of Notre Dame (University of Notre Dame 2020). Generally speaking, however, this analysis is not found in commercial laboratories as it requires the use of a particle accelerator which is beyond the budget, space and operating capabilities of commercial laboratories.

Combustion ion chromatography (CIC) is a technique that is amenable to commercial environmental laboratories and is capable of capturing Total Organofluorines with slightly better sensitivity than the PIGE analysis, resulting in reporting limits in the single digit or less ppb range. This is intended to be a rapid screening tool and would be most useful at a contaminated site with concentrations at or above the ppb range. Many AFFF-impacted sites would fall into this category. There are three CIC analysis options: total organic fluorine (TOF), adsorbable organic fluorine (AOF), or extractable organic fluorine (EOF). As seen from the example summarized in Figure 5.8, TOF and EOF results are within experimental measurement uncertainty but demonstrate a large difference between the conventional targeted LC/MS/MS analysis, meaning that there are considerable amounts of unknown PFAS in this example. These unknown PFAS are most likely precursor compounds which have the potential to transform in the environment into shorter-chain, stable perfluorinated chemicals. These precursors make up many of the unknown chemicals in this class of over 4700 PFAS.

PFAS chemicals are mostly unknown. The ability to test for unknowns, otherwise known as non-targeted analysis, has become very attractive. A traditional laboratory analysis requires the laboratory to be specific about what they are looking for in a sample. This requires setting up a method and calibrating an instrument with reference materials to look for those specific constituents. The reference materials are man-made versions of the exact compounds of interest in the environment. It is with these standards that the laboratory can identify the specific target compounds of concern in a sample and report a concentration for them. With advanced technology, certain instruments can operate in different modes, no longer using reference materials to detect specific compounds. In this mode, the instrument scans for all chemicals

Variance between Analytical Methods for Total PFAS

Total Organic Fluorine (TOF)	410 mg F/kg
Extractable Organic Fluorine (EOF)	390 mg F/kg
LC-MS/MS ΣPFAS (n=28)	120 mg/kg

CIC: **Combustion Ion Chromatography**

FIGURE 5.8 Data demonstrating the difference in mass identified by various analytical approaches to measuring Total PFAS. As demonstrated by this case study, the difference in mass captured by the traditional 537 method and Total or Extractable Organic Fluorine indicates the presence of many unknown PFAS compounds in this sample.

and the complexity lies in sorting through the data to identify PFAS and which PFAS appear to be more relevant than others.

Non-target analysis has promise across a wide range of PFAS-related applications, ranging from discovery of additional analytes of interest, to elucidation of environmental transformation pathways, to unique characterization of product formulations. Multiple academic institutions and EPA's Office of Research and Development (ORD) have made notable advancements in terms of discovery of next-generation PFAS chemicals and characterization of product formulations. Since it is challenging to obtain government and academic resources to support commercial work, commercial laboratories are developing these methods directly to support the industry as a whole. Although commercial laboratories are capable of contributing to the research efforts previously mentioned, it is more likely that their services will be called upon to assist in forensics applications where there is a lack of clarity regarding responsible parties.

5.6 CONCLUSIONS

Establishing formal guidance, finalizing methods, or promulgating rules in the face of rapidly evolving science is a tall order but one that the EPA will have to tackle sooner rather than later. When they do not, Congress and the States have shown over the last few years they will move forward with their own efforts while drawing their own conclusions about the state of the science. What we are left with is a mismatch of actionable parameters from one state to the next or laws from Congress being enacted where the technology does not yet exist to comply with them. The EPA has developed the PFAS Action Plan for the purposes of moving all of these efforts forward, but it appears their timeline for doing so is not well aligned with what communities and elected officials are demanding of the environmental industry. There is a race to develop technologies and methodologies capable of supporting these demands but the toxicology lags behind, so we are left with testing initiatives that are in front of the science needed to guide us with where to look next.

REFERENCES

American Society for Testing and Materials, 2017. *ASTM D7968: Standard Test Method for Determination of Per- and Polyfluoroalkyl Substances in Soil by Liquid Chromatography Tandem Mass Spectrometry (LC/MS/MS)*. Available at: https://www.astm.org/Standards/D7968.htm (accessed October 19, 2020).

Busch, J., Ahrens, L., Sturm, R., and Ebinghaus, R. 2010. Polyfluoroalkyl compounds in landfill leachates. *Environmental Pollution*, 158, 1467–1471.

American Society for Testing and Materials, 2020. *ASTM D7979: Standard Test Method for Determination of Per- and Polyfluoroalkyl Substances in Water, Sludge, Influent, Effluent, and Wastewater by Liquid Chromatography Tandem Mass Spectrometry (LC/MS/MS)*. Available at: https://standards.globalspec.com/std/14326571/ASTM%20D7979 (accessed October 19, 2020).

Department of Defense, 2019. *Department of Defense (DoD) and Department of Energy (DOE) Consolidated Quality Systems Manual (QSM) for Environmental Laboratories*. Available at: https://denix.osd.mil/edqw/documents/manuals/qsm-version-5-3-final/ (accessed October 19, 2020).

International Organization for Standardization, 2009. *Water Quality – Determination of Perfluorooctanesulfonate (PFOS) and Perfluorooctanoate (PFOA) – Method for Unfiltered Samples Using Solid Phase Extraction and Liquid Chromatography/Mass Spectrometry*. Available at: https://www.iso.org/standard/42742.html (accessed October 19, 2020).

Rosenblum, L., Wendelken, S.C., 2019. *Method 533: Determination of Per- and Polyfluoroalkyl Substances in Drinking Water by Isotope Dilution Anion Exchange Solid Phase Extraction and Liquid Chromatography/Tandem Mass Spectrometry*. Available at: https://www.epa.gov/sites/production/files/2019-12/documents/method-533-815b19020.pdf (accessed October 19, 2020).

Shoemaker, J.A., Grimmett, P.E., Boutin, B.K., 2009. Method 537: Determination of selected perfluorinated alkyl acids in drinking water by solid phase extraction and liquid chromatography/tandem mass spectrometry (LC/MS/MS). *EPA/600/R-08/092*. Available at: http://file:///C:/Users/racet/AppData/Local/Temp/MicrosoftEdgeDownloads/dcb8fc5b-610c-4cd1-911b-1a05ea719dfd/METHOD%20537_FINAL_REV1.1.PDF (accessed October 19, 2020).

Shoemaker, J.A., Tettenhorst, D.R., 2018. Method 537.1: Determination of selected per- and polyfluorinated alkyl substances in drinking water by solid phase extraction and liquid chromatography/tandem mass spectrometry (LC/MS/MS). *EPA/600/R-18/352*. Available at: https://cfpub.epa.gov/si/si_public_record_Report.cfm?Lab=NERL&dirEntryId=343042 (accessed October 19, 2020).

University of Notre Dame, 2020. *Department of Chemistry and Biochemistry*. Available at: https://chemistry.nd.edu/people/graham-peaslee/ (accessed October 19, 2020).

US Environmental Protection Agency, 1994. *Method 8321: Solvent Extractable Non-volatile Compounds by High Performance Liquid Chromatography/Thermospray/Mass Spectrometry (HPLC/TSP/MS) or Ultraviolet (UV) Detection*. Available at: http://legismex.mty.itesm.mx/secc_inter/SW-846/8321.pdf (accessed October 19, 2020).

US Environmental Protection Agency, 2019. *Method 8327: Per-and Polyfluoroalkyl Substances (PFAS) Using External Standard Calibration and Multiple Reaction Monitoring (MRM) Liquid Chromatography/Tandem Mass Spectrometry (LC/MS/MS)*. Available at: https://www.epa.gov/sites/production/files/2019-06/documents/proposed_method_8327_procedure.pdf (accessed October 19, 2020).

US Environmental Protection Agency, 2020. *Status of EPA research and development on PFAS*. Available at: https://www.epa.gov/chemical-research/status-epa-research-and-development-pfas (accessed October 19, 2020).

6 Landfills as Sources of PFAS Contamination of Soil and Groundwater

Nanthi Bolan, Son A. Hoang
University of Newcastle, Australia

Yubo Yan
Huaiyin Normal University, China

Sammani Ramanayaka
Lancaster University, United Kingdom

P. Koliyabandara
University of Sri Jayewardenepura, Sri Lanka

Gayathri Chamanee, Hasintha Wijesekara
Sabaragamuwa University, Sri Lanka

Raj Mukhopadhyay
ICAR-Central Soil Salinity Research Institute, India

Binoy Sarkar
Lancaster University, United Kingdom

Meththika Vithanage
University of Sri Jayewardenepura, Sri Lanka

M. B. Kirkham
Kansas State University, United States of America

CONTENTS

6.1 Introduction 120
6.2 Landfills for Waste Management 121
6.3 Accumulation of PFAS in Landfills 123
6.4 Leaching of PFAS from Landfills 126
6.5 Water and Air Pollution by PFAS 129
6.6 Treatment of PFAS in Landfill Leachate 131
6.7 Conclusions 135
References 135

6.1 INTRODUCTION

Per- and poly-fluoroalkyl substances (PFAS) are a group of synthetic chemicals that are often found on consumer and industrial products such as non-stick cookwares, textiles, furniture, carpet stain protectors, food packaging, plastic materials, and firefighting foams. Aqueous film forming foam (AFFF) used in firefighting is one of the major point sources of PFAS input to terrestrial (i.e., soil) and aquatic (i.e., groundwater) environments. Application of biosolids from wastewater treatment and leachate from landfill sites are considered as the major nonpoint sources of PFAS in these environments (Buck et al., 2011; Kannan, 2011; Bolan et al., 2021).

There are thousands of various PFAS, and more than 600 compounds are used for industrial applications and domestic purposes (US EPA, 2020). In addition to the domestic products listed above, PFAS are included in food contact papers, microwave popcorn bags, carpets, outdoor clothings, cosmetics, and water-resistant upholstery. Textiles, surfactants, insecticides, aqueous film-forming foams, lubricants, semiconductors, and metal plating are industrial products or processes that use PFAS at varying contents and concentrations (Herzke et al., 2012; Kotthoff et al., 2015; Benskin et al., 2012a; Wei et al., 2019). High PFAS levels have been recorded in ski waxes (up to ~2000 μg/kg PFOA) (perfluorooctanoic acid), leather samples (up to ~200 μg/kg PFBA and ~ 120 μg/kg PFBS) (perfluorobutanoic acid and perfluorobutanesulfonic acid), outdoor textiles (up to ~19 $\mu g/m^2$ PFOA), and baking papers (up to ~15 $\mu g/m^2$ PFOA) (Kotthoff et al., 2015). Fluorotelomer alcohols (FTOHs) have also been recorded in outdoor textiles (maximum levels as 379.9 $\mu g/m^2$ FTOH 8:2) (an 8:2 fluorotelomer alcohol is a molecule with 8 fluorinated carbons and a 2 carbon ethyl alcohol group), cleaning agents (up to 73,000 μg/kg 8:2 FTOH), impregnating sprays (up to 719,000 μg/kg 8:2 FTOH), and non-stick cookwares (436 μg/kg) (Kotthoff et al., 2015; Herzke et al., 2012). Many of these consumer products are landfilled after their primary uses, thereby becoming a source for accumulation of PFAS in landfills.

PFAS can enter the terrestrial and aquatic environments from landfill sites that receive waste containing PFAS, because the landfills leach the PFAS to soil and the leachate is discharged to surface and groundwaters. PFAS reaches landfill sites through the deliberate disposal of products containing PFAS to landfills at the end of their lifetimes. Although a number of studies have examined the redistribution and groundwater mobility of PFAS resulting from AFFF use in legacy firefighting sites and biosolid application, there are only limited reports on the distribution and fate of these compounds in soils and groundwater following discharge of landfill leachate. For example, more than 1200 landfill sites in Australia receive around 20 million tons of domestic and industrial waste each year. It is estimated that around 50 GL (1×10^9 L) of leachate is produced from these landfill sites, which need to be treated before being discharged to wastewater treatment plants for subsequent treatment and safe disposal (Bolan et al., 2013).

Most PFAS are chemically and thermally stable, and the remediation of PFAS contamination in both solid and aqueous media is challenging. In the case of solid media such as soil and wastes including biosolids derived from wastewater treatment plants, remediation of PFAS can be achieved through abiotic mineralization or immobilization using adsorbents. Three of the most common abiotic PFAS mineralization remediation technologies include: (i) oxidation processes including electrochemical oxidation, photolysis, and photocatalysis; (ii) reduction processes involving the use of zero-valent metals/bimetals with clay interlayers; and (iii) thermal decomposition aimed at breaking C-F bonds under high temperature. Immobilization of PFAS using adsorbents such as activated carbon reduces the mobility and bioavailability of these compounds (Ross et al., 2018; Mahinroosta and Senevirathna, 2020; Bolan et al., 2021).

For aqueous media including groundwater, landfill leachates, and stormwater sources, the relatively successful remediation approach is to use a pump-and-treat method with adsorbents like activated carbon followed by off-site incineration or safe disposal of the spent activated carbon. However, activated carbon has a relatively low capacity to adsorb PFAS, particularly when shorter-chain compounds are present. Other methods for *ex situ* PFAS removal include expensive high-pressure membrane treatment using nanofiltration or reverse osmosis (Kannan, 2011; Ross et al., 2018; Bolan et al., 2021). Often pre-treatment technologies are incorporated ahead of the adsorbents or membranes, primarily to minimize the impact of other organic compounds such as dissolved organic carbon in landfill leachates on the removal of PFAS compounds. This chapter covers the sources, distribution, and treatment of PFAS in landfill leachate.

6.2 LANDFILLS FOR WASTE MANAGEMENT

Landfills provide the most economical and safe avenue for the disposal of waste, and, in many countries, domestic and industrial wastes are managed primarily using landfills (Bolan et al., 2013; Lamb et al., 2014; Ferronato and Torretta, 2019). Although globally there has been a major shift in the reduction, reuse, and recycling of solid waste, safe disposal to landfill remains the most commonly practiced sustainable waste management method. While landfilling provides an economic practice of waste disposal, it can lead to environmental degradation through the emission of various contaminants, especially when they are not managed properly (Bolan et al., 2013).

The major environmental challenges linked to landfills include surface water and groundwater contamination, release of greenhouse gases (GHGs), and emission of odorous compounds (Albright et al., 2006; Bolan et al., 2013; Lamb et al.,

2014). Landfill leachates are enriched with a range of organic and inorganic contaminants which can impact terrestrial and aquatic environments when landfill leachates reach these sources of water (Scott et al., 2005; Bolan et al., 2013; Lamb et al., 2014). Similarly, the microbial decomposition of putrescible organic waste in landfills can result in the release of GHGs, such as methane (CH_4) and carbon dioxide (CO_2), and odorous compounds, such as volatile fatty acids (Lou and Nair, 2009; Bolan et al., 2013).

Microbial decomposition of organic matter in landfills tends to occur when water comes in contact with the buried waste material, and the release of leachates and GHGs are promoted due to an increase in moisture content in landfills (Lamb et al., 2014). As the water received through rainfall or irrigation infiltrates through landfill waste, it becomes contaminated with dissolved and suspended components originating from the decomposing waste material. The discharged leachate from landfills can contaminate groundwater, surface waters, and soil, potentially impacting the terrestrial and aquatic environments and health. Even closed landfills can continue to release leachates and gases for a long period of time after they have ceased to operate, making the sustainable management of emissions from landfills a long-term challenge for landfill operators and environmental regulators (Brand et al., 2018).

Landfill leachates contain a range of contaminants, including dissolved gases, inorganic compounds, odorous organic compounds, and dissolved organic carbon compounds (i.e., with biological/chemical oxygen demand) (Brand et al., 2018; Kulikowska and Klimiuk, 2008; Bolan et al., 2013). The most common inorganic compounds include heavy metal(loid)s, such as cadmium, lead, and arsenic, and nutrient ions, such as nitrate and phosphate. The common organic contaminants in landfill leachates include pesticides, BTEX (benzene, toluene, ethylbenzenes, and xylenes), total petroleum hydrocarbons, chlorinated aliphatic hydrocarbons, and chlorinated benzene compounds (Slack et al., 2005). However, a range of emerging compounds, including flame retardants (i.e., PFAS), pharmaceuticals and perfluroinated compounds, and nanomaterials have been reported at elevated concentrations in landfill leachate, leading to contamination of surface waters and the underlying groundwater (Lang et al., 2017; Brand et al., 2018; Scheutz and Kjeldsen, 2005).

Microbial decomposition of organic wastes in landfills also results in the generation of a range of gases, which include CH_4, CO_2, nitrogen gases (N_2, N_2O), sulfur gases (SO_2 and H_2S), and a range of other odorous sulfurous trace gases (Zhang et al., 2019; Penza et al., 2010). However, the predominant gas emitted in most landfill sites is CH_4, produced through microbial methanogenic process involving *Methanobacterium* (Zhang et al., 2019; Lou and Nair, 2009), which is captured and used as an energy source. For example, in recent times many local governments have introduced engineered landfills with gas recovery systems and odor management; thereby, landfills potentially provide a significant source of CH_4 as a fuel source (Bolan et al., 2013; Andriani and Atmaja, 2019). Furthermore, increasingly revegetation (i.e., phytocapping) is practiced at traditionally managed landfill sites to reduce water infiltration, thereby mitigating leachate generation and GHG emissions (Lamb et al., 2014).

6.3 ACCUMULATION OF PFAS IN LANDFILLS

Industrial sources of PFAS include primary and secondary manufacturing facilities. In primary manufacturing facilities, PFAS-containing products are synthesized. PFAS-based materials are used in the production process in secondary manufacturing facilities, such as the application of a coating to finished products. PFAS are used for the production of car interior materials and electrical and electronic equipment (PFOS range from 0.07–0.43 µg/kg). High PFAA levels have been recorded in firefighting agents (PFBS, 253,700 µg/kg; PFBA, 27,647 µg/kg) (Bečanová et al., 2016; Herzke et al., 2012). Chemical attributes of PFAS are used in building and construction materials for product strength and durability. PFAS are used for manufacturing lightweight concrete blocks and concrete sandwich panels (Bečanová et al., 2016). Perfluorinated carboxylic acids (PFCAs) and PFOA have been detected in composite wood samples and oriented stand boards with a range of 1.38–13.9 µg/kg. Wood fiber insulations also have shown high PFCAs (12.3 and 5.8 µg/kg) and PFHpA (perfluoroheptanoic acid) (20.6 and 28.4 µg/kg) (Bečanová et al., 2016). PFHpA concentrations are found in wall insulations (61.5 µg m^{-2}) and floor insulations (181.8 µg m^{-2}). The construction and demolition waste that could be derived from these woods can become potential sources of PFAS after disposal in landfills. Uses of PFAS are summarized in Table 6.1. Many landfills receive PFAS contaminated sewage sludge from wastewater treatment plants (Gallen et al., 2017). Wastewater treatment plants (WWTPs) are unable to degrade or remove PFAS entirely, and can even cause the conversion of PFAS precursors to more stable PFAS. For example, fluorotelomer polymers (FTPs), which are commonly used as surface protection agents, were degraded into FTCAs (fluorotelomer carboxylic acids) and FTUCAs (fluorotelomer unsaturated carboxylic acid), and subsequently to PFCAs (Perfluoroalkyl carboxylic acids) under both aerobic and anaerobic conditions in activated sludge from WWTPs (Liu and Avendano, 2013). Microbial degradation of PAPs (polyfluorinated alkyl phosphate esters), which are used in paper and synthetic fibers, produced a mixture of FTCAs and PFCAs in activated sludge (Lee et al., 2010). Concentrations of PFOA and PFOS in sludge ranged from 8300–21,900 pg/g and 8200–993,000 pg/g, respectively, in two WWTPs in the US (Loganathan et al., 2007). In Germany, 61 WWTPs sludge samples had concentrations of PFOS ranging from 14,000 to 2,615,000 pg/g (Omotayo et al., 2013). Decades of consumer usage and the subsequent disposal of PFAS-containing products and disposal of PFAS-containing industrial waste and sludge have led to the PFAS accumulation in both active and old (i.e., aged) landfills (Figure 6.1).

The fate of accumulated PFAS inside landfills is regulated by a combination of biological (biodegradation) and abiotic (desorption) processes (Hamid et al., 2018). A study of the physical and biological release of PFAS from municipal solid waste in anaerobic type model landfill reactors showed the average PFAS leaching measured in biological reactors was 16.7 nmol/kg dry-refuse, which exceeded the average for abiotic reactors (2.83 nmol/kg dry-refuse). This indicated that biological processes are primarily responsible for releasing PFAS from disposed waste (Allred et al., 2015). Biodegradation of PFAS is difficult due to their unique molecular structures and strong C–F bond energy (536 kJ/mol) (Wei et al., 2019). Degradation of the PFAS chain is usually limited to molecules or regions of

TABLE 6.1
Possible sources of perfluoroalkyl polyfluoroalkyl substances (PFAS) in landfills.

Substance	Family	Class	Example	Use	Reference
Perfluoroalkyl substances	Perfluoroalkyl acids (PFAA)	Perfluoroalkyl carboxylic acids (PFCA)	Perfluorooctanoic acid (PFOA)	Ski waxes, outdoor textiles, baking papers, non-stick cookware, water proofing agents, lubricants, surfactant, composite wood, oriented stand boards	Bečanová et al., 2016; Buck et al., 2011; Hamid et al., 2018; Herzke et al., 2012; Matthias et al., 2015
		Perfluoroalkyl sulfonic acids (PFSA)	Perfluorooctane sulfonic acid (PFOS)	Paint, AFFF, leather samples, carpets, printed circuit boards	Bečanová et al., 2016; Herzke et al., 2012
			Perfluorobutane sulfonate (PFBS)	Leather samples, AFFF, semiconductors	Bečanová et al., 2016; Herzke et al., 2012
Polyfluoroaalkyl substances	Perfluoroalkane sulfonamido derivatives		N-Methyl perfluorooctane sulfonamide (MeFOSA)	Surfactant, surface protection agents of paper and packaging products	Buck et al., 2011; Matthias et al., 2015; Wei et al., 2019
		Fluorotelomer iodides (FTI)	8:2Fluorotelomer iodide (8:2) FTI	Surfactant, surface protection agents	Buck et al., 2011
	Fluoroterometer substances	Fluorotelomer alcohols (FTOH)	10:2 Fluorotelomer alcohol (10:2) FTOH	Outdoor textiles, cleaning agents, impregnating sprays, lubricants, carpets, upholstery, papers	Matthias et al., 2015
		Polyfluoroalkyl phosphates (PAP)	8:2 Fluorotelomer phosphate monoester (8:2 monoPAP)	Surfactants, paper and food packaging materials, synthetic fibers, semiconductor materials, personal care products	Bečanová et al., 2016; Herzke et al., 2012; Matthias et al., 2015

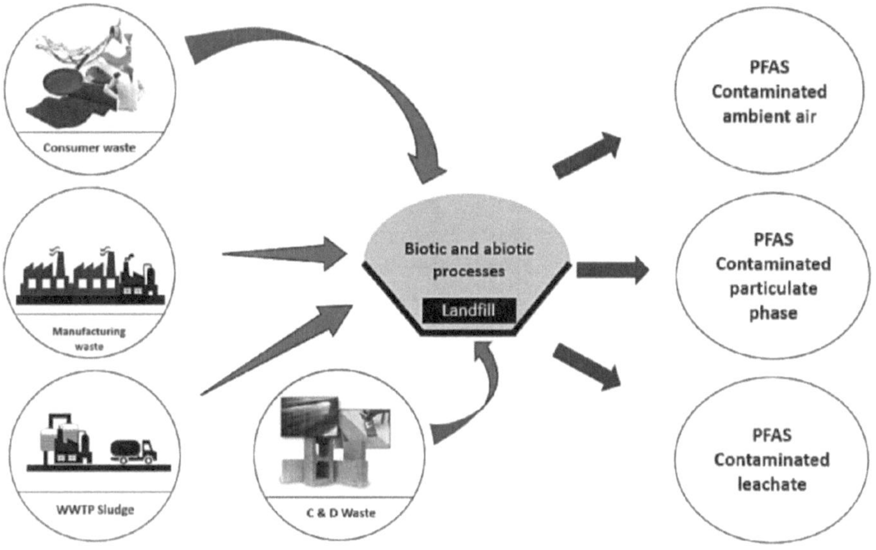

FIGURE 6.1 Sources of perfluoroalkyl polyfluoroalkyl substances (PFAS) in landfills (WWTP: wastewater treatment plant; C&D: construction and demolition waste).

molecules that are not fully fluorinated (Liu and Avendano, 2013). FTCAs and FTUCAs have known degradation products of FTOHs (Buck et al., 2011). PFAA precursors [FTOH, n:2 fluorotelomer carboxylic acids (n:2 FTCA) and n:2 unsaturated fluorotelomer carboxylic acids (n:2 FTUCAs)] present in consumer products degrade to PFAAs after being disposed in landfills (Allred et al., 2015, Lang et al., 2016). Concentrations of known biodegradation intermediates of PFAA precursors, such as methylperfluorobutane sulfonamido acetic acid (MeFBSAA), and the n:2 and n:3 fluorotelomer carboxylates, have increased gradually after the onset of methanogenesis in model landfill studies.

Allred et al. (2015) observed that the single most concentrated PFAS is the 5:3 fluorotelomer carboxylate (9.53 nmol/kg dry-refuse). Fluorotelomer sulfonates (FTSA)-containing surface protectors and complex fluorotelomer-based substances used in food packaging applications are degraded to n:2 FTSAs (Lang et al., 2016, Allred et al., 2015). Biodegradation of ethyl-perfluorooctane sulfonamidoethanol (Et-FOSE), which is a primary raw material of packaging and papers, produces C8-based ethyl perfluorooctane sulfonamido acetic acid (EtFOSAA) (Rhoads et al., 2008). These transformations are influenced by the physio-chemical properties of the PFAS and the landfill leachate (Hamid et al., 2018). Landfilled waste passes through aerobic, acetogenic, and methanogenic stabilization stages. Significant changes occur in the physicochemical properties of landfill leachates such as pH, organic, and inorganic constituents, which affect the mobility and degradation of PFAS in landfills (Renou et al., 2008). Several studies have observed the greater mobility of PFAA with increasing pH (>7) (Gallen et al., 2017, Benskin et al., 2012a) due to protonation of the adsorbent surface leading to fewer positive sites on the sorbent and decreased sorption ability (Wang and Shih, 2011). Furthermore, the fate and transformation of PFAS in landfills are affected by climatic factors that change the moisture content inside landfills and operating conditions, such as compaction of the waste, waste filling techniques, and leachate recirculation (Hamid et al., 2018).

The estimate of total PFAS mass release in US landfill leachate is 600 kg/yr^{-1}. PFAS release from waste is slow compared to the release from manufacturing processes. It is difficult to apply a rate of release to consumer products annually due to sorption, low PFAS solubility, bonding to polymers that are slowly degradable, low infiltration rates, and subsequent waste flushing. However, buried waste has been found to release PFAS at similar rates for over a ten-year period (Johnsie et al., 2017). PFAS have been widely detected in landfills with a large range of concentrations worldwide. PFOA concentration has ranged from 0.15 to 9.2 μg/L in 13 US landfills, and, for Chinese landfills, it has been reported as high as 214 μg/L (Johnsie et al., 2017). PFAS with lower than eight carbons tend to predominate in the landfill leachate because they are less hydrophobic and more likely to partition to the aqueous phase (Huset et al., 2011). Short-chain PFAS, defined as PFCAs with six or fewer perfluorinated carbons or PFSAs with five or fewer perfluroinated carbons (OECD 2013), are the frequently detected PFAS in landfills (Ahrens and Bundschuh, 2014). PFAS have been recorded in municipal landfill leachate, the gaseous phase of ambient air of landfills, and in the particulate phase (Ahrens et al., 2011, Weinberg et al., 2011).

6.4 LEACHING OF PFAS FROM LANDFILLS

PFAS in landfill leachate depend on factors like the type of waste, age of waste, climatic conditions, practices carried out in waste landfilling, and methanogenesis (Benskin et al., 2012b; Yan et al., 2015a; Gallen et al., 2017). PFAS concentrations tend to become reduced with an increase of landfill age because of biological activities (Li et al., 2020). Countries receiving frequent precipitation have higher PFAS leaching compared to countries in arid regions (Lang et al., 2017).

PFAS are released from residential areas (sources include carpets, microwave popcorn bags, and non-stick cookwares) and commercial and industrial areas (sources include textiles, surfactants, firefighting foams, and photography). Landfills, manufacturing plants, wastewater and sewage treatment plants, and houses are initial sources of PFAS, while runoff from lands and roads, where biosolids are applied, and wet and dry atmospheric deposition are nonpoint sources (Davis et al., 2007, Shoeib et al., 2011, Ahrens et al., 2011).

Ethylperfluorooctane sulfonamido acetic acid (EtFOSAA) is generated from ethylperfluorooctane sulfonamidoethanol (EtFOSE), which is used as a raw material for the production of packaging materials and paper (Buck et al., 2011). The production of C8-based chemicals generates PFBS more than that of PFOS (Paul et al., 2009). PFAA are even used for polymerization in manufacturing of fluoropolymers and for formulations in aqueous media used in firefighting foam (Konwick et al., 2008). Fluorotelomer alcohols (6:2 and 8:2) have been the most abundant extractable PFAS in consumer products in Norway (Vestergren et al., 2015).

PFAS, like perfluoroalkyl acids, perfluoroalkyl sulfonamide derivatives, polyfluoroalkyl phosphate esters, perfluoroalkyl carboxylic acids, and perfluoroalkyl sulfonic acids, have been recorded in landfill leachate (Hamid et al., 2018; Rahman et al., 2014).

Perfluoroalkyl sulfonamide derivatives such as methyl- and ethyl-perfluoroalkane sulfonamido acetic acids (FASAAs) have been observed in landfill leachate (Allred et al., 2014; Lang et al., 2017; Benskin et al., 2012b). Certain classes of polyfluoroalkyl phosphate esters, such as di-substituted fluorotelomer phosphate esters and EtFOSE-based polyfluoroalkyl phosphate diester (EtFOSE stands for ethyl perfluorooctane sulfonamido ethanol), have been detected in leachate (Allred et al., 2014). Microbially degraded polyfluoroalkyl phosphate esters of FTCA and PFCA have been found in activated sludge and in soils that are under aerobic conditions (Lee et al., 2010; Liu and Liu, 2016). PFCAs, which are short in length and even in length, have shown to be dominant in leachates tested from 22 sites in Germany (Busch et al., 2010).

The fate of PFAS and their transport depends on factors such as landfill age, climate, type of landfilled waste, and discharges from industrial sources (Ahrens and Bundschuh, 2014; Hamid et al., 2018; Schwarzbauer et al., 2002). Leachate recirculation can trigger leaching, and dilution of contaminants occurs with elevated moisture levels (Huset et al., 2011). Allred et al. (2015) found short-chain PFAS when waste carpets were fed to reactors, and long- and short-chain PFAS were concentrated in bioreactors fed with clothing waste. Biodegradation intermediates, such as methyl perfluorobutane sulfonamido acetic acid (MeFBSAA) and fluorotelomer carboxylates (or fluorotelomer carboxylic acids) (FTCAs) increase under methanogenic conditions (Xiao et al., 2012; Allred et al., 2015). Leached PFAS concentration from a municipal-solid-waste-filled anaerobic reactor was ten times higher than that in a reactor that was biologically inactive (Allred et al., 2015; Gallen et al., 2016). During the methanogenic phase of a landfill, intermediates from biodegradation increase, such as n:2 and n:3 fluorotelomer carboxylates, including methyl perfluorobutane sulfonamido acetic acid. In live reactors, the 5:3 fluorotelomer carboxylate was found to be the most concentrated PFAS. The presence of PFCAs less than eight carbons in length causes a reduction of leaching in reactors that are controlled abiotically (Allred et al., 2015). Landfills where wastes with fluorochemicals are deposited have high levels of them in their leachates, and their concentrations range between 48,000 to 82,000 ng/L (Kriens and Kessler, 2006; Huset et al., 2011). Huset et al. (2011) quantified fluorochemicals at sites in the US and found that perfluoroalkyl carboxylates accounted for the highest percentage (67 ± 4%); next were the perfluoroalkyl sulfonates at 22 ± 2%, and their concentrations ranged between 16 to 2300 ng/L. Among the perfluoroalkyl sulfonates, PFBS ranged in concentration between 280 and 2300 ng/L. Concentrations of PFBA (1700 ng/L) and PFHpA (2800 ng/L) were recorded in the study. These concentrations are beyond or equal to values reported in landfill leachates generated at sites without fluorochemical waste (Bossi et al., 2008; Company, 2001; Kallenborn, 2004). PFAS that are water-soluble, like PFAA, can be leached into the environment by landfill leachate (Yan et al., 2015b). PFAS, like fluorotelomer alcohols (FTOHs), which are neutral, less water soluble, and have high vapor pressures can be released with landfill gas (Hamid et al., 2018). In a multicity analysis of PFAA conducted by the 3 M Company in the US, the highest values of PFOS were found in the effluent, sludge, and landfill leachate of the Publicly Owned Treatment Works in Decatur, Illinois (3 M Company, 2001). Table 6.2 shows concentrations of PFAS in landfill leachates, as found in different studies.

TABLE 6.2
The global distribution of PFAS in landfill leachate (<LOD = less than limit of detection).

Compound	Regions (ng/L)						
	North America	Europe	Australia	China	Singapore	Germany	Northern European countries
PFBS(C4)	28–3200	<0.39–1356	<840	1600–41,600	161.6–1916.3		
PFPeS(C5)	–	–	–	–	–		
PFHxS(C6)	45–1100	<0.20–1800	<1900	<479	89.7–307.5		
PFHpS(C7)	–	-	–	–	–		
PFOS(C8)	<9.5–4400	0.01–1500	<1100	1150–6020	–		
PFDcS(C10)	–	<1–0.28	<3	–	<0.028–0.72		
PFPrA(C3)	–	–	–	638–10,000	–		
PFBA(C4)	69–660	<3.36–2968	<1600	1100–9270	–		
PFPeA(C5)	54–3200	<829	-	609–6530	–		
PFHxA(C6)	190–8900	<2900	12–5700	146–4430	128.0–867.8		
PFHpA(C7)	62–3100	<600	2.2–3500	75.4–5830	41.8–485.5		
PFOA(C8)	42–5000	<4200	19–2100	281–214,000	537.9–3457.6		
PFNA(C9)	11–450	<680	<89	<381	14.7–100.4		
PFDA(C10)	0.3–1100	<410	<57	<18.8	7.1–26.9		
PFUnA(C11)	<120	<430	<18	–	0.59–3.12		
PFDoA(C12)	<1.4–16	<25	<28	–	<0.11–3.89		
PFOSA	3.4–220	<0.15–14.0	–	–	–		
FTCA	10–15,000	–	–	–	20.4–215.4		
FTSA	0.3–300	–	–	–	7.1–54.8		
Sum PFSAs	<82.5–8700	<1.6–4656	<3843	<2750–48,099	<251.328–2224.52	1889–12,819 in active landfills; 31 in closed landfills (1979)	1.6–4656
PFCAs	<433.1–22,766	<3.51–13,056	<33.2–13,092	<2849.4–250,460	<730.2–4945.21	31–12,819 in untreated leachate; 4–8060 in treated leachate	3.51–13,056
Reference	Allred et al., 2014; Benskin et al., 2012a, 2012b; Fuertes et al., 2017; Gewurtz et al., 2013;Huset et al., 2011; Lang et al., 2017; Li et al., 2012; Schwarzbauer et al., 2002	Busch et al., 2010; Eggen et al., 2010; Bossi et al., 2008; Fuertes et al., 2017; Kallenborn, 2004; Perkola and Sainio, 2013;Woldegiorgis et al., 2006; Eschauzier et al., 2013	Fuertes et al., 2017; Gallen et al., 2016	Fuertes et al., 2017; Yan et al., 2015b	Yin et al., 2017	Busch et al., 2010	Ahrens and Bundschuh, 2014; Busch et al., 2010; Bossi et al., 2008; Kallenborn, 2004

Lang et al. (2017) found high values of standard deviations for PFAS concentrations due to climate variation (wet, temperate, arid). In their study, 5:3 FTCA was the largest contributor to PFAS release from landfill leachate. Becker et al. (2008) found that cumulative discharge of PFAS in all landfill sites in Germany was 240 g/day. For effluents from German wastewater treatment plants, they found discharge rates of 30 to 2200 mg/day for PFOA and 20 to 2800 mg/day for PFOS. Busch et al. (2010) estimated that the release of 44 PFAS from treated leachate in nearly 1700 landfills in Germany was 90 kg/yr. Schultz et al. (2006) measured eight PFAS from wastewater treatment plants in the US and found that the cumulative discharge was 2000 mg/day. Buck et al. (2011) showed that wastes that release PFAS is low compared to the production of consumer products in a year. The slow release can be due to binding to slowly biodegrading polymers or to low PFAS solubility. Freshly landfilled waste will continue to release PFAS at the same frequency for more than 10 years (Allred et al., 2015). Lang et al. (2017) studied US landfill-generated leachate that was directed to wastewater treatment plants during the year 2013, and estimated that the PFAS contributions for the year ranged from 563 to 638 kg. The main PFAS in the leachate was 5:3 FTCA. Compounds like n:2 FTCAs, n:2 FTUCAs, n:3 FTCAs, n:2 FTSAs, which are fluorotelomer-based compounds, have been detected in landfill leachate (Lang et al., 2017; Allred et al., 2014). Long-chain PFOA and PFOS have been observed in landfill leachate. China has restricted the production of long-chain PFAS, which are leached from landfilled waste (Wei et al., 2019). Lang et al. (2017) estimate that leaching of PFAS will continue at the same rate from both fresh and stabilized waste with similar concentrations of PFAS.

6.5 WATER AND AIR POLLUTION BY PFAS

It is not known whether the dominant transport of ionizable PFAS is through the water phase or the atmospheric phase. However, in remote regions, neutral, volatile PFAA are more likely to be heavily transported through the atmosphere (Ahrens et al., 2011). Temperature, organic content, salinity, level of atmospheric oxidants, and physical and chemical characteristics (long chain-hydrophobic or short chain-hydrophilic) of PFAS control the environmental cycling of them (Inoue et al., 2012). Globally, sediments and oceans are considered to be the largest reservoir of PFAS (Yamashita et al., 2008).

Pollution of water bodies by PFAS occurs near industrial sites. Konwick et al. (2008) studied PFAS in waters near the largest carpet manufacturing site in North America located in Whitfield County in northwest Georgia. Concentrations of PFAS, including PFOA (253–1150 ng/L), PFUA (perfluoroundecanoic acid) (58.0–99.2 ng/L), PFDA (perfluorodecanoic acid) (30.1–113 ng/L), and PFOS (192–318 ng/L), were recorded in the Conasauga River, which runs through the southeast Tennessee and northwest Georgia. Downstream of the land application system of the effluent from a carpet manufacturing site, pond waters had PFOA concentrations of 49.9 to 299 ng/L and PFOS concentrations of 15.8 to 120 ng/L. These values are the highest concentrations recorded when there has been no spill or direct release.

PFAS can be leached from landfills to nearby waterbodies (Lang et al., 2017). Leachate-impacted groundwaters have increased values for total organic carbon, ammonia, bicarbonate, and dissolved methane (Eschauzier et al., 2013). Hepburn et al. (2019) measured concentrations of PFAS in groundwater close to legacy landfill sites in Australia and found that they ranged between 26 and 5200 ng/L. The leachate-contaminated groundwater was dominated by PFOA. The levels varied between 25 and 45% of the sum of the detected PFAS. Ammonia N levels were strongly correlated with PFOA and short-chain PFAS levels. The relatively conservative behavior during leaching was suggested by the consistent correlation coefficients between PFAS and leachate indicators. The results of Hepburn et al. (2019) suggested that the ratio of PFOA/short-chain PFAS can be used as a tracer for leachate that has come from PFAS-contaminated sites.

The presence of short-chain PFAS have been detected in both the gaseous phase of ambient air of landfills and in the particulate phase (Weinberg et al., 2011; Ahrens et al., 2011). In air near landfill sites in Germany, PFOS was observed in low concentrations below 5 pg/m^3. Trapping PFOS during landfill gas collection, sorption by landfill solids, and partitioning of PFOS to the leachate can be reasons for the observed low concentration of PFOS (Ahrens et al., 2011). Airborne volatile compounds can be transported to far away locations. These compounds include FTOH and PFOS (Martin et al., 2006; Ellis et al., 2004). Studies have recorded PFAS concentrations 2 to 30 times higher in the landfill ambient air compared to control sites (Ahrens et al., 2011; Weinberg et al., 2011; Hamid et al., 2018). Recorded values in ambient air from active and closed landfill sites are tabulated in Table 6.3. Of all the volatile precursors to perfluorinated carboxylic acids, FTOHs have been the most commonly detected ones and they have occurred in more than 90% of ambient air samples tested at landfills (Ahrens et al., 2011; Weinberg et al., 2011).

The ample presence of PFOA and PFCAs in landfilled waste is thought to be the reason for elevated levels of PFBA, PFHxA (perfluorohexanoic acid), and PFOA in the gas phase (Ahrens et al., 2011). Arvaniti et al. (2012) found that detection of long-chain PFCAs was minimal in ambient air of landfills due to their affinity to solid particles. Volatile PFAS, such as perfluoroalkane sulphonamide (FASA) and FTOHs, can enter air and be degraded into intermediates by atmospheric oxidation or be transformed into PFSAs or PFCAs, which are persistent PFAS. They can finally accumulate in the aquatic environment (Young and Mabury, 2010). Polyfluoroalkyl phosphate esters, FTUCAs, and FTSAs are known to be major sources of PFSAs and PFCAs in water environments, and these precursors are normally transported by the atmosphere due to their volatile or semi-volatile nature (Ahrens and Bundschuh, 2014). A variety of transformations occurs with FTUCAs and FTSAs, both under aerobic and anaerobic conditions (Butt et al., 2014). Acute and chronic toxic effects caused by reactive intermediate degradation products of PFAS have been detected in both aquatic

TABLE 6.3
PFAS concentrations from landfill ambient air (pg/m^3).

PFAAS	FTOHs	FOSAs, FOSEs	Remarks	Reference
130–320	2500–26,000	60–120	Active landfill	Ahrens et al., 2011
5–10	70–100	6–20	Closed landfill	Weinberg et al., 2011

invertebrates and green algae. These reactive products are formed during biotransformations and atmospheric transformations (Phillips et al., 2010, Hoke et al., 2012).

6.6 TREATMENT OF PFAS IN LANDFILL LEACHATE

In addition to sites that manufacture PFAS and firefighting training sites, landfills are major sources of PFAS in the environment (Allred et al. 2015). Contamination of landfill leachates with PFAS is increasing rapidly. Hence, disposal of PFAS-enriched wastes in landfills should be stopped in order to protect the surrounding environment (Masoner et al., 2020). Landfill-leachate treatment is a crucial step to remediate PFAS-contaminated landfills. The efficacy of treatment of landfill leachate depends on various factors, including the composition of the varying types of PFAS-containing wastes that are placed in landfill, climatic and environmental conditions, landfilling practices, waste composition (mainly the moisture content), and degree of compaction, which affects aeration and liquid-flow patterns (Yan et al., 2015b). Landfill leachate treatment is challenging due to the complexity of the system and strong matrix effects. Treatment requirements for landfill leachate are based upon several organic and inorganic monitoring parameters, such as biological and chemical oxygen demand (COD and BOD), pH, electrical conductivity, NH_4^+-N, and potentially toxic elements (Al-Yaqout and Hamoda, 2003). Landfill leachate can be treated by methods such as: (a) biological treatment, which involves aerobic (aerated lagoons, activated sludge) and anaerobic degradation (anaerobic lagoons, reactors) (Wei et al., 2019); (b) physicochemical treatment, including sonochemical treatment, membrane technology, photochemical and electrochemical oxidation, air stripping, coagulation using lime, alum, and ferric chloride, and chemical precipitation (Yin et al., 2017); and (c) advanced techniques involving adsorption, nanofiltration, and ion exchange processes (Yin et al., 2017) (Figure 6.2, Table 6.4). Selected studies examined the treatment of PFAS in landfill leachate are listed in Table 6.5. Conventional biological

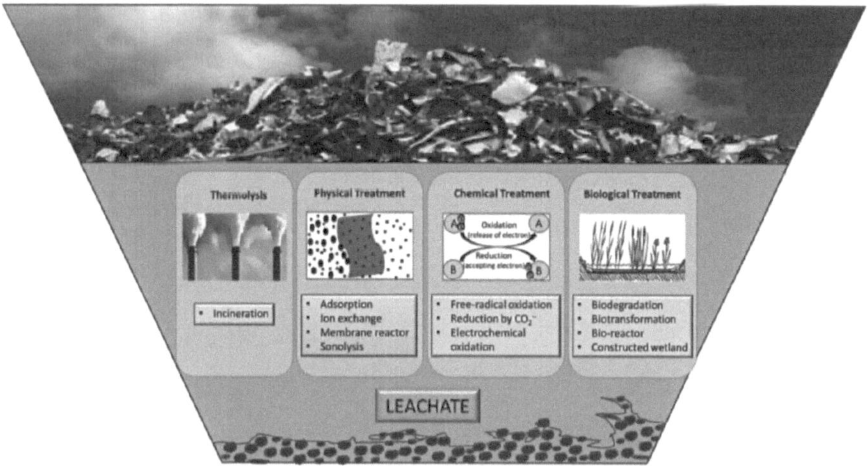

FIGURE 6.2 Treatment options for PFAS in landfill leachates.

TABLE 6.4
Overview of PFAS treatment methods for landfill leachates (adapted from Wei et al., 2019).

Method	Process	Sources	Mechanisms	Advantages	Disadvantages	References
Physical treatment	Adsorption	Perfluorooctanesulfonyl fluoride (PFOSF) washing wastewater, surface water, drinking water	Hydrophobic interaction, ligand exchange, π-π interaction, and electrostatic attraction	Easy to perform, inexpensive process	Low adsorption and high desorption capacity of the adsorbents	Du et al., 2014, Du et al., 2015; Rahman et al., 2014
	Ion exchange resin	Industrial wastewater, surface and underground water	Anion-exchange reactions	Quick and target specific	Expensive resin polymers, and difficulty in regeneration	Chularueangaksorn et al., 2013; Deng et al., 2010; Schuricht et al., 2017; Zaggia et al., 2016
	Membrane filtration	Semiconductor wastewater, semiconductor and metal plating, food packaging products	Size exclusion, Coulomb's force interaction	High removal efficiency, easy to perform	Expensive maintenance and difficulties in disposal	Appleman et al., 2013; Tang et al., 2006, 2007
Oxidation (through free radicals)	•OH	Emulsion, fire retardants, paper coatings, carpet cleaners	e^- transfer, decarboxylate, CF_2 elimination	Quick generation of radicals by precursors	Ineffective	Mitchell et al., 2013; Tang et al., 2012
	$SO_4^{\bullet -}$	Surface waters, polymer synthesis	Strong sulphate oxidizer, Electron transfer, CF_2 elimination	Quick generation of radicals by precursors, effective degradation	Slow degradation, high production of degradation intermediates, high energy consumption depending upon source of light	Hori et al., 2005; Qian et al., 2015; Xiao et al., 2015
	h^+	Fire retardants, firefighting foams, surfactants, polymer additives, lubricants, emulsifiers	Decarboxylation, unimolecular decomposition	Low energy consumption	Ineffective	Chen et al., 2015; Estrellan et al., 2010; Gomez-Ruiz et al., 2018
Reduction	e^-, $CO_2^{\bullet -}$	Wastewater discharge, industry and consumer products	Aqueous electron generation, unimolecular decomposition	Quick degradation	High required amount, easily scavenged	Gu et al., 2016; Wang and Zhang, 2011, 2016
Thermal treatment	Incineration	Paper and textiles, oxidative protective coatings on metals, inert surfactants for semi-conductor etching, aqueous film forming foams, thermally stable lubricants, groundwater beneath a landfill	Thermal decomposition	Complete destruction	High energy consumption, problems in disposal	Cheng et al., 2008; Vecitis et al., 2009

TABLE 6.5
Methods and their efficiency to treat PFAS in landfill leachate.

Nature of landfill	Country	PFAS compounds (concentration)	Treatment methods	Removal rate	References
Municipal solid waste	Poland	PFOA (1350 ± 320 ng/L); PFOS (3280 ± 940 ng/L)	Electrochemical oxidation	80 and 78% for PFOA and PFOS, respectively	Pierpaoli et al., 2020
Municipal solid waste	Spain	PFAS (1379 ng/L)	Membrane bioreactor (MBR)	PFAS concentration increased to 3162 ng/L from 1379 ng/L due to possible degradation of PFAS precursors such as fluorotelomer alcohols and/or fluorotelomer sulfonates	Fuertes et al., 2017
Municipal landfill	Canada	6:2 fluorotelomer sulfonate (FTS)	Microbial transformation (heterotrophic + ammonia + nitrite oxidizing bacteria)	20% of the 6:2 FTS biotransformation within 10 days	Hamid et al., 2020a
Municipal landfill	China	Perfluoro alkyl acid (PFAA) (Raw leachate: 7280 to 292,000 ng/L)	MBR + reverse osmosis or nanofiltration	Treated leachate: 98.4–282,000 ng/L	Yan et al., 2015b
Primarily municipal and commercial waste	Australia	Perfluorohexanoic acid (PFHxA) (12–5700 ng/L)	Off-site wastewater treatment plant	-	Gallen et al., 2016
Municipal solid, construction and demolition waste	Singapore	PFAS (1259–7661 ng/L)	Constructed wetland system	61% removal for total PFAS, and 50–96% removal of individual PFAS	Yin et al., 2017
Municipal solid waste, construction and demolition waste	USA	PFAS (>13,000 ng/L)	Incineration (930–980 °C)	PFAS <3400 ng/L after treatment in municipal solid waste ash	Solo-Gabriele et al., 2020
Municipal solid waste	Canada	8:2 fluorotelomer alcohol (FTOH)	Aerobic biotransformation	>20% volatilization in 90 days	Hamid et al., 2020b
Municipal solid waste	Germany	PFAS (30.5 ng/L - 12,922 ng/L)	Adsorption by activated carbon	PFAS in treated landfill: 9.26–4079 ng/L	Busch et al., 2010
Dumping site	China	PFAS (~5000 ng/L)	Adsorption by magnetic activated carbon	72.8–89.6% removal efficiency	Zhang et al., 2018

and chemical treatment technologies are ineffective in breaking the strong C–F bonds in PFAS. Adsorption, in general, has found wider acceptability than other techniques for PFAS remediation, and activated carbon and ion exchange resins have been shown to adsorb substantial amounts of PFAS. However, major drawbacks of the adsorption technique are chances of PFAS desorption following adsorption and production of large volumes of spent adsorbent that incurs a huge challenge for environmentally friendly disposal (Du et al., 2014; Ateia et al., 2018).

A few conventional and advanced treatment technologies have been evaluated for PFAS removal from landfill leachate. The most widely accepted method used for PFAS remediation in landfill leachate is the pump-and-treat method in which the leachate is pumped out and treated using granular activated carbon (GAC), followed by off-site incineration of the spent adsorbent (McGregor, 2020). GAC is a non-selective adsorbent of PFAS with relatively low adsorption capacity toward short-chain PFAS compounds (Ateia et al., 2019). Other than adsorption, constructed wetland treatment of PFAS-containing landfill leachate has been tried in the recent past. This technique uses an equalization tank, aerated lagoons, sedimentation tank, reed beds, and polishing ponds to eliminate PFAS compounds at a rather slow rate, but it shows enormous promise in developing an environmentally benign PFAS remediation technology for landfill leachates (Yin et al., 2017). However, careful consideration is needed through regular monitoring that PFAS do not leak through the linings of the lagoons and ponds, which would result in contamination of the groundwater underneath.

Other *ex situ* PFAS removal methods include membrane technology using nanofiltration (NF) and reverse osmosis (RO). But these methods have more feasibility for drinking-water treatment than landfill leachate remediation due to the high sophistication and expense of the techniques. Both NF and RO show successful rejection of PFAS compounds. In general, NF might show a better practical utility for PFAS treatment in landfill leachate than RO because NF involves a lower organic fouling potential and energy consumption than RO (Appleman et al., 2013). Photocatalysis, advanced oxidation, oxidation–reduction reactions, and sonolysis were found successful in transforming some long-chain PFAS (e.g., PFOA and PFOS) with >90% transformation in drinking water, but the methods have limited mineralization capacity (Wei et al., 2019). Background electrolytes and natural organic matter (NOM) of various molecular weights substantially affect the degradation of PFAS compounds in natural or wastewater (Gagliano et al., 2020). Photolysis and photocatalytic degradation of PFAS can be successful in landfill leachate, but the background matrix complexity is a big challenge. Heterogeneous adsorptive photocatalysts also hold potential to degrade PFAS in landfill leachate via a 'hook and destroy' approach in two mechanistic steps: (1) a functionalized surface attracts PFAS on reactive sites of the photocatalyst anchored on environmentally benign adsorbents; and (2) the photocatalyst then degrades the concentrated PFAS *in situ* by producing radicals and aqueous electrons (Liu et al., 2016).

No singular conventional biological, chemical, or advanced technique has been effective in eliminating PFAS completely from wastewater and landfill leachate. Future research should be concentrated on the integration of physicochemical treatments with biological treatments in order to make the technologies effective, energy-neutral, and inexpensive. Some intermediate products of photocatalytic PFAS degradation might be difficult to biodegrade, making the success of biological treatment challenging. Selective physicochemical treatments (e.g., sonolysis) could be applied as a priming treatment, making the metabolites conveniently degradable by microorganisms (e.g., in microbial lagoons). All combinations of compatible technologies need to be tested for landfill leachate treatment in and outside the laboratory for complete validation of the treatment effectiveness and economic feasibility (Wei et al., 2019). Continuous monitoring of treated landfill leachates for any potential leaking or secondary release should also be included in the PFAS remediation program.

6.7 CONCLUSIONS

Landfill leachate is one of the most important point sources of PFAS reaching surface and groundwater due to disposal of PFAS-containing products in landfills, such as carpet and food storage containers. The occurrence, distribution, and transformation of PFAS in landfill leachate are dependent on a number of factors that include the nature of waste disposed, age of the landfill site, the type of landfill (i.e., engineered or conventional), and climatic conditions (i.e., temperature and rainfall). The fate of accumulated PFAS inside the landfills is regulated by a combination of biological (biodegradation) and abiotic (desorption) processes. While various technologies have been tested for treating firefighting foam-derived PFAS in groundwater, our knowledge is limited on the treatability of PFAS in landfill leachate, in particular, as it pertains to the effect of the organic solute-rich leachate matrix. The organic solutes in landfill leachates, especially dissolved organic matter, interfere with the treatment of PFAS. It is important to undertake pretreatment of landfill leachates that are rich in organic solutes in order to achieve reasonable treatment of PFAS.

Adsorption of PFAS using carbon-based and ion exchange-based materials has been found effective in surface and groundwater sources that are not enriched with organic solutes. However, regeneration of spent conventional adsorbents requires costly follow-up treatment (e.g., thermal or incineration). The issue with the regeneration of spent adsorbents is exacerbated in the case of treating PFAS in landfill leachate, where the adsorbents are subjected to competitive adsorption and organic fouling by other organic solutes such as dissolved organic carbon. Therefore, there is a need to develop selective adsorbents such as functionalized natural clay minerals that inhibit the competition from other organic contaminants.

REFERENCES

Ahrens, L., and Bundschuh, M. 2014. Fate and effects of poly-and perfluoroalkyl substances in the aquatic environment: a review. *Environmental Toxicology and Chemistry*, 33, 1921–1929.

Ahrens, L., Shoeib, M., Harner, T., Lee, S. C., Guo, R., and Reiner, E. J. 2011. Wastewater treatment plant and landfills as sources of polyfluoroalkyl compounds to the atmosphere. *Environmental Science and Technology*, 45, 8098–8105.

Albright, W.H., Benson, C.H., Gee, G.W., Abichou, T., Tyler, S.W., and Rock, S.A. 2006. Field performance of three compacted clay landfill covers. *Vadose Zone Journal*, 5, 1157–1171.

Allred, B. M., Lang, J. R., Barlaz, M. A., and Field, J. A. 2014. Orthogonal zirconium diol/C18 liquid chromatography–tandem mass spectrometry analysis of poly and perfluoroalkyl substances in landfill leachate. *Journal of Chromatography A*, 1359, 202–211.

Allred, B. M., Lang, J. R., Barlaz, M. A., and Field, J. A. 2015. Physical and biological release of poly-and perfluoroalkyl substances (PFASs) from municipal solid waste in anaerobic model landfill reactors. *Environmental Science and Technology*, 49, 7648–7656.

Al-Yaqout, A. F., and Hamoda, M. F. 2003. Evaluation of landfill leachate in arid climate—a case study. *Environment International*, 29, 593–600.

Andriani, D., and Atmaja, T. D. 2019. The potentials of landfill gas production: a review on municipal solid waste management in Indonesia. *Journal of Material Cycles and Waste Management*, 21, 1572–1586.

Appleman, T. D., Dickenson, E. R. V., Bellona, C., and Higgins, C. P. 2013. Nanofiltration and granular activated carbon treatment of perfluoroalkyl acids. *Journal of Hazardous Materials*, 260, 740–746.

Arvaniti, O. S., Ventouri, E. I., Stasinakis, A. S., and Thomaidis, N. S. 2012. Occurrence of different classes of perfluorinated compounds in Greek wastewater treatment plants and determination of their solid–water distribution coefficients. *Journal of Hazardous Materials*, 239, 24–31.

Ateia, M., Arifuzzaman, M., Pellizzeri, S., Attia, M. F., Tharayil, N., Anker, J. N., and Karanfil, T. 2019. Cationic polymer for selective removal of GenX and short-chain PFAS from surface waters and wastewaters at ng/L levels. *Water Research*, 163, 114874.

Ateia, M., Attia, M. F., Maroli, A. S., Tharayil, N., Alexis, F.,Whitehead, D. C., and Karanfil, T. 2018. Rapid removal of poly-and perfluorinated alkyl substances by polyethyleni-mine-functionalized cellulose microcrystals at environmentally relevant conditions. *Environmental Science & Technology Letters*, 5, 764–769.

Bečanová, J., Melymuk, L., Vojta, Š., Komprdová, K., and Klánová, J. 2016. Screening for perfluoroalkyl acids in consumer products, building materials and wastes. *Chemosphere*, 164, 322–329.

Becker, A. M., Gerstmann, S., and Frank, H. 2008. Perfluorooctane surfactants in waste waters, the major source of river pollution. *Chemosphere*, 72, 115–121.

Benskin, J. P., Ikonomou, M. G., Woudneh, M. B., and Cosgrove, J. R. 2012a. Rapid characterization of perfluoralkyl carboxylate, sulfonate, and sulfonamide isomers by high-performance liquid chromatography–tandem mass spectrometry. *Journal of Chromatography A*, 1247, 165–170.

Benskin, J. P., Li, B., Ikonomou, M. G., Grace, J. R., and Li, L. Y. 2012b. Per- and polyfluoroalkyl substances in landfill leachate: patterns, time trends, and sources. *Environmental Science and Technology*, 46, 11532–11540.

Bolan, N. S., Thangarajan, R., Seshadri, B., Jena, U., Das, K. C., Wang, H., and Naidu, R. 2013. Landfills as a bio refinery to produce biomass and capture biogas. *Bioresource Technology*, 135, 578–587. doi:10.1016/j.biortech.2012.08.135.

Bolan, N. S., Sarkar, B., Yan, Y., Qiao, L., Wijesekara, H., et al. (2021). Remediation of poly- and perfluoroalkyl substances (PFAS) contaminated soils – to mobilize or to immobilize or to destroy?. *Journal of Hazardous Materials*, 401, 123892.

Bossi, R., Strand, J., Sortkjær, O., and Larsen, M. M. 2008. Perfluoroalkyl compounds in Danish wastewater treatment plants and aquatic environments. *Environment International*, 34, 443–450.

Brand, J. H., Kate L. Spencer, Francis T. O'Shea, and John E. Lindsay (2018). Potential pollution risks of historic landfills on low-lying coasts and estuaries. *WIREs Water 2018*, 5, e1264.

Buck, R. C., Franklin, J., Berger, U., Conder, J. M., Cousins, I. T., de Voogt, P., Jensen, A. A., Kannan, K., Mabury, S. A., and van Leeuwen, S. P. 2011. Perfluoroalkyl and polyfluoroalkyl substances in the environment: terminology, classification, and origins. *Integrated Environmental Assessment and Management*, 7, 513–541.

Busch, J., Ahrens, L., Sturm, R., and Ebinghaus, R. 2010. Polyfluoroalkyl compounds in landfill leachates. *Environmental Pollution*, 158, 1467–1471.

Butt, C. M., Muir, D. C., and Mabury, S. A. 2014. Biotransformation pathways of fluorotelomer-based polyfluoroalkyl substances: a review. *Environmental Toxicology and Chemistry*, 33, 243–267.

Chen, M. J., Lo, S. L., Lee, Y. C., and Huang, C. C. 2015. Photocatalytic decomposition of perfluorooctanoic acid by transition-metal modified titanium dioxide. *Journal of Hazardous Materials*, 288, 168–175.

Cheng, J., Vecitis, C. D., Park, H., Mader, B. T., and Hoffmann, M. R. 2008. Sonochemical degradation of perfluorooctane sulfonate (PFOS) and perfluorooctanoate (PFOA) in landfill groundwater: environmental matrix effects. *Environmental Science and Technology*, 42, 8057–8063.

Chularueangaksorn, P., Tanaka, S., Fujii, S., and Kunacheva, C. 2013. Regeneration and reusability of anion exchange resin used in perfluorooctane sulfonate removal by batch experiments. *Journal of Applied Polymer Science*, 130, 884–890.

Company, 3 M. 2001. Environmental Monitoring – Multi-City Study Water, Sludge, Sediment, POTW Effluent and Landfill Leachate Samples.

Davis, K. L., Aucoin, M. D., Larsen, B. S., Kaiser, M. A., and Hartten, A. S. 2007. Transport of ammonium perfluorooctanoate in environmental media near a fluoropolymer manufacturing facility. *Chemosphere*, 67, 2011–2019.

Deng, S., Yu, Q., Huang, J., and Yu, G. 2010. Removal of perfluorooctane sulfonate from wastewater by anion exchange resins: effects of resin properties and solution chemistry. *Water Research*, 44, 5188–5195.

Du, Z., Deng, S., Bei, Y., Huang, Q., Wang, B., Huang, J., and Yu, G. 2014. Adsorption behavior and mechanism of perfluorinated compounds on various adsorbents—A review. *Journal of Hazardous Materials*, 274, 443–454.

Du, Z., Deng, S., Chen, Y., Wang, B., Huang, J., Wang, Y., and Yu, G. 2015. Removal of perfluorinated carboxylates from washing wastewater of perfluorooctanesulfonyl fluoride using activated carbons and resins. *Journal of Hazardous Materials*, 286, 136–143.

Eggen, T., Moeder, M., and Arukwe, A. 2010. Municipal landfill leachates: a significant source for new and emerging pollutants. *Science of the Total Environment*, 408, 5147–5157.

Ellis, D., Martin, J., De Silva, A., Mabury, S., and Hurley, M. 2004. Degradation of fluorotelomer alcohols: a likely atmospheric source of perfluorinated carboxylic acids. *Environmental Science and Technology*, 38, 3316–3321.

Eschauzier, C., Raat, K. J., Stuyfzand, P. J., and De Voogt, P. 2013. Perfluorinated alkylated acids in groundwater and drinking water: identification, origin and mobility. *Science of the Total Environment*, 458, 477–485.

Estrellan, C. R., Salim, C., and Hinode, H. 2010. Photocatalytic decomposition of perfluorooctanoic acid by iron and niobium co-doped titanium dioxide. *Journal of Hazardous Materials*, 179, 79–83.

Ferronato, N., and Torretta, V. 2019. Waste mismanagement in developing countries: a review of global issues. *International Journal of Environmental Research and Public Health*, 16, 1060.

Fuertes, I., Gómez-Lavín, S., Elizalde, M., and Urtiaga, A. 2017. Perfluorinated alkyl substances (PFASs) in northern Spain municipal solid waste landfill leachates. *Chemosphere*, 168, 399–407.

Gagliano, E., Sgroi, M., Falciglia, P. P., Vagliasindi, F. G. A., and Roccaro, P. 2020. Removal of poly- and perfluoroalkyl substances (PFAS) from water by adsorption: role of PFAS chain length, effect of organic matter and challenges in adsorbent regeneration. *Water Research*, 171, 115381.

Gallen, C., Drage, D., Eaglesham, G., Grant, S., Bowman, M., and Mueller, J. 2017. Australia-wide assessment of perfluoroalkyl substances (PFASs) in landfill leachates. *Journal of Hazardous Materials*, 331, 132–141.

Gallen, C., Drage, D., Kaserzon, S., Baduel, C., Gallen, M., Banks, A., Broomhall, S., and Mueller, J. 2016. Occurrence and distribution of brominated flame retardants and perfluoroalkyl substances in Australian landfill leachate and biosolids. *Journal of Hazardous Materials*, 312, 55–64.

Gewurtz, S. B., Backus, S. M., De Silva, A. O., Ahrens, L., Armellin, A., Evans, M., Fraser, S., Gledhill, M., Guerra, P., Harner, T., Helm, P. A., Hung, H., Khera, N., Kim, M. G., King, M., Lee, S. C., Letcher, R. J., Martin, P., Marvin, C., McGoldrick, D. J., Myers, A. L., Pelletier, M., Pomeroy, J., Reiner, E. J., Rondeau, M., Sauve, M. C., Sekela, M., Shoeib, M., Smith, D. W., Smyth, S. A., Struger, J., Spry, D., Syrgiannis, J., and Waltho, J. 2013. Perfluoroalkyl acids in the Canadian environment: multi-media assessment of current status and trends. *Environment International*, 59, 183–200.

Gomez-Ruiz, B., Ribao, P., Diban, N., Rivero, M. J., Ortiz, I., and Urtiaga, A. 2018. Photocatalytic degradation and mineralization of perfluorooctanoic acid (PFOA) using a composite TiO_2-rGO catalyst. *Journal of Hazardous Materials*, 344, 950–957.

Gu, Y., Dong, W., Luo, C., and Liu, T. 2016. Efficient reductive decomposition of perfluorooctanesulfonate in a high photon flux UV/sulfite system. *Environmental Science and Technology*, 50, 10554–10561.

Hamid, H., Li, L. Y., and Grace, J. R. 2018. Review of the fate and transformation of per-and polyfluoroalkyl substances (PFASs) in landfills. *Environmental Pollution*, 235, 74–84.

Hamid, H., Li, L. Y., and Grace, J. R. 2020a. Formation of perfluorocarboxylic acids from 6:2 fluorotelomer sulfonate (6:2 FTS) in landfill leachate: role of microbial communities. *Environmental Pollution*, 259, 113835.

Hamid, H., Li, L. Y., and Grace, J. R. 2020b. Aerobic biotransformation of fluorotelomer compounds in landfill leachate-sediment. *Science of the Total Environment*, 713, 136547.

Hepburn, E., Madden, C., Szabo, D., Coggan, T. L., Clarke, B., and Currell, M. 2019. Contamination of groundwater with per-and polyfluoroalkyl substances (PFAS) from legacy landfills in an urban re-development precinct. *Environmental Pollution*, 248, 101–113.

Herzke, D., Olsson, E., and Posner, S. 2012. Perfluoroalkyl and polyfluoroalkyl substances (PFASs) in consumer products in Norway—a pilot study. *Chemosphere*, 88, 980–987.

Hoke, R. A., Bouchelle, L. D., Ferrell, B. D., and Buck, R. C. 2012. Comparative acute freshwater hazard assessment and preliminary PNEC development for eight fluorinated acids. *Chemosphere*, 87, 725–733.

Hori, H., Yamamoto, A., Hayakawa, E., Taniyasu, S., Yamashita, N., Kutsuna, S., Kiatagawa, H., and Arakawa, R. 2005. Efficient decomposition of environmentally persistent perfluorocarboxylic acids by use of persulfate as a photochemical oxidant. *Environmental Science and Technology*, 39, 2383–2388.

Huset, C. A., Barlaz, M. A., Barofsky, D. F., and Field, J. A. 2011. Quantitative determination of fluorochemicals in municipal landfill leachates. *Chemosphere*, 82, 1380–1386.

Inoue, Y., Hashizume, N., Yakata, N., Murakami, H., Suzuki, Y., Kikushima, E., and Otsuka, M. 2012. Unique physicochemical properties of perfluorinated compounds and their bioconcentration in common carp *Cyprinus carpio* L. *Archives of Environmental Contamination and Toxicology*, 62, 672–680.

Johnsie, R. Lang, McKay, B. A., Jennifer, A. Field, James, W. Levis, and Morton, A. B. 2017. National Estimate of Per- and Polyfluoroalkyl Substance (PFAS) Release to U.S. Municipal Landfill Leachate. *Environmental Science and Technology*, 51, 2197–2205.

Kallenborn, R. 2004. *Perfluorinated alkylated substances (PFAS) in the Nordic environment*, Nordic Council of Ministers, Copenhagen 2005.

Kannan, K., 2011. Perfluoroalkyl and polyfluoroalkyl substances: current and future perspectives. *Environmental Chemistry*, 8, 333–338. doi: 10.1071/EN11053.

Konwick, B. J., Tomy, G. T., Ismail, N., Peterson, J. T., Fauver, R. J., Higginbotham, D., and Fisk, A. T. 2008. Concentrations and patterns of perfluoroalkyl acids in Georgia, USA surface waters near and distant to a major use source. *Environmental Toxicology and Chemistry: An International Journal*, 27, 2011–2018.

Kotthoff, M., Müller, J., Jürling, H., Schlummer, M., and Fiedler, D. 2015. Perfluoroalkyl and polyfluoroalkyl substances in consumer products. *Environmental Science and Pollution Research International*, 22(19), 14546–14559. doi:10.1007/s11356-015-4202-7

Kriens, D., and Kessler, K. 2006. Investigation of perfluorochemical (PFC) contamination in Minnesota phase one. Report to Senate Environment Committee, Minnesota 2006.

Kulikowska, D., and Klimiuk, E., 2008. The effect of landfill age on municipal leachate composition. *Bioresource Technology*, 99, 5981–5985.

Lamb, D. T., Venkatraman, K., Bolan, N., Ashwath, N., Choppala, G., and Naidu, R. 2014. Phytocapping: an alternative technology for the sustainable management of landfill sites. *Critical Reviews in Environmental Science and Technology*, 44(6), 561–637.

Lang, J. R., Allred, B. M., Field, J. A., Levis, J. W., and Barlaz, M. A. 2017. National estimate of per-and polyfluoroalkyl substance (PFAS) release to US municipal landfill leachate. *Environmental Science and Technology*, 51, 2197–2205.

Lang, J. R., Allred, B. M., Peaslee, G. F., Field, J. A., Barlaz, M. A. 2016. Release of per-and polyfluoroalkyl substances (PFASs) from carpet and clothing in model anaerobic landfill reactors. *Environmental Science & Technology*. 50, 5024–5032.

Lee, H., D'eon, J., and Mabury, S. A. 2010. Biodegradation of polyfluoroalkyl phosphates as a source of perfluorinated acids to the environment. *Environmental Science and Technology*, 44, 3305–3310.

Li, B., Danon-Schaffer, M. N., Li, L. Y., Ikonomou, M. G., and Grace, J. R. 2012. Occurrence of PFCs and PBDEs in landfill leachates from across Canada. *Water, Air, and Soil Pollution*, 223, 3365–3372.

Li, F., Duan, J., Tian, S., Ji, H., Zhu, Y., Wei, Z., and Zhao, D. 2020. Short-chain per-and polyfluoroalkyl substances in aquatic systems: occurrence, impacts and treatment. *Chemical Engineering Journal*, 380, 122506.

Liu, C., and Liu, J. 2016. Aerobic biotransformation of polyfluoroalkyl phosphate esters (PAPs) in soil. *Environmental Pollution*, 212, 230–237.

Liu, J. X., and Avendano, S. M. 2013. Microbial degradation of polyfluoroalkyl chemicals in the environment. *Environment International*, 61, 98–114.

Liu, W., Cai, Z., Zhao, X., Wang, T., Li, F., and Zhao, D. 2016. High capacity and photoregenerable composite material for efficient adsorption and degradation of phenanthrene in water. *Environmental Science and Technology*, 50, 11174–11183.

Loganathan, B. G., Sajwan, K. S., Sinclair, E., Senthil Kumar, K., and Kannan, K. 2007. Perfluoroalkyl sulfonates and perfluorocarboxylates in two wastewater treatment facilities in Kentucky and Georgia. *Water Reseach*, 41 (20), 4611–4620.

Lou, X., and Nair, J., 2009 The impact of landfilling and composting on greenhouse gas emissions – a review. *Bioresource Technology*, 100, 3792–3798.

Mahinroosta, R., and Senevirathna, L., 2020. A review of the emerging treatment technologies for PFAS contaminated soils. *Journal of Environmental Management*, 255, 109896. doi:10.1016/j.jenvman.2019.109896.

Martin, J. W., Ellis, D. A., Mabury, S. A., Hurley, M., and Wallington, T. 2006. Atmospheric chemistry of perfluoroalkanesulfonamides: kinetic and product studies of the OH radical and Cl atom initiated oxidation of N-ethyl perfluorobutanesulfonamide. *Environmental Science and Technology*, 40, 864–872.

Masoner, J. R., Kolpin, D. W., Cozzarelli, I. M., Smalling, K. L., Bolyard, S. C., Field, J. A., Furlong, E. T., Gray, J. L., Lozinski, D., Reinhart, D., Rodowa, A., and Bradley, P. M. 2020. Landfill leachate contributes per–/poly-fluoroalkyl substances (PFAS) and pharmaceuticals to municipal wastewater. *Environmental Science: Water Resource Technology*, 6, 1300–1311.

Matthias, K., Josef. M., Heinrich, J., Martin, S., and Dominik, F. 2015. Perfluoroalkyl and polyfluoroalkyl substances in consumer products. *Environmental Science and Pollution Research*, doi: 10.1007/s11356-015-4202-7.

McGregor, R. 2020. Six pilot-scale studies evaluating the in situ treatment of PFAS in groundwater. *Remediation Journal*, 30, 39–50.

Mitchell, S. M., Ahmad, M., Teel, A. L., and Watts, R. J. 2013. Degradation of perfluorooctanoic acid by reactive species generated through catalyzed H_2O_2 propagation reactions. *Environmental Science & Technology Letters*, 1, 117–121.

OECD 2013. OECD/UNEP Global PFC Group, Synthesis paper on per- and polyfluorinated chemicals (PFCs), Environment, Health and Safety, Environment Directorate, OECD.

Omotayo, S., Francis, Orata, Roland, W., and Oladele, O., 2013 Per- and polyfluoroalkyl substances in selected sewage sludge in Nigeria. *Chemosphere*, 92, 329–335.

Paul, A. G., Jones, K. C., and Sweetman, A. J. 2009. A first global production, emission, and environmental inventory for perfluorooctane sulfonate. *Environmental Science and Technology*, 43, 386–392.

Penza, M., Rossi, R., Alvisi, M., and Serra, E., 2010. Metal-modified and vertically aligned carbon nanotube sensors array for landfill gas monitoring applications. *Nanotechnology*, 21, 105–501.

Perkola, N., and Sainio, P. 2013. Survey of perfluorinated alkyl acids in Finnish effluents, storm water, landfill leachate and sludge. *Environmental Science and Pollution Research*, 20, 7979–7987.

Phillips, M. M., Dinglasan-Panlilio, M. J., Mabury, S. A., Solomon, K. R., and Sibley, P. K. 2010. Chronic toxicity of fluorotelomer acids to Daphnia magna and Chironomus dilutus. *Environmental Toxicology and Chemistry*, 29, 1123–1131.

Pierpaoli, M., Szopińska, M., Wilk, B. K., Sobaszek, M., Łuczkiewicz, A., Bogdanowicz, R., and Fudala-Książek, S. 2020. Electrochemical oxidation of PFOA and PFOS in landfill leachates at low and highly boron-doped diamond electrodes. *Journal of Hazardous Materials*, 123, 606.

Qian, Y., Guo, X., Zhang, Y., Peng, Y., Sun, P., Huang, C. -H., Niu, J., Zhou, X., and Crittenden, J. C. 2015. Perfluorooctanoic acid degradation using UV–persulfate process: modeling of the degradation and chlorate formation. *Environmental Science and Technology*, 50, 772–781.

Rahman, M. F., Peldszus, S., and Anderson, W. B. 2014. Behaviour and fate of perfluoroalkyl and polyfluoroalkyl substances (PFASs) in drinking water treatment: a review. *Water Research*, 50, 318–340.

Renou, S., Givaudan, J. G., Poulain, S., Dirassouyan, F., and Moulin, P. 2008. landfill leachate treatment: review and opportunity. *Journal of Hazardous Materials*, 150, 468–493.

Rhoads, K. R., Janssen, E. M., Luthy, R. G., and Criddle, C. S. 2008. Aerobic biotransformaton and fate of N-ethyl perfluorooctane sulfonamidoethanol (N-EtFOSE) in actactivated sludge. *Enironmental Science & Technology*, 42, 2873–2878.

Ross, I., McDonough, J., Miles, J., Storch, P., Thelakkat Kochunarayanan, P., Kalve, E., Hurst, J., Dasgupta, S. S., and Burdick, J., 2018. A review of emerging technologies for remediation of PFASs. *Remediation Journal*, 28, 101–126. doi: 10.1002/rem.21553.

Scheutz, C., and Kjeldsen, P., 2005. Biodegradation of trace gases in simulated landfill soil cover systems. *Journal of the Air & Waste Management Association*, 55, 878–885.

Schultz, M. M., Higgins, C. P., Huset, C. A., Luthy, R. G., Barofsky, D. F., and Field, J. A. 2006. Fluorochemical mass flows in a municipal wastewater treatment facility. *Environmental Science and Technology*, 40, 7350–7357.

Schuricht, F., Borovinskaya, E. S., and Reschetilowski, W. 2017. Removal of perfluorinated surfactants from wastewater by adsorption and ion exchange—Influence of material properties, sorption mechanism and modeling. *Journal of Environmental Sciences*, 54, 160–170.

Schwarzbauer, J., Heim, S., Brinker, S., and Littke, R. 2002. Occurrence and alteration of organic contaminants in seepage and leakage water from a waste deposit landfill. *Water Research*, 36, 2275–2287.

Scott, J., Beydoun, D., Amal, R., Gary, L., and Cattle, J., 2005. Landfill management, leachate generation, and leach testing of solid wastes in Australia and overseas. *Critical Reviews in Environmental Science and Technology*, 35, 239–332.

Shoeib, M., Harner, T. M. Webster, G., and Lee, S. C. 2011. Indoor sources of poly-and perfluorinated compounds (PFCS) in Vancouver, Canada: implications for human exposure. *Environmental Science and Technology*, 45, 7999–8005.

Slack, R., Gronow, J., and Voulvoulis, N., 2005. Household hazardous waste in municipal landfills: contaminants in leachate. *Science of the Total Environment*, 337, 119–137.

Solo-Gabriele, H. M., Jones, A. S., Lindstrom, A. B., and Lang, J. R. 2020. Waste type, incineration, and aeration are associated with per- and polyfluoroalkyl levels in landfill leachates. *Waste Management*, 107, 191–200.

Tang, C. Y., Fu, Q. S., Criddle, C. S., and Leckie, J. O. 2007. Effect of flux (transmembrane pressure) and membrane properties on fouling and rejection of reverse osmosis and nanofiltration membranes treating perfluorooctane sulfonate containing wastewater. *Environmental Science and Technology*, 41, 2008–2014.

Tang, C. Y., Fu, Q. S., Robertson, A., Criddle, C. S., and Leckie, J. O. 2006. Use of reverse osmosis membranes to remove perfluorooctane sulfonate (PFOS) from semiconductor wastewater. *Environmental Science and Technology*, 40, 7343–7349.

Tang, H., Xiang, Q., Lei, M., Yan, J., Zhu, L., and Zou, J. 2012. Efficient degradation of perfluorooctanoic acid by UV–Fenton process. *Chemical Engineering Journal*, 184, 156–162.

USEPA 2016a. Drinking water health advisory for perfluorooctane sulfonate (PFOS). Office of Water (4304 T), Health and Ecological Criteria Division, Washington, DC 20460.

USEPA 2016b. Drinking Water Health Advisory for Perfluorooctanoic Acid (PFOA). Office of Water (4304 T), Health and Ecological Criteria Division, Washington, DC 20460.

USEPA 2020, *EPA PFAS Action Plan: Program Update* https://www.epa.gov/sites/production/files/202001/documents/pfas_action_plan_feb2020.pdf.

Vecitis, C. D., Park, H., Cheng, J., Mader, B. T., and Hoffmann, M. R. 2009. Treatment technologies for aqueous perfluorooctanesulfonate (PFOS) and perfluorooctanoate (PFOA). *Frontiers of Environmental Science & Engineering in China*, 3, 129–151.

Vestergren, R., Herzke, D., Wang, T., and Cousins, I. T. 2015. Are imported consumer products an important diffuse source of PFASs to the Norwegian environment? *Environmental Pollution*, 198, 223–230.

Wang, F., and Shih, K. 2011. Adsorption of perfluorooctanesulfonate (PFOS) and perfluorooctanoate (PFOA) on alumina: influence of solution pH and cations. *Water Research*, 45,2925–2930.

Wang, Y., and Zhang, P. 2011. Photocatalytic decomposition of perfluorooctanoic acid (PFOA) by TiO_2 in the presence of oxalic acid. *Journal of Hazardous Materials*, 192, 1869–1875.

Wang, Y., and Zhang, P. 2016. Enhanced photochemical decomposition of environmentally persistent perfluorooctanoate by coexisting ferric ion and oxalate. *Environmental Science and Pollution Research*, 23, 9660–9668.

Wei, Z., Xu, T., and Zhao, D. 2019. Treatment of per-and polyfluoroalkyl substances in landfill leachate: status, chemistry and prospects. *Environmental Science: Water Research & Technology*, 5, 1814–1835.

Weinberg, I., Dreyer, A., and Ebinghaus, R. 2011. Landfills as sources of polyfluorinated compounds, polybrominated diphenyl ethers and musk fragrances to ambient air. *Atmospheric Environment*, 45, 935–941.

Woldegiorgis, A., Andersson, J., Remberger, M., Kaj, L., Ekheden, Y., Blom, L., Brorstrom-Lunden, E., Borgen, A., and Schlabach, M. 2006. Results from the Swedish National Screening Programme2005, Subreport 1: Antibiotics, Anti-inflammatory Substances, and Hormones. B1689, Swedish Environmental Research Institute, Stockholm.

Xiao, F., Davidsavor, K. J., Park, S., Nakayama, M., and Phillips, B. R. 2012. Batch and column study: sorption of perfluorinated surfactants from water and cosolvent systems by Amberlite XAD resins. *Journal of Colloid and Interface Science*, 368, 505–511.

Xiao, R., Ye, T., Wei, Z., Luo, S., Yang, Z., and Spinney, R. 2015. Quantitative structure–activity relationship (QSAR) for the oxidation of trace organic contaminants by sulfate radical. *Environmental Science and Technology*, 49, 13394–13402.

Yamashita, N., Taniyasu, S., Petrick, G., Wei, S., Gamo, T., Lam, P. K., and Kannan, K. 2008. Perfluorinated acids as novel chemical tracers of global circulation of ocean waters. *Chemosphere*, 70, 1247–1255.

Yan, C., Zhang, W., Zhang, Z., Liu, Y., Deng, C., and Nie, N. 2015a. Assessment of water quality and identification of polluted risky regions based on field observations & GIS in the honghe river watershed, *China PloS One*, 10, e0119130.

Yan, H., Cousins, I. T., Zhang, C., and Zhou, Q. 2015b. Perfluoroalkyl acids in municipal landfill leachates from China: occurrence, fate during leachate treatment and potential impact on groundwater. *Science of the Total Environment*, 524, 23–31.

Yin, T., Chen, H., Reinhard, M., Yi, X., He, Y., and Yew-Hoong Gin, K. 2017. Perfluoroalkyl and polyfluoroalkyl substances removal in a full-scale tropical constructed wetland system treating landfill leachate. *Water Research*, 125, 418–426.

Young, C. J., and Mabury, S. A. 2010. Atmospheric perfluorinated acid precursors: chemistry, occurrence, and impacts. *Reviews of Environmental Contamination and Toxicology*, 208, 1–109. Springer.

Zaggia, A., Conte, L., Falletti, L., Fant, M., and Chiorboli, A. 2016. Use of strong anion exchange resins for the removal of perfluoroalkylated substances from contaminated drinking water in batch and continuous pilot plants. *Water Research*, 91, 137–146.

Zhang, C., Jiang, S., Tang, J., Zhang, Y., Cui, Y., Su, C., Qu, Y., Wei, L., Cao, H., and Quan, J. 2018. Adsorptive performance of coal based magnetic activated carbon for perfluorinated compounds from treated landfill leachate effluents. *Process Safety and Environmental Protection*, 117, 383–389.

Zhang, C., Xu, T., Feng, H., and Chen, S., 2019. Greenhouse gas emissions from landfills: a review and bibliometric analysis. *Sustainability*, 11, 2282.

Section II

Toxicology and Epidemiology

7 Exposure to PFAS
Biomonitoring Insights

Kayoko Kato, Julianne Cook Botelho, and Antonia M. Calafat

Division of Laboratory Sciences, National Center for Environmental Health, Centers for Disease Control and Prevention, USA

CONTENTS

7.1 Introduction ... 145
7.2 The National Health and Nutrition Examination Survey (NHANES) ... 147
7.3 Biomonitoring Concentration Profiles ... 150
7.4 Biomonitoring Matrices for Assessing PFAS Exposure ... 152
7.5 Alternative PFAS ... 153
7.6 NHANES Limitations: Relevance of other Biomonitoring Investigations ... 153
Acknowledgement ... 154
Disclaimer ... 154
References ... 154

7.1 INTRODUCTION

Per- and polyfluoroalkyl substances (PFAS) encompass thousands of man-made chemicals, all containing the perfluoroalkyl moiety (C_nF_{2n+1}–), with carbon alkyl chains of variable length and diverse chemical functionalities (Buck et al. 2011). Because of the strength of the C–F bond, PFAS resist environmental degradation and remain intact in the environment for decades. In addition, long-alkyl chain "legacy" PFAS also persist in humans with elimination half-lives of years. This includes three of the most studied PFAS: perfluorooctane sulfonate (PFOS), perfluorooctanoate (PFOA), and perfluorohexane sulfonate (PFHxS) (ATSDR 2018; Dewitt 2015). By contrast, some short-alkyl chain PFAS (e.g., perfluorobutane sulfonate [PFBS], perfluorohexanoate [PFHxA]), and other "alternative" (or substitute) PFAS with different chemical functional groups such as perfluoroalkyl ether carboxylic acids or fluoroethers (e.g., GenX, or ammonium salt of 2,3,3,3-tetrafluoro-2-(1,1,2,2,3,3,3-heptafluoropropoxy)-propanoic acid) have much shorter biological elimination half-lives (Gannon et al. 2011; Gannon et al. 2016; Nilsson et al. 2010; Olsen et al. 2009).

Manufacturing of long-alkyl chain PFAS (including PFOS and PFOA, their precursors, and related compounds) changed in the United States in the early 2000s

(ATSDR 2018; Dewitt 2015). In 2002, 3M, the primary worldwide manufacturer of PFOS, discontinued the production of PFOS precursors and related compounds in the United States (Paul et al. 2009). In 2006, the US Environmental Protection Agency (EPA) invited eight major leading companies in the PFAS industry to join a global stewardship program to eliminate emissions and product content of PFOA and related chemicals by 2015 (EPA 2006). As a result of these changes, production and use of other PFAS, including short-alkyl chain PFAS and "alternative" PFAS, may be on the rise (Brendel et al. 2018).

PFAS have been used since the middle of the twentieth century in a wide range of industrial applications such as (ATSDR 2018; Dewitt 2015):

- soil-, stain-, grease-, and water-resistant coatings on textiles, paper, and cookware;
- automotive, mechanical, aerospace, chemical, electrical, medical, and building/construction industries;
- personal care products; and
- aqueous film-forming foams.

Because PFAS can migrate to soil and water, and contaminate food and drinking water supplies, exposure to PFAS in populations around the world is widespread (CDC 2019; Duffek et al. 2020; Haines et al. 2017).

Like traditional persistent organic pollutants (POPs), PFAS persist in the environment, distribute widely, and can negatively impact human health and the environment. Unlike traditional POPs, however, PFAS do not accumulate in fat tissues, instead binding tightly to blood proteins (Jones et al. 2003; Liu et al. 2018). The Stockholm Convention, a global treaty adopted in 2001 to protect human health and the environment from POPs, has considered several PFAS for inclusion (UNEP 2019). PFOS, its salts, and its precursor perfluorooctanesulfonyl fluoride (POFS) were added to Annex B (i.e., restricted use) of the POPs list in 2009 (and amended acceptable purposes and specific exemptions in 2019). PFOA, its salts and PFOA-related compounds were added to Annex A (i.e., to eliminate production and use) in 2019. The listing of PFHxS, its salts and PFHxS-related compounds is currently under review (UNEP 2019). Furthermore, the International Agency for Research on Cancer (IARC) classified PFOA "as possibly carcinogenic to humans" (Group 2B) (IARC 2017).

Public health concerns regarding PFAS originate in part from universal exposure, the aforementioned assessments made by the Stockholm Convention and IARC, and a growing body of experimental and epidemiological data suggesting adverse health outcomes (ATSDR 2018; Dewitt 2015; Gomis et al. 2018). Further, in 2016, the US EPA released lifetime health advisories for PFOA and PFOS (EPA 2016). State, tribal, and local officials as well as people who operate drinking-water systems can use the information in these advisories to protect their residents from health risks. Of late, the above facts and detection of PFAS in surface and drinking waters used for human consumption (sometimes at levels above EPA's health advisory), and in the blood of people exposed to contaminated drinking water has heightened scientific and public interest in better understanding the factors that affect human exposure to PFAS (Gebbink et al. 2017; Heydebreck et al. 2015; Kabore et al. 2018; Pan et al. 2017; Sun et al. 2016; Wei et al. 2018; ATSDR 2019; Barton et al. 2020; Daly et al. 2018;

Fromme et al. 2017; Graber et al. 2019; Hoffman et al. 2011; Holzer et al. 2009; Ingelido et al. 2018; Kotlarz et al. 2020; Landsteiner et al. 2014; Stubleski et al. 2017).

Historically, scientists have used several approaches to assess environmental exposures (NRC 2012), including surveys of product use and food consumption, environmental monitoring (e.g., measuring target chemicals in food or environmental media), and biomonitoring (i.e., quantifying concentrations of chemicals, their metabolic products, or adducts in human specimens). Indirect measures of exposure relied on combining environmental monitoring data with exposure history and questionnaire data, estimates of human contact, and pharmacokinetic assumptions based on animal data. Recent advances in analytical chemistry allow the accurate quantification of trace levels of environmental chemicals in select biospecimens. These biomonitoring data (i.e., actual concentrations of chemicals in the body), which represent an integrative measure of chemical exposures from multiple sources and routes, are increasingly used to support exposure assessment. In this chapter, we focus on PFAS biomonitoring.

7.2 THE NATIONAL HEALTH AND NUTRITION EXAMINATION SURVEY (NHANES)

General population programs are useful for investigating human exposure to environmental chemicals using biomonitoring (Angerer et al. 2006; Casteleyn et al. 2015; Schulz et al. 2007). In the United States, the Centers for Disease Control and Prevention (CDC) conducts one such program: the National Health and Nutrition Examination Survey (NHANES) (CDC 2016, 2017). The first NHANES, NHANES I, took place in 1971–1975, followed by NHANES II (1976–1980), Hispanic HANES (1982–1984), and NHANES III (1988–1994). In 1999, NHANES became an annual survey (Calafat 2012). Collection of biomonitoring data began with the measurement of blood lead levels in NHANES II participants three years and older.

NHANES relies on a complex multistage area probability design, which includes selection of primary sampling units (counties), household segments within the counties, and sample persons from select households (CDC 2016, 2017). Analyses of NHANES results incorporate sample weights and design variables to account for unequal selection probabilities caused by the complex, clustered design of the survey and to account for the oversampling of certain demographic groups. Therefore, because NHANES data estimates are probability-based, they are representative of the United States' civilian noninstitutionalized population. Since 1999, each year, approximately 5000 randomly chosen residents across the United States participate in the survey. Participants undergo a physical examination in a mobile examination center; they also provide sociodemographic and health behavior data, detailed personal and family medical history, a blood sample, and a urine specimen. These biospecimens are analyzed to evaluate participants' clinical and nutritional status (CDC 2016, 2017). Biospecimens from a representative subset of participants, often a one-third subsample, are used to assess exposure to metals and hundreds of other environmental chemicals, including PFAS (CDC 2019). NHANES data are unique among biomonitoring surveys in that they provide a representative assessment of Americans' exposures to potentially hazardous chemicals.

NHANES biomonitoring data integrate environment and health. For instance, as discussed below, NHANES results are critical to establish reference ranges for select environmental chemicals, set research priorities, evaluate effectiveness of public health interventions, and monitor exposure trends.

TABLE 7.1
PFAS measured in NHANES from 1999–2000 through 2015–2016.

PFAS in serum		NHANES Cycle[a]			
		1999–2000	2003–2004 through 2011–2012	2013–2014	2015–2016
Short-alkyl chain	PFBS		X	X	
	PFHpA	X	X	X	
Long-alkyl chain	PFHxS	X	X	X	X
	PFOS[b]	X	X	X	X
	PFOA[b]	X	X	X	X
	PFNA	X	X	X	X
	PFDA	X	X	X	X
	PFUnDA	X	X	X	X
	PFDoDA	X	X	X	X
	FOSA	X	X		
	EtFOSAA	X	X		
	MeFOSAA	X	X	X	X

[a] Pools analyzed in NHANES 2001–2002 because of unavailability of serum for individual biomonitoring measurements. When concentrations were largely undetectable for previous survey periods, measurements of certain PFAS (e.g., FOSA, EtFOSAA, PFBS, PFHpA) were discontinued (https://www.cdc.gov/exposurereport/faq.html).

[b] PFOS and PFOA measured as separate isomers (i.e., linear and branched) starting with 2013–2014.

Starting with NHANES 1999–2000, biomonitoring data exist for select long- and short-alkyl chain PFAS in the serum of random one-third subsamples of NHANES participants 12 years of age and older (Table 7.1) (CDC 2019). To provide nationally representative PFAS exposure data for young children in the United States, during NHANES 2013–2014 PFAS were also measured in the serum of a random one-third subsample of 639 participants 3–11 years of age (CDC 2019; Ye et al. 2018). Serum concentrations of PFOS, PFOA, PFHxS, and perfluorononanoate (PFNA) in these young children were similar to those among NHANES 2013–2014 adolescents and adults (Figure 7.1) (Ye et al. 2018). Together, these data suggest widespread exposure to several "legacy" long-alkyl chain PFAS in the US general population (CDC 2019), even among children born after the changes in production in the early 2000s (CDC 2019; Ye et al. 2018). NHANES data also show PFAS serum concentration differences by sex, race/ethnicity, and age (CDC 2019) perhaps related, at least in part, to lifestyle differences (e.g., diet) that may impact exposure to these compounds.

NHANES biomonitoring data (CDC 2019) suggest a downward exposure trend for several long-alkyl chain PFAS (Figure 7.2). For example, geometric mean serum concentrations of PFOS decreased by more than 84% from 30.4 μg/L in 1999–2000 to 4.72 μg/L in 2015–2016. Similarly, during the same time period, geometric mean

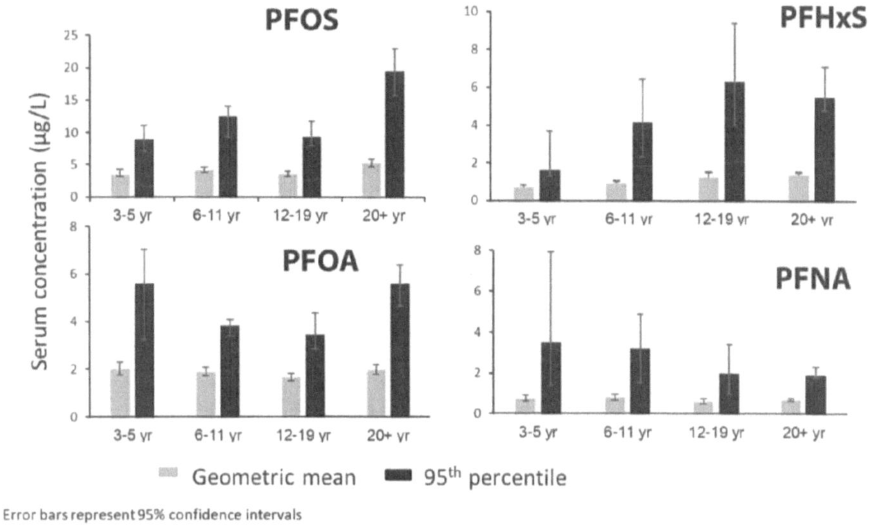

FIGURE 7.1 PFAS detected in 3–11 year old children (NHANES 2013–2014) (Data from CDC 2019, Ye et al. 2018).

serum concentrations of PFOA and PFHxS decreased by approximately 70% and 45%, respectively. By contrast, concentrations of other PFAS did not appreciably change (e.g., 5% increase for PFNA). These trends likely reflect discontinued production of PFOS-based materials (Paul et al. 2009), including PFOS, PFOA, and related compounds in the United States since 2002. The trends may also reflect efforts from United States industry and the EPA to reduce emissions and product content of PFOA and related chemicals by 95% by 2010 and toward elimination by 2015 (EPA 2006). For PFNA, which was present as a reaction by-product in PFOS-based materials no longer produced (Prevedouros et al. 2006), the increasing PFNA concentration trends may relate to degradation of volatile precursors such as fluoro telomer alcohols (Ellis et al. 2004).

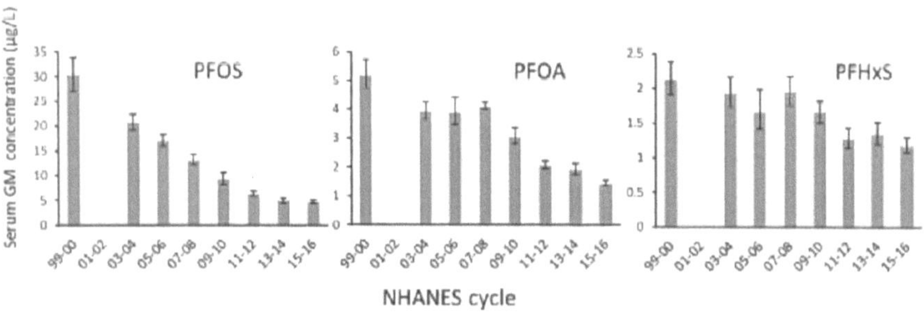

FIGURE 7.2 Temporal trends: long alkyl chain PFAS (Data from CDC 2019).

The fact that some PFAS concentrations are changing faster than others can be explained, in part, by differences in toxicokinetics. PFHxS, which has the longest estimated elimination half-life in humans (4.7–15.5 years) compared to PFOS (3.1–7.4 years) and PFOA (2.1–8.5 years) (ATSDR 2018), showed the smallest concentration decrease. Further, although manufacturing of some PFAS stopped in the United States approximately two decades ago, PFAS persist in the environment. So, we cannot rule out that certain exposures, such as those related to the consumption of PFAS-contaminated drinking water (as discussed later in the chapter), may still occur.

Although regulatory and other initiatives intended to reduce environmental emissions of PFAS also exist in Canada and the European Union, other countries still produce long-alkyl chain PFAS (Buck et al. 2011). Forthcoming NHANES biomonitoring data will provide critical information to evaluate whether PFAS manufactured abroad may impact Americans' future exposures.

7.3 BIOMONITORING CONCENTRATION PROFILES

NHANES reference ranges can be used to assist epidemiologic investigations in identifying segments of the population with exposures that exceed those of the civilian noninstitutionalized United States population. NHANES (and other general population-based studies) have commonly evaluated four PFAS: PFOS, PFOA, PFHxS, and PFNA. Remarkably, for background general population exposures, the relative concentrations of various PFAS (i.e., biomonitoring profiles) are quite consistent over time. These profiles, both at the median and at the high end (e.g., 95th percentile) serum concentrations, show PFOS with the highest concentrations followed by comparable concentrations of PFOA and PFHxS; PFNA has the lowest concentrations (Figure 7.3). However, in certain cases (e.g., occupational settings, populations accidentally exposed to specific PFAS), concentration patterns may differ from those reported among the general population (ATSDR 2019; Barton et al. 2020; Brede et al. 2010; Daly et al. 2018; Emmett et al. 2006; Graber et al. 2019; Hoffman et al.

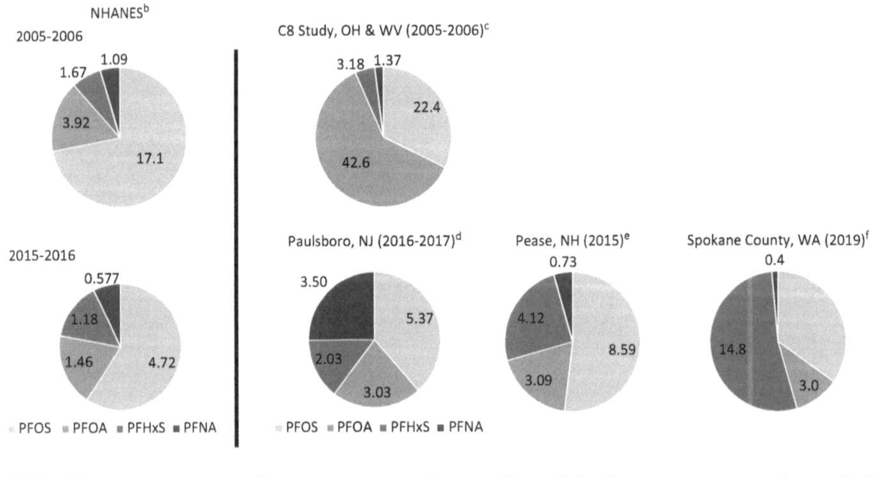

FIGURE 7.3 Examples of serum concentration profiles of the four most commonly studied PFAS: PFOS, PFOA, PFHxS and PFNA.[a] [a]Geometric mean concentrations shown (µg/L). [b](CDC 2019). [c](Gallo et al. 2013). [d](Graber et al. 2019). [e,f](Barton et al. 2020; Daly et al. 2018).

2011; Holzer et al. 2008; Holzer et al. 2009; Ingelido et al. 2018; Landsteiner et al. 2014; Stubleski et al. 2017; Wilhelm et al. 2009).

The differences between population-specific and NHANES profiles during a similar time period suggest that the population under study experienced unique exposures at some time that may not apply to the general population. More importantly, concentration profiles in combination with other data (e.g., environmental levels of PFAS, proximity to manufacturing sites or military facilities) can provide useful insight to identify exposure sources and inform public health efforts to reduce or stop exposures. Figure 7.3 illustrates examples in which higher-than-background exposures may have resulted from consumption of contaminated drinking water from industrial pollution (Graber et al. 2019; Hoffman et al. 2011; Ingelido et al. 2018; Landsteiner et al. 2014; Panikkar et al. 2019; Stubleski et al. 2016; Wilhelm et al. 2009) or from use of aqueous film-forming foam (AFFF), a highly effective fire suppressant agent used to extinguish petroleum-based fires (ATSDR 2019; Barton et al. 2020; Daly et al. 2018). Interestingly, individuals or communities with identical exposure sources may have concentration profiles that are quite different (Figure 7.4). For example, in the case of contamination of drinking water supplies from AFFF use (ATSDR 2019), differences in serum concentration profiles reflect differences in the magnitude of exposure, length of time exposure took place, composition of the AFFF used, design and complexity of the drinking water system, and use or installation of treatment systems, among other reasons. Biomonitoring profiles can help increase the understanding of factors that affect exposure to PFAS, which, in turn, can inform communities on ways to reduce their exposures.

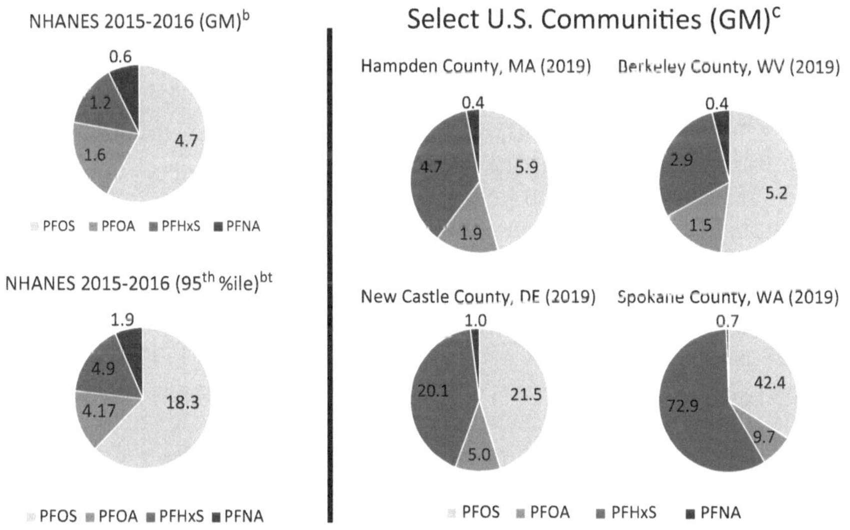

FIGURE 7.4 Differences in PFAS concentration profiles are evident across four communities in which the source of contamination (i.e., aqueous film-forming foam) is presumed to be the same. [a]Geometric mean (GM) or 95th percentile (95th %ile) serum concentrations shown (μg/L). [b](CDC 2019). [c](ATSDR 2019).

7.4 BIOMONITORING MATRICES FOR ASSESSING PFAS EXPOSURE

Blood (or its components) and urine are the most common matrices for biomonitoring. Blood is widely used to measure chemicals with long biological elimination half-lives (i.e., years), while chemicals with shorter elimination half-lives (e.g., hours) are most frequently measured in urine (Needham and Sexton 2000). Therefore, for PFAS, exposure has been largely evaluated from measuring PFAS in components of blood, mainly serum or plasma.

NHANES serum data suggest universal exposure to several long-alkyl chain legacy PFAS (CDC 2019). However, manufacturers are replacing select chemistries ("legacy" PFAS) with PFAS with shorter biological elimination half-lives (e.g., GenX, perfluorobutanoate [PFBA]) which may be eliminated efficiently in urine. Information regarding exposure to these other PFAS are limited.

To fill this data gap, CDC quantified 14 C_4-C_{11} PFAS and three fluorinated PFAS alternatives or substitutes in the urine of a random one-third subsample (N = 2682) of NHANES 2013–2014 participants ≥6 years (Calafat et al. 2019). NHANES estimates showed that about two-thirds of the population (67.5%) did not have detectable urinary concentrations (i.e., above the limit of detection of 0.1 μg/L) of any of the 17 PFAS examined, 27.2% had detectable concentrations of at least one PFAS, 4.7% of at least two, and 0.6% of three. Furthermore, the estimated detection frequencies in urine were 0–1.2% for 15 of the 17 target PFAS. Although estimated frequencies were higher for two short-alkyl chain PFAS (PFBA 13.3%, PFHxA 22.6%), concentrations of these chemicals were at or around the limit of detection. Based on these results, the general population's background exposure to short-alkyl chain or to alternative PFAS appear to be infrequent or at levels below those that would result in detectable concentrations (Calafat et al. 2019).

Furthermore, 2273 of these NHANES participants had their urine and serum analyzed for two short-chain PFAS and eight long-alkyl chain PFAS (Calafat et al. 2019). The detection frequency of the two short-alkyl chain PFAS was higher in serum (0.6% [PFBS], 12.6% [PFHpA] (CDC 2019, Ye et al. 2018) than in urine (0% [PFBS], 0–1.2%, depending on the age group [PFHpA]). Similarly, although the detection frequency in 2013–2014 NHANES participants' serum was >98% for the four most studied long-alkyl chain PFAS, namely PFHxS, PFOS, PFOA and PFNA, the corresponding frequency was <0.1% in urine.

These concentrations of PFAS in serum and urine in NHANES mirror those from occupational studies in which PFAS concentrations are much higher than those in NHANES participants. For example, median concentrations of PFHpA and PFBS were five to seven times higher in serum than in urine among people working in the proximity of fluorochemical manufacturing facilities (Zhou et al. 2014). In a separate study of six PFBS-manufacturing workers, PFBS concentrations were also higher in serum than in urine (Olsen et al. 2009). Similarly, concentrations of legacy PFAS are much higher in serum than in urine in:

- biomonitoring surveys of general population groups around the world (Genuis et al. 2013; Kim et al. 2014; Kim et al. 2019; Li et al. 2013; Zhang et al. 2015; Zhang et al. 2013),

- highly exposed populations including workers (Fu et al. 2016; Gao et al. 2015; Zhou et al. 2014), and
- people accidentally exposed to PFAS through contaminated water or dust (ATSDR 2019; Beesoon et al. 2012; Pritchett et al. 2019; Worley et al. 2017).

The above findings highlight the importance of selecting the appropriate biomonitoring matrix and being able to achieve the needed analytical sensitivity. For experimental research and pharmacokinetic studies, matrices other than serum (e.g., urine) may provide relevant information to increase understanding of the biological fate of PFAS. For exposure assessment of PFAS using biomonitoring, however, serum is far superior to urine regardless of the exposure type (e.g., background vs. occupational or accidental) and the PFAS biological elimination half-lives (which can range from several days to years).

7.5 ALTERNATIVE PFAS

PFAS exposure can vary considerably from region to region due to differences in demographics, exposure sources, and other factors, which leads to disparities in exposure. Of late, environmental detection of legacy and alternative PFAS, many of which are not part of NHANES, has highlighted the importance of conducting focused studies (e.g., investigations of residents of communities exposed through drinking water). These studies can examine whether localized exposures to PFAS, particularly short-alkyl chain and alternative PFAS (e.g., GenX, PFBA, PFHxA), in specific populations who live or work nearby exposure sources occur. One of these alternatives, GenX, introduced as a PFOA substitute, was detected with other PFAS in the Cape Fear River Basin in North Carolina (Sun et al. 2016). However, area residents whose drinking water private wells had an average 660 ng/L of GenX or whose tap water had a median concentration of GenX of 50 ng/L did not have detectable concentrations of GenX in urine or serum in two recent investigations (Kotlarz et al. 2020; Pritchett et al. 2019). Although residents might have consumed these water sources for years, they consumed bottled water for several months before providing their serum or urine; whether GenX would have been detectable before residents' switch to bottled water for drinking is unknown.

The results of one of these investigations (Kotlarz et al. 2020) support the observation from prior studies that the biological elimination half-life of GenX is much shorter than of legacy PFAS with similar carbon-chain lengths (on the order of years), and closer to estimates of other alternatives and short-alkyl chain PFAS (e.g., PFBA two to four days (Chang et al. 2008) and PFHxA 32 days (Russell et al. 2013)). Despite their relatively rapid elimination from the body, short-chain and alternative PFAS can persist in the environment, thus providing an ongoing source of exposure (Ahearn 2019; Gomis et al. 2015; Wang et al. 2015).

7.6 NHANES LIMITATIONS: RELEVANCE OF OTHER BIOMONITORING INVESTIGATIONS

Notwithstanding its many strengths, NHANES also has important limitations. First, the survey is cross-sectional, excludes persons under one year of age, and does not collect data on fetal exposures. Second, NHANES biomonitoring data provide national estimates (i.e., reference ranges) and not regional or local estimates, and generally do not allow the identification of seasonal or specific exposures, or of exposure source(s) or route(s). Third, because the amount of biospecimen available for biomonitoring is limited, quantifying trace concentrations of relatively non-biologically persistent PFAS is challenging because they are detected in drinking water at or below part-per-trillion levels (ng/L). Increasing the matrix volume can improve method sensitivity for environmental samples. For example, EPA method 537.1 uses 250 mL water to quantify 18 PFAS with estimated laboratory lowest concentration minimum reporting levels of 0.53–6.3 ng/L, depending on the analyte (Shoemaker and Tettenhorst 2018). Yet this approach is impractical for biomonitoring studies where the available sample volume is sometimes limited to fractions of a milliliter (e.g., CDC's current NHANES method uses 0.05 mL serum to achieve limits of detection of 100 ng/L (Kato et al. 2018)). Lastly, only a few hundred of the tens of thousands of chemicals in commerce are included in the NHANES biomonitoring program. Coupling the traditional targeted biomonitoring of NHANES with the nonspecific approaches of untargeted analyses and high-resolution metabolomics (Dennis et al. 2017) will help increase the understanding of environmental exposures and their public health relevance. Therefore, studies specifically designed to identify the relative contribution of exposure sources and routes, as well as additional chemicals of potential concern, can complement NHANES findings and provide relevant information to reduce exposures.

In summary, biomonitoring plays a key role in and has strengthened the exposure assessment process. Americans' exposure to PFAS is widespread, as shown by decades of NHANES data, and continuation of NHANES will provide critical information to understand the effectiveness of regulatory and voluntary efforts to reduce PFAS emissions. NHANES data should be augmented by additional studies designed to improve understanding of the determinants of PFAS exposure; delineate PFAS toxicokinetics (particularly for PFAS that are not as well studied as legacy PFAS); and characterize exposure, identify health effects, and assess usefulness of intervention strategies to reduce exposures among specific populations.

ACKNOWLEDGEMENT

We acknowledge CDC researchers for their tireless dedication throughout the years to the success of NHANES.

DISCLAIMER

The findings and conclusions in this report are those of the authors and do not necessarily represent the official position of the CDC. Use of trade names is for identification only and does not imply endorsement by the CDC, the Public Health Service, or the U.S. Department of Health and Human Services. The authors declare no competing financial interest.

REFERENCES

Ahearn, A. 2019. A regrettable substitute: The story of GenX. *Environmental Health Perspectives* doi: 10.1289/EHP5134.

Angerer, J., Bird, M.G., Burke, T.A., Doerrer, N.G., Needham, L., Robison, S.H., et al. 2006. Strategic biomonitoring initiatives: Moving the science forward. *Toxicological Sciences* 93: 3–10.

ATSDR. 2018. *Draft toxicological profile for perfluoroalkyls*. Available at: https://www.atsdr.cdc.gov/ToxProfiles/tp.asp?id=1117&tid=237 (accessed July 26, 2020.

ATSDR. 2019. PFAS *exposure assessment sites*. Available at: https://www.atsdr.cdc.gov/pfas/activities/assessments/sites.html (accessed July 29, 2020).

Barton, K.E., Starling, A.P., Higgins, C.P., McDonough, C.A., Calafat, A.M., Adgate, J.L. 2020. Sociodemographic and behavioral determinants of serum concentrations of per- and polyfluoroalkyl substances in a community highly exposed to aqueous film-forming foam contaminants in drinking water. *International Journal of Hygiene and Environmental Health* 223:256–266.

Beesoon, S., Genuis, S.J., Benskin, J.P., Martin, J.W. 2012. Exceptionally high serum concentrations of perfluorohexanesulfonate in a Canadian family are linked to home carpet treatment applications. *Environmental Science & Technology* 46:12960–12967.

Brede, E., Wilhelm, M., Goen, T., Muller, J., Rauchfuss, K., Kraft, M., et al. 2010. Two-year follow-up biomonitoring pilot study of residents' and controls' PFC plasma levels after PFOA reduction in public water system in Arnsberg, Germany. *International Journal of Hygiene and Environmental Health* 213:217–223.

Brendel, S., Fetter, E., Staude, C., Vierke, L., Biegel-Engler, A. 2018. Short-chain perfluoroalkyl acids: Environmental concerns and a regulatory strategy under REACH. *Environmental Sciences Europe* 30:9.

Buck, R.C., Franklin, J., Berger, U., Conder, J.M., Cousins, I.T., de Voogt, P., et al. 2011. Perfluoroalkyl and polyfluoroalkyl substances in the environment: Terminology, classification, and origins. *Integrated Environmental Assessment and Management* 7:513–541.

Calafat, A.M. 2012. The U.S. National Health and Nutrition Examination Survey and human exposure to environmental chemicals. *International Journal of Hygiene and Environmental Health* 215:99–101.

Calafat, A.M., Kato, K., Hubbard, K., Jia, T., Botelho, J.C., Wong, L.Y. 2019. Legacy and alternative per- and polyfluoroalkyl substances in the U.S. General population: Paired serum-urine data from the 2013–2014 National Health and Nutrition Examination Survey. *Environment International* 131:105048.

Casteleyn, L., Dumez, B., Becker, K., Kolossa-Gehring, M., Den, H.E., Schoeters, G., et al. 2015. A pilot study on the feasibility of european harmonized human biomonitoring: Strategies towards a common approach, challenges and opportunities. *Environmental Research* 141:3–14.

CDC. 2016. *National Health and Nutrition Examination Survey 1999–2016 survey content brochure*. Available at: http://www.cdc.gov/nchs/data/nhanes/survey_content_99_16.pdf (accessed July 23, 2020).

CDC. 2017. *About the National Health and Nutrition Examination Survey*. Available at: https://www.cdc.gov/nchs/nhanes/about_nhanes.htm#content (accessed July 22, 2020).

CDC. 2019. *Fourth national report on human exposure to environmental chemicals*. Updated tables, january 2019. Available at: http://www.cdc.gov/exposurereport (accessed July 20, 2020).

Chang, S.C., Das, K., Ehresman, D.J., Ellefson, M.E., Gorman, G.S., Hart, J.A., et al. 2008. Comparative pharmacokinetics of perfluorobutyrate in rats, mice, monkeys, and humans and relevance to human exposure via drinking water. *Toxicological Sciences* 104:40–53.

Daly, E.R., Chan, B.P., Talbot, E.A., Nassif, J., Bean, C., Cavallo, S.J., et al. 2018. Per- and polyfluoroalkyl substance (PFAS) exposure assessment in a community exposed to contaminated drinking water, New Hampshire, 2015. *International Journal of Hygiene and Environmental Health* 221:569–577.

Dennis, K.K., Marder, E., Balshaw, D.M., Cui, Y., Lynes, M.A., Patti, G.J., et al. 2017. Biomonitoring in the era of the exposome. *Environmental Health Perspectives* 125:502–510.

Dewitt, J.C.. 2015. *Toxicological effects of perfluoroalkyl and polyfluoroalkyl substances.* Basel, Switzerland:Humana Press.

Duffek, A., Conrad, A., Kolossa-Gehring, M., Lange, R., Rucic, E., Schulte, C., et al. 2020. Per- and polyfluoroalkyl substances in blood plasma - results of the german environmental survey for children and adolescents 2014–2017 (GerES V). *International Journal of Hygiene and Environmental Health* 228:113549.

Ellis, D.A., Martin, J.W., De Silva, A.O., Mabury, S.A., Hurley, M.D., Sulbaek Andersen, M.P., et al. 2004. Degradation of fluorotelomer alcohols: A likely atmospheric source of perfluorinated carboxylic acids. *Environmental Science & Technology* 38:3316–3321.

Emmett, E.A., Shofer, F.S., Zhang, H., Freeman, D., Desai, C., Shaw, L.M. 2006. Community exposure to perfluorooctanoate: Relationships between serum concentrations and exposure sources. *Journal of Occupational and Environmental Medicine* 48:759–770.

EPA. 2006. *PFOA stewardship program.* Available at: https://www.epa.gov/assessing-and-managing-chemicals-under-tsca/risk-management-and-polyfluoroalkyl-substances-pfass#tab-3 (accessed July 31, 2020).

EPA. 2016. *Drinking water health advisories for PFOA and PFOS.* Available at: https://www.epa.gov/ground-water-and-drinking-water/drinking-water-health-advisories-pfoa-and-pfos (accessed July 30, 2020).

Fromme, H., Wockner, M., Roscher, E., Volkel, W. 2017. ADONA and perfluoroalkylated substances in plasma samples of German blood donors living in south Germany. *International Journal of Hygiene and Environmental Health* 220:455–460.

Fu, J, Gao, Y., Cui, L., Wang, T., Liang, Y., Qu, G., et al. 2016. Occurrence, temporal trends, and half-lives of perfluoroalkyl acids (PFAAS) in occupational workers in China. *Scientific Reports* 6:38039.

Gallo, V., Leonardi, G., Brayne, C., Armstrong, B., Fletcher, T. 2013. Serum perfluoroalkyl acids concentrations and memory impairment in a large cross-sectional study. *BMJ Open* 3:e002414.

Gannon, S.A., Johnson, T., Nabb, D.L., Serex, T.L., Buck, R.C., Loveless, S.E. 2011. Absorption, distribution, metabolism, and excretion of [1-C-14]-perfluorohexanoate ([C-14]-PFHx) in rats and mice. *Toxicology* 283:55–62.

Gannon, S.A., Fasano, W.J., Mawn, M.P., Nabb, D.L., Buck, R.C., Buxton, L.W., et al. 2016. Absorption, distribution, metabolism, excretion, and kinetics of 2,3,3,3-tetrafluoro-2-(heptafluoropropoxy)propanoic acid ammonium salt following a single dose in rat, mouse, and cynomolgus monkey. *Toxicology* 340:1–9.

Gao, Y., Fu, J., Cao, H., Wang, Y., Zhang, A., Liang, Y., et al. 2015. Differential accumulation and elimination behavior of perfluoroalkyl acid isomers in occupational workers in a manufactory in China. *Environmental Science & Technology* 49:6953–6962.

Gebbink, W.A., van Asseldonk, L., van Leeuwen, S.P.J. 2017. Presence of emerging per- and polyfluoroalkyl substances (PFASs) in river and drinking water near a fluorochemical production plant in the Netherlands. *Environmental Science & Technology* 51:11057–11,065.

Genuis, S.J., Beesoon, S., Birkholz, D.. 2013. Biomonitoring and elimination of perfluorinated compounds and polychlorinated biphenyls through perspiration: Blood, urine, and sweat study. *ISRN Toxicology* 2013:483832.

Gomis, M.I., Wang, Z., Scheringer, M., Cousins, I.T. 2015. A modeling assessment of the physicochemical properties and environmental fate of emerging and novel per- and polyfluoroalkyl substances. *The Science of the Total Environment* 505:981–991.

Gomis, M.I., Vestergren, R., Borg, D., Cousins, I.T. 2018. Comparing the toxic potency in vivo of long-chain perfluoroalkyl acids and fluorinated alternatives. *Environment International* 113:1–9.

Graber, J.M., Alexander, C., Laumbach, R.J., Black, K., Strickland, P.O., Georgopoulos, P.G., et al. 2019. Per and polyfluoroalkyl substances (PFAS) blood levels after contamination of a community water supply and comparison with 2013–2014 NHANES. *Journal of Exposure Science & Environmental Epidemiology* 29:172–182.

Haines, D.A., Khoury, C., Saravanabhavan, G., Werry, K., Walker, M., Malowany, M. 2017. Human biomonitoring reference values derived for persistent organic pollutants in blood plasma from the Canadian Health Measures Survey 2007–2011. *International Journal of Hygiene and Environmental Health* 220:744–756.

Heydebreck, F., Tang, J., Xie, Z., Ebinghaus, R. 2015. Alternative and legacy perfluoroalkyl substances: Differences between European and Chinese river/estuary systems. *Environmental Science & Technology* 49:8386–8395.

Hoffman, K., Webster, T.F., Bartell, S.M., Weisskopf, M.G., Fletcher, T., Vieira, V.M. 2011. Private drinking water wells as a source of exposure to perfluorooctanoic acid (PFOA) in communities surrounding a fluoropolymer production facility. *Environmental Health Perspectives* 119:92–97.

Holzer, J., Midasch, O., Rauchfuss, K., Kraft, M., Reupert, R., Angerer, J., et al. 2008. Biomonitoring of perfluorinated compounds in children and adults exposed to perfluorooctanoate-contaminated drinking water. *Environmental Health Perspectives* 116:651–657.

Holzer, J., Goen, T., Rauchfuss, K., Kraft, M., Angerer, J., Kleeschulte, P., et al. 2009. One-year follow-up of perfluorinated compounds in plasma of German residents from Arnsberg formerly exposed to PFOA-contaminated drinking water. *International Journal of Hygiene and Environmental Health* 212:499–504.

IARC. 2017. *Some chemicals used as solvents and in polymer manufacture. IARC monographs on the evaluation of carcinogenic risks to humans volume 110*. Available at: https://publications.iarc.fr/547 (accessed July 29, 2020).

Ingelido, A.M., Abballe, A., Gemma, S., Dellatte, E., Iacovella, N., De, A.G., et al. 2018. Biomonitoring of perfluorinated compounds in adults exposed to contaminated drinking water in the Veneto region, Italy. *Environment International* 110:149–159.

Jones, P.D., Hu, W.Y., De Coen, W., Newsted, J.L., Giesy, J.P. 2003. Binding of perfluorinated fatty acids to serum proteins. *Environmental Toxicology and Chemistry* 22:2639–2649.

Kabore, H.A., Vo, D.S., Munoz, G., Meite, L., Desrosiers, M., Liu, J., et al. 2018. Worldwide drinking water occurrence and levels of newly-identified perfluoroalkyl and polyfluoroalkyl substances. *The Science of the Total Environment* 616–617:1089–1100.

Kato, K., Kalathil, A.A., Patel, A.M., Ye, X., Calafat, A.M. 2018. Per- and polyfluoroalkyl substances and fluorinated alternatives in urine and serum by on-line solid phase extraction-liquid chromatography–tandem mass spectrometry. *Chemosphere* 209:338–345.

Kim, D.H., Lee, M.Y., Oh, J.E. 2014. Perfluorinated compounds in serum and urine samples from children aged 5–13 years in South Korea. *Environmental Pollution* 192:171–178.

Kim, D.H., Lee, J.H., Oh, J.E. 2019. Perfluoroalkyl acids in paired serum, urine, and hair samples: Correlations with demographic factors and dietary habits. *Environmental Pollution* 248:175–182.

Kotlarz, N., McCord, J., Collier, D., Lea, C.S., Strynar, M., Lindstrom, A.B., et al. 2020. Measurement of novel, drinking water-associated PFAS in blood from adults and children in Wilmington, North Carolina. *Environmental Health Perspectives* 128:77005.

Landsteiner, A., Huset, C., Williams, A., Johnson, J. 2014. Biomonitoring for perfluorochemicals in a Minnesota community with known drinking water contamination. *Journal of Environmental Health* 77:14–19.

Li, J.G., Guo, F.F., Wang, Y.X., Zhang, J.L., Zhong, Y.X., Zhao, Y.F., et al. 2013. Can nail, hair and urine be used for biomonitoring of human exposure to perfluorooctane sulfonate and perfluorooctanoic acid? *Environment International* 53:47–52.

Liu, H.S., Wen, L.L., Chu, P.L., Lin, C.Y. 2018. Association among total serum isomers of perfluorinated chemicals, glucose homeostasis, lipid profiles, serum protein and metabolic syndrome in adults: NHANES, 2013–2014. *Environmental Pollution* 232:73–79.

Needham, L.L., Sexton, K. 2000. Assessing children's exposure to hazardous environmental chemicals: An overview of selected research challenges and complexities. *Journal of Exposure Analysis and Environmental Epidemiology* 10:611–629.

Nilsson, H., Karrman, A., Westberg, H., Rotander, A., van Bavel, B., Lindstrom, G. 2010. A time trend study of significantly elevated perfluorocarboxylate levels in humans after using fluorinated ski wax. *Environmental Science & Technology* 44:2150–2155.

NRC. 2012. *Exposure science in the 21st century: A vision and a strategy*. National Research Council.

Olsen, G.W., Chang, S.C., Noker, P.E., Gorman, G.S., Ehresman, D.J., Lieder, P.H., et al. 2009. A comparison of the pharmacokinetics of perfluorobutanesulfonate (PFBS) in rats, monkeys, and humans. *Toxicology* 256:65–74.

Pan, Y., Zhang, H., Cui, Q., Sheng, N., Yeung, L.W.Y., Guo, Y., et al. 2017. First report on the occurrence and bioaccumulation of hexafluoropropylene oxide trimer acid: An emerging concern. *Environmental Science & Technology* 51:9553–9560.

Panikkar, B., Lemmond, B., Allen, L., DiPirro, C., Kasper, S. 2019. Making the invisible visible: Results of a community-led health survey following PFAS contamination of drinking water in Merrimack, New Hampshire. *Environmental Health* 18:79.

Paul, A.G., Jones, K.C., Sweetman, A.J. 2009. A first global production, emission, and environmental inventory for perfluorooctane sulfonate. *Environmental Science & Technology* 43:386–392.

Prevedouros, K., Cousins, I.T., Buck, R.C., Korzeniowski, S.H. 2006. Sources, fate and transport of perfluorocarboxylates. *Environmental Science & Technology* 40:32–44.

Pritchett, J.R., Rinsky, J.L., Dittman, B., Christensen, A., Langley, R., Moore, Z., et al. 2019. Targeted biomonitoring for GenX and other per- and polyfluoroalkyl substances (PFAS)—North Carolina, 2018. *MMWR. Morbidity and Mortality Weekly Report* 68:647–648.

Russell, M.H., Nilsson, H., Buck, R.C. 2013. Elimination kinetics of perfluorohexanoic acid in humans and comparison with mouse, rat and monkey. *Chemosphere* 93:2419–2425.

Schulz, C., Angerer, J., Ewers, U., Kolossa-Gehring, M. 2007. The German Human Biomonitoring Commission. *International Journal of Hygiene and Environmental Health* 210:373–382.

Shoemaker, J., Tettenhorst, D. 2018. *Method 537.1: Determination of selected per- and polyfluorinated alkyl substances in drinking water by solid phase extraction and liquid chromatography/tandem mass spectrometry (LC/MS/MS)*. Washington, DC: U.S. Environmental Protection Agency, Office of Research and Development, National Center for Environmental Assessment.

Stubleski, J., Salihovic, S., Lind, L., Lind, P.M., van, B.B., Karrman, A. 2016. Changes in serum levels of perfluoroalkyl substances during a 10-year follow-up period in a large population-based cohort. *Environment International* 95:86–92.

Stubleski, J., Salihovic, S., Lind, P.M., Lind, L., Dunder, L., McCleaf, P., et al. 2017. The effect of drinking water contaminated with perfluoroalkyl substances on a 10-year longitudinal trend of plasma levels in an elderly Uppsala cohort. *Environmental Research* 159:95–102.

Sun, M., Arevalo, E., Strynar, M., Lindstrom, A., Richardson, M., Kearns, B., et al. 2016. Legacy and emerging perfluoroalkyl substances are important drinking water contaminants in the Cape Fear River watershed of North Carolina. *Environmental Science & Technology Letters* 3:415–419.

UNEP. 2019. *The new POPs under the Stockholm Convention* Available at: http://chm.pops.int/TheConvention/ThePOPs/TheNewPOPs/tabid/2511/Default.aspx [accessed July 28, 2020].

Wang, Z., Cousins, I.T., Scheringer, M., Hungerbuehler, K. 2015. Hazard assessment of fluorinated alternatives to long-chain perfluoroalkyl acids (PFAAs) and their precursors: Status quo, ongoing challenges and possible solutions. *Environment International* 75:172–179.

Wei, C., Wang, Q., Song, X., Chen, X., Fan, R., Ding, D., et al. 2018. Distribution, source identification and health risk assessment of PFASs and two PFOS alternatives in groundwater from non-industrial areas. *Ecotoxicology and Environmental Safety* 152:141–150.

Wilhelm, M., Holzer, J., Dobler, L., Rauchfuss, K., Midasch, O., Kraft, M., et al. 2009. Preliminary observations on perfluorinated compounds in plasma samples (1977–2004) of young German adults from an area with perfluorooctanoate-contaminated drinking water. *International Journal of Hygiene and Environmental Health* 212:142–145.

Worley, R.R., Moore, S.M., Tierney, B.C., Ye, X., Calafat, A.M., Campbell, S, et al. 2017. Per- and polyfluoroalkyl substances in human serum and urine samples from a residentially exposed community. *Environment International* 106:135–143.

Ye, X., Kato, K., Wong, L.Y., Jia, T., Kalathil, A., Latremouille, J., et al. 2018. Per- and polyfluoroalkyl substances in sera from children 3 to 11 years of age participating in the National Health and Nutrition Examination Survey 2013–2014. *International Journal of Hygiene and Environmental Health* 221:9–16.

Zhang, T., Sun, H., Qin, X., Gan, Z., Kannan, K. 2015. PFOS and PFOA in paired urine and blood from general adults and pregnant women: Assessment of urinary elimination. *Environmental Science and Pollution Research* 22:5572–5579.

Zhang, Y.F., Beesoon, S., Zhu, L.Y., Martin, J.W. 2013. Biomonitoring of perfluoroalkyl acids in human urine and estimates of biological half-life. *Environmental Science & Technology* 47:10619–10627.

Zhou, Z., Shi, Y.L., Vestergren, R., Wang, T., Liang, Y., Cai, Y.Q. 2014. Highly elevated serum concentrations of perfluoroalkyl substances in fishery employees from Tangxun Lake, China. *Environmental Science & Technology* 48:3864–3874.

8 State of the Science for Risk Assessment of PFAS at Contaminated Sites

Jeanmarie Zodrow, Jennifer Arblaster, and Jason Conder
Geosyntec Consultants

CONTENTS

8.1 Introduction 161
8.2 General Overview of Fate and Exposures at PFAS Contaminated Sites 163
8.3 Human Health Risk Assessment 163
8.3.1 Overview of Human Health Effects and Exposures 163
8.3.2 Human Health Assessment of PFAS in Drinking Water 164
8.3.3 Human Health Assessment of PFAS in the Diet 168
8.3.3.1 PFAS Exposure via Fish Consumption 168
8.3.3.2 Calculation of Site-specific Fish Tissue Criteria 169
8.3.3.3 Other Dietary Exposure Routes – Game, Agriculture, and Homegrown Produce 170
8.3.4 Human Health Assessment of PFAS Associated with Other Exposure Pathways 171
8.3.4.1 Incidental Soil Ingestion and Dust Inhalation 171
8.3.4.2 Dermal Contact with Soils and Water 172
8.3.4.3 Vapor Inhalation 172
8.4 Ecological Risk Assessment 173
8.4.1 Ecological Health Assessment of PFAS for Aquatic Life 173
8.4.2 Ecological Health Assessment of PFAS for Aquatic-dependent and Terrestrial Vertebrate Wildlife 175
8.4.3 Ecological Health Assessment of PFAS for Soil Life 177
8.5 Conclusions 178
References 180

8.1 INTRODUCTION

The widespread detection of per- and polyfluoroalkyl substances (PFAS) at thousands of sites worldwide will require the application of human health and ecological risk assessments to enable the effective and efficient assessment and management of potentially impacted areas by stakeholders. Detectable concentrations of PFAS are ubiquitous in the environment because of the high persistence and water solubility of many of the PFAS used in various applications and subsequent high volume/mass usage, release, and disposal of materials containing PFAS (Hekster et al. 2003; Buck et al. 2011; International Training and Research Center [ITRC] 2020). In particular, a variety of site settings can exhibit concentrations of PFAS that are elevated with respect to ambient levels in the environment (although not every site exhibits detectable and/or actionable concentrations of PFAS, and other types of sites may also exhibit elevated levels), including: (1) facilities that have manufactured PFAS; (2) military facilities; (3) airports; (4) firefighting training areas; (5) chemical facilities; (6) chrome/metal plating facilities; (7) textile/carpet manufacturing facilities; (8) wastewater treatment plants; (9) sewage sludge land application areas; and (10) landfills.

Analytical chemistry methods for PFAS are now sufficiently sensitive to yield detectable concentrations of at least a few PFAS in a typical environmental sampling program conducted in areas impacted by anthropogenic uses. The widespread detection of PFAS in soils in remote locations of North America (Rankin et al. 2016; Strynar et al. 2012) has indicated PFAS are widespread in the environment. Using analytical detections as a trigger for active management of PFAS at sites will not be logistically feasible give the large volumes of environmental media likely to yield a detection using current (and improving) analytical techniques. The "how clean is clean?" question should be answered with the underlying dictum that "the dose makes the poison" rather than by the presence of concentrations that can be detected by the ever-improving abilities of our analytical chemists. Risk assessment of PFAS at contaminated sites is needed to facilitate remediation decisions about PFAS at sites, offering key input on the approaches used to investigate, the potential adverse impacts to humans and ecological resources, and the optimization of our efforts to manage unacceptable risks in a timely, resource efficient, and realistic manner.

This goal of this chapter is to present an overview of the current state of human health and ecological risk assessment at contaminated sites impacted by PFAS, particularly sites impacted with aqueous film forming foam (AFFF) and other point sources of PFAS releases. It will not discuss the considerable body of work and ongoing focus on the consumer product, occupational, industrial, and epidemiological assessments that have evaluated human exposures and effects. Our review focuses on risk assessments that primarily aim to address human health risks and ecological risks that are often evaluated at the point of releases, as well as human health and ecological risks that occur off-site as a result of PFAS transport away from the initial point of release. The remainder of this chapter is organized into the following sections:

Section 8.1: General Overview of Fate and Exposures at PFAS Contaminated Sites
Section 8.2: Human Health Risk Assessment
Section 8.3: Ecological Risk Assessment.
Section 8.4: Conclusions

8.2 GENERAL OVERVIEW OF FATE AND EXPOSURES AT PFAS CONTAMINATED SITES

Environmentally relevant PFAS exhibit a wide range of physical and chemical properties that greatly influence their human health and ecological hazard profiles (Hekster et al. 2003; Conder et al. 2008; Ahrens 2011; Buck et al. 2011; Guelfo and Higgins 2013; ITRC 2020). The fate of the perfluoroalkyl acids (PFAAs), particularly the perfluoroalkyl carboxylic acids (PFCAs) and perfluoroalkyl sulfonic acids (PFSAs) like perfluorooctanoate (PFOA) and perfluorooctane sulfonic acid (PFOS), have received the most study, as PFCAs and PFSAs are often the most prevalent transformation products for many PFAS precursors in the environment (Houtz et al. 2013). Under normal environmental conditions, PFCAs and PFSAs cannot transform any further and thus are persistent in the environment (Buck et al. 2011; ITRC 2020). PFCAs and PFSAs are organic anions at common environmental pH values (Buck et al. 2011) and are relatively water-soluble and mobile in the environment compared to other persistent organic chemicals of concern at contaminated sites. PFCAs and PFSAs are not volatile (i.e., they do not evaporate to the atmosphere readily [Field et al. 2017]). PFCAs and PFSAs can sorb to the organic carbon present in soil or sediment such that organic carbon-normalized distribution coefficients (i.e., K_{OC} values) can be useful parameters for evaluating transport potential in soil, sediment, and water (Higgins and Luthy 2006; Guelfo and Higgins, 2013). Given their sorptive properties, PFCAs and PFSAs, in addition to other neutral, cationic, and zwitterionic PFAS (ITRC 2020), can accumulate in soils and sediments. Some PFAS have a proclivity to bioaccumulate in animals and plants, and persistent PFAS such as PFCAs and PFSAs do not undergo significant biodegradation or biotransformation once present in a biological system (Conder et al. 2008). Given the wide range of solubility, sorption, and bioaccumulation properties, PFCAs and PFSAs (and other PFAS) can be prevalent in a wide variety of environmental media, including groundwater, surface water, soil, sediment, biosolids, landfill leachate, plants, fish, invertebrates, and wildlife, both at contaminated sites where PFAS are initial released to the environment, and in off-site areas to which PFAS has been transported by natural physical processes (Lau 2012).

8.3 HUMAN HEALTH RISK ASSESSMENT

8.3.1 Overview of Human Health Effects and Exposures

The characterization of effects in health risk assessments is derived from toxicological studies of mammalian exposures to PFAS, primarily studies involving PFOS and PFOA. The most sensitive non-cancer adverse effects have been observed to involve the liver and kidney, immune system, and various developmental and reproductive endpoints (USEPA, 2016a, 2016b, 2016c, 2016d; Agency for Toxic Substances and Disease Registry (ATSDR), 2018). There is limited information on whether PFAS are carcinogenic in laboratory test mammals and humans (ATSDR 2018). Some increases in cancers have been found in workers and in community members living near a facility that produced PFOA (USEPA 2016b), although more research is being conducted to elucidate the potential causal links between PFOA and cancer.

The USEPA currently considers carcinogenicity potential as "suggestive" for both PFOA and PFOS (USEPA 2016a, 2016b) and several US states have concluded carcinogenicity for PFOA and PFOS (e.g., Office of Environmental Health Hazard Assessment [OEHHA] 2019). The International Agency for Research on Cancer considers PFOA "possibly" carcinogenic (USEPA 2016a, 2016b).

Given the current understanding of PFAS releases, fate, exposure, and effects, the most important human health PFAS risk issues associated with most PFAS sites are for direct exposure pathways associated with drinking water ingestion and dietary intake (Table 8.1). Given the chronic, long-term nature of these pathways, particularly for off-site receptors impacted by long-range transport of PFAS in groundwater and surface water, these pathways are expected to result in the highest risk levels and will tend to drive decision-making at many sites. Incidental soil ingestion and/or dust inhalation are typically of moderate concern and are recommended for inclusion into human health risk assessments, but compared to drinking water and dietary ingestion, generally result in lower exposures for most receptors. Exposures via dermal contact with soils and water, and inhalation of vapors (due to volatilization of PFAS), are generally of low concern for most sites. The remainder of this section provides detail on the state of human health risk assessment approaches for each of these exposure pathways.

8.3.2 Human Health Assessment of PFAS in Drinking Water

Human health exposures via drinking water have been the focus in risk assessments and investigations at PFAS impacted sites (Post et al. 2012; Guelfo et al. 2018; USEPA 2019). PFAS investigation areas in groundwater aquifers and surface water bodies are often large (e.g., several square miles) and may extend several miles from PFAS release locations, thus presenting exposures via drinking water pathways.

Risk assessment approaches for PFAS for drinking water follow typical, well-established chemical drinking water risk assessment approaches (Table 8.1). Several international agencies, the US federal government, and states have established health-based regulatory values and criteria for drinking water (ITRC 2020; USEPA 2016a, 2016b), primarily for the stable PFAAs such as PFOA and PFOS. In May 2016, the USEPA released established lifetime health advisory (LHA) levels for PFOS and PFOA in drinking water at 70 nanograms per liter (ng/L) or 0.07 micrograms per liter (µg/L) (USEPA 2016a, 2016b). Based on observations that PFOS and PFOA have similar effects in animal studies and may exert toxic effects additively when present in a mixture, USEPA suggested that when both chemicals (PFOS and PFOA) occur at the same location in a drinking water source, the sum of the concentrations should not exceed the LHA of 70 ng/L (USEPA 2016a, 2016b). In 2019, USEPA (2019) noted individual screening levels of 40 ng/L (PFOA or PFOS) as thresholds for additional investigation where impacted groundwater is encountered.

Many US states have adopted the USEPA (2016a, 2016b) LHA for PFOA and PFOS or have derived their own risk-based values, and some state agencies have calculated risk-based thresholds for other PFAAs (ITRC 2020). In addition, as of early 2020, the USEPA was following the Safe Drinking Water Act process that may culminate in the development of a Maximum Contaminant Level (MCL) for PFOA

TABLE 8.1
An overview of human health risk assessment approaches for PFAS at contaminated sites.

Exposure Pathway	Relative Importance as Risk-based Management Driver at Contaminated Sites	Screening Level Availability	Relative Maturity of Site Investigation Tools	Relative Maturity of Risk Assessment Tools
Drinking Water	**Higher**: PFAS in drinking water sources due to long-range transport	**Higher**: Many screening values, but limited to PFAAs	**Higher**: Standard measurement and sampling methods available	**Higher**: Standard drinking water risk assessment models apply
Dietary Items	**Higher**: PFAS in water and soil can result in uptake by homegrown produce, wild-caught fish, and other game	**Moderate**: Limited screening level values for food and abiotic media are available or are under development	**Moderate**: Standard measurement methods available, but optimal sampling methods and strategies needed	**Higher**: Standard dietary ingestion risk assessment models apply
Incidental Soil Ingestion and Dust Inhalation	**Moderate**: Exposures in past/present industrial/military areas	**Moderate**: Some regulatory authorities have values	**Higher**: Standard measurement and sampling methods available	**Higher**: Standard soil ingestion risk assessment models apply
Dermal Contact with Soils and Water	**Lower**: Less important than direct exposure pathways	**Lower**: Few/no values	**Higher**: Standard measurement and sampling methods available	**Lower**: Standard dermal uptake risk assessment models apply, but limited dermal uptake parameters
Vapor Inhalation	**Lower**: Most exposures relatively insignificant due to low volatility from water and soil	**Lower**: No/ limited values	**Lower**: Standard measurement and sampling methods not fully developed for many PFAS	**Lower**: Standard inhalation models apply, but inhalation toxicity data for many PFAS are limited or unavailable

and PFOS. For PFOS and PFOA, alternate risk-based concentrations for drinking water for US states and international organizations vary widely from the USEPA LHA values. For example, the California Environmental Protection Agency's OEHHA (2019) has calculated a risk-based level of 0.4 ng/L in drinking water for PFOS, a value that is 100 to 175 times lower than that of the USEPA's 40 ng/L screening level and 70 ng/L LHA. The risk-based concentrations for PFOS in drinking water calculated independently by other US state agencies and those of other countries currently spans two orders-of-magnitude (ITRC 2020).

The large discrepancies among the various regulatory organizations are largely due to the different effect endpoints and exposure doses being used to calculate risk-based levels in drinking water. Put simply, the calculations are executed to determine the concentration in drinking water that is expected to reach an exposure associated with an effect or lack of effect (for non-cancer effects) or a particular risk level associated with development of cancer, as determined from laboratory studies with standard test animals, such as rodents. The calculation process begins with review of animal laboratory studies, and the evaluation of the adverse effects in animals at the various doses received by the test animals. Often the product of this review is a benchmark dose level that is associated with a lack of effects for a variety of relevant endpoints (developmental abnormalities, growth reductions, cancer effects, etc.). The various toxicologists that have independently conducted these exercises do not necessarily agree on the most applicable study, the adverse effect endpoint, or the particular PFAS benchmark doses at which adverse effects appear or doses at which adverse effects are absent. This results in the selection of benchmark dose values that can vary widely. Additionally, because of the differences in PFAS behavior in laboratory animals studies and humans, the doses identified from the laboratory studies are mathematically converted to values for humans that are then used as reference doses (*RfDs*) and cancer slope factors (CSFs) for calculating site-specific risks or deriving risk-based concentrations in drinking water or other environmental media. Once again, a variety of different laboratory-to-human exposure dose extrapolation approaches have been used by different groups of toxicologists (USEPA 2016a, 2016b; ITRC 2020). Given the discordance in the interpretation of animal toxicity studies and various approaches to convert the resulting data for use with humans, the result is a wide range of *RfDs* and CSFs that result in a wide range of acceptable levels in drinking water.

In terms of application for site-specific risk-based decision-making, the USEPA LHA and US state/international screening levels are intended to provide technical guidance on health effects and treatment technologies. They are suggested as preliminary remediation goals (PRGs) to inform final cleanup levels for PFOA and/or PFOS contamination of groundwater that is a current or potential source of drinking water (USEPA 2019). Other state-promulgated risk-based levels in drinking water have been used in a similar fashion to address groundwater and drinking water resources that indicate PFAS concentrations above the respective criteria. Measurement of drinking water is fairly straightforward (USEPA Method 537 and Method 533 for drinking water), and a variety of guidance documents and technical resources are available to address challenges associated with sampling drinking water and groundwater (Michigan Department of Environmental Quality [MDEQ] 2018,; ITRC 2020; Rodowa et al. 2020).

PFAS levels for drinking water include a Relative Source Contribution (RSC), which adds additional conservatism and protectiveness to the drinking water value under the assumption that drinking water consumers are exposed to other non-drinking water PFAS sources. Instead of calculating the water concentration at which a drinking water consumer may be exposed at a dose associated with a lack of effects or a dose associated with a particular cancer risk level, an RSC value is applied such that only a fraction of that critical dose is allowable to be reached via drinking water consumption. For example, the USEPA (2016a, 2016b) and many US states assume an RSC value of 20%, thus allowing for up to 80% of PFAS exposures to occur via other media (indoor dust, diet, etc.) before an acceptable exposure is reached. In USEPA (2019), the risk-based level (calculated without the application of the RSC) associated with a lack of non-cancer effects is approximately 400 ng/L for PFOA or PFOS. As more is learned regarding the daily ambient PFAS exposures to the general public, it may be possible to explicitly quantify these values and either include them explicitly in site-specific risk assessments or modify RSC values. For example, ToxConsult (2016) conducted dose estimation for ambient levels of PFSAs for members of the Australian public, and, rather than application of RSC values, those exposure values have been explicitly included in human health risk calculations for site-specific risk assessments in Australia (EnRisk 2017; AECOM Technical Services [AECOM] 2018a, 2018b). ToxConsult (2016) reports that 0.0008 and 0.0014 μg/kg/day are appropriate average and high-end estimates for background PFOS exposures for Australians.

In addition to accounting for ambient exposures via RSC or other approaches, perhaps the largest challenge for addressing drinking water risks is the absence of risk-based levels (or associated *RfD* and CSF values) for many of the PFAS that can be detected in drinking water. As of mid-2020, fewer than 20 PFAS have available toxicity data sufficient for site-specific risk assessments or screening level calculation, leaving no clear approach for understanding the risks of the other 10 to 20 PFAS that can be detected by commercial analytical laboratories in drinking water. Additionally, analytical methods cannot currently measure all PFAS, and the presence of PFAS that cannot be currently detected remains a huge uncertainty. Additionally, mixture toxicity remains a key data gap, making even the assessment of PFAS with toxicity information problematic. Various jurisdictions (e.g., USEPA, several US states, Australia, Germany) have indicated that some PFAS should be summed for a risk-based evaluation (e.g., PFOS + PFOA, PFOS + perfluorohexane sulfonate [PFHxS]). The technical and regulatory approaches for understanding risk of multiple PFAS are likely to continue to evolve over the next few years.

Beyond addressing PFAS risks in drinking water directly, another key challenge will be understanding and managing the contribution of PFAS from sources in contaminated soils to drinking water aquifers. For example, the USEPA's Regional Screening Levels (RSLs) online calculator (USEPA 2020a) provides approaches that can be used to predict a concentration of a chemical in soil that could result in a particular risk-based concentration in groundwater, assuming leaching of the chemical from the soil. As detailed in ITRC (2020), the concentration of PFOS in soil predicted by the RSL calculator to be protective of drinking water uses is 0.4 micrograms per kilogram (μg/kg). This value is lower than ambient or background concentrations of PFOS in soils obtained from remote locations in North America (Rankin et al. 2016; Strynar et al. 2012). Given that the majority of drinking water systems in the US do not exceed the USEPA PFOS LHA level of 70 ng/L (as tested by Hu et al. 2016), an RSL value of 0.4 μg/kg for soil does not appear to be accurate in identifying and prioritizing soils that should be investigated or managed to reduce their potential to impact aquifers used as drinking water. The evaluation of the soil to groundwater transport pathway may be only appropriate using site-specific data and assumptions at this time. Much work needs to be done on fate assessment and modeling approaches to relate concentrations in soils to concentrations in groundwater, as the management of PFAS-impacted soils and source areas at many sites will be a key focus of managing PFAS drinking water risks.

8.3.3 Human Health Assessment of PFAS in the Diet

As noted in the previous section, due to the prevalence of PFAS in groundwater, the initial regulatory focus for PFAS has been on establishing drinking water criteria for the protection of human health. However, at PFAS-impacted contaminated sites, the bioaccumulation potential of many PFAS results in dietary exposure pathways that can be complete and significant. Singularly evaluating drinking water does not account for, and may not be protective of, these pathways which can include the consumption of vegetable produce irrigated with, and livestock products fed with PFAS-impacted groundwater or the consumption of fish or shellfish living in PFAS-impacted water bodies. For example, the risk-based threshold concentration of PFOS in water associated with exposure of fishermen consuming fish from Pool 2 of the Mississippi River in Minnesota was set at 14 ng/L (Minnesota Pollution Control Agency [MPCA] 2013), almost five times lower than the USEPA 70 ng/L drinking water LHA.

A site-specific Conceptual Site Model should inform the determination of complete exposure pathways. At PFAS-impacted aquatic sites, bioaccumulation of PFAS into shellfish and fish from surface water and sediment results in complete exposure pathways to both human and ecological consumers. At PFAS-impacted terrestrial sites, PFAS in soil and soil porewater, including from irrigation with PFAS-impacted water or application of biosolids, can bioaccumulate into plants, food crops, and livestock or other agricultural products, resulting in complete exposure pathways to consumers.

8.3.3.1 PFAS Exposure via Fish Consumption

Field and laboratory studies have shown that some PFCAs and PFSAs bioaccumulate in fish and other aquatic life at rates that could result in relevant dietary PFAS exposures for consumers of fish and other seafood (Martin et al. 2003a, 2003b; Chen et al. 2016; Fang et al. 2016; Bertin et al. 2014, 2016; Dai et al. 2013; Prosser et al. 2016; Muller et al. 2011; Rich et al. 2015). Due to this bioaccumulation potential and the widespread nature of PFAS in US waterways, PFAS are commonly detected in samples of fish and other aquatic animals. For example, Stahl et al. (2014) performed a large study of urban lakes, rivers, and the Great Lakes and reported that 73% of all fish samples had detectable concentrations of PFOS. PFOS is typically the most frequently detected PFAS and can account for greater than 80% of the total PFAS detected in fish (Ye et al. 2008; Stahl et al. 2014). Although PFOS is typically the dominant PFAS detected in fish tissue, other long-chain PFAS (perfluorodecanoic acid [PFDA], perfluoroundecanoic acid [PFUnA], perfluorododecanoic acid [PFDoA]) are also prevalent in fish (Stahl et al. 2014). As a result of the high bioaccumulation potential and widespread detections of PFAS in fish tissues, PFAS-impacted sites near water bodies often require site-specific fish tissue criteria as a risk management tool.

8.3.3.2 Calculation of Site-specific Fish Tissue Criteria

Consumption-based fish tissue criteria (*FTC*) are calculated to represent the concentrations of a chemical in edible fish tissues that is safe for consumption at a specific frequency. Often, *FTC* are calculated to indicate if a fish is safe to eat in unlimited amounts, one meal per week, or one meal per month, while accounting for regional fish consumption rates. *FTC* are typically calculated using a specified *RfD* (often selected based on the jurisdiction) following USEPA (2000a, 2000b) guidance. The general equation for calculating fish tissue concentrations is shown in Equation 8.1. A similar approach can also be used to derive a criterion for other aquatic dietary items, such as shellfish.

$$FTC = \frac{BW(kg)\,x\left(RfD\left[mg/kg/day\right] * RSC\left[\%\right]\right)}{FIR\left(kg/day\right)} \tag{8.1}$$

where:

BW = body weight.
RfD = selected human health Reference Dose.
RSC = Relative Source Contribution.
FIR = Fish ingestion Rate.

Human health exposure factors (body weight, fish ingestion rate) in the above equation are obtained from national guidance documents (USEPA 2011), state or regional guidance documents, or are based on site-specific surveys of fishing use and fish consumption for a specific water body. When bioaccumulation between water or sediment has been characterized at a site, these FTC can be back-calculated to a sediment- or surface-water-specific threshold. General site-specific human health risk assessment methods as outlined by USEPA (1989) are applicable to risk assessments for dietary exposures for PFAS for which a human health *RfD* has been developed.

Several US states have developed consumption-based FTC, often following the detection of PFAS in fish tissues during contaminated site investigations. Since anglers do not have the ability to measure PFAS in fish when caught, state-wide criteria are applied to monitoring programs to identify specific water bodies where Fish Consumption Advisories are needed; these water bodies are consistently found in close proximity to PFAS-contaminated sites. Minnesota, Maine, New Jersey, and Michigan are among the states that have developed state-wide criteria to identify potentially impacted water bodies. Maine has set action levels for PFOS in fish tissues for sensitive and the general populations (Maine PFAS Task Force 2020), while Minnesota, as part of the Great Lakes Consortium (GLC), and New Jersey have developed consumption advisories based on ranges of concentrations that correspond to recommended limits on frequency of consumption (New Jersey Department of Environmental Protection [NJDEP] 2019; GLC 2019). Michigan provides a listing of recommended consumptions frequencies for individual water bodies (Michigan PFAS Action Response Team [MPART] 2020), while New York and Wisconsin have developed Fish Consumption Advisories for specific water bodies based on regionally elevated concentrations of PFAS near a known contaminated site (New York

State Department of Health 2020) or by applying FTCs developed by other states (Minnesota Department of Natural Resources 2020). Most states have developed *FTCs* for PFOS only, though some have included PFOA, PFNA, and PFBS. Fish tissue criteria and fish consumption limits vary as a result of different selections of *RfD* and fish consumption rates.

As noted above, standard human health risk assessment (HHRA) methodologies are appropriate to evaluate potential risks from fish or shellfish consumption when analytical data are available. Analysis of PFAS in tissues is available from many commercial laboratories, though tissues are a more challenging matrix. When tissue data are unavailable for use in HHRAs, concentrations of PFAS in fish can be estimated from concentrations in water and sediment under most scenarios. Available approaches for modeling concentrations of PFAS in fish currently rely on empirical food web modeling that use laboratory developed measures of bioaccumulation from abiotic to biotic media (Larson et al. 2018). The uptake of some PFAS (e.g., PFOS) by aquatic food webs may be originating from sediment-associated PFOS rather than water-associated PFOS (Larson et al. 2018); therefore, modeling approaches that rely solely on water may misidentify the most optimal abiotic source for potential investigation and management. In addition, for predatory fish, uptake from diet (aquatic plants, invertebrates, other fish) should be considered. Much work is needed to refine our approaches, but guidance on current best practices is available (Conder et al. 2020; Zodrow et al. 2020).

8.3.3.3 Other Dietary Exposure Routes – Game, Agriculture, and Homegrown Produce

In addition to fish, terrestrial wildlife can accumulate contaminants from impacted sites, resulting in potential exposures to consumers of wild game. In these cases, specific game consumption advisories may be needed to protect hunters and consumers of wild game. Game consumption advisories are typically site-specific and reflect local hunting practices, though that can be challenging to quantify (Conder and Arblaster 2016).

The last important dietary exposure route to consider is consumption of homegrown produce or agricultural products that originate from areas irrigated with PFAS-impacted groundwater, or are amended with biosolids that contain PFAS, or contain soils that were directly affected by PFAS releases. Multiple studies have found PFAS can be taken up by plants from soil porewater (Blaine et al. 2013, 2014; Ghisi et al. 2019), and livestock can accumulate PFAS from drinking water and/or feed (van Asselt et al. 2013). As noted in Blaine et al. (2013), PFAS with shorter perfluorinated chain lengths are more bioaccumulative in plants, and PFSAs are less bioaccumulative than PFCAs. This is thought to be because of the higher water solubility of shorter PFSAs and PFCAs, which can result in their accumulation in leaf structures as the water taken up by a plant evaporates (leaving the relatively nonvolatile PFCAs and PFSAs behind in the tissue). With regards to agriculture, dairy farms and other livestock have also been found to accumulate elevated levels of PFAS following exposure from nearby PFAS-impacted sites (van Asselt et al. 2013; Sunderland et al. 2019). Dairy farms and many other agricultural systems have historically land-applied biosolids, which has resulted in cases of elevated PFAS in both cows and milk products (Sepulvado et al. 2011). Additionally, some PFAS have been shown to transfer from maternal tissues to eggs in poultry (Zafeiraki et al. 2016), indicating that poultry exposure to PFAS via water or feed may result in detectable concentrations in eggs.

There are currently only a few widely available agency-promulgated screening levels or criteria to evaluate these pathways (Food Standards Australia New Zealand 2017). The calculation of screening levels for food (i.e., risk-based concentrations in food items) is possible given traditional risk assessment models. Calculation of risk-based concentration in abiotic media, such as would be applied to soils in gardens or water used for agricultural purposes, is likely to be more difficult, as these models would require a variety of uptake factors into plants and agricultural animals. Standard human health risk assessment methods are available and appropriate for estimating risks from a variety of food products on a site-specific basis (AECOM 2018a, 2018b; EnRisk 2017; Scher et al. 2018). While this exposure pathway may not be complete at all areas near PFAS-impacted sites, studies indicate that when complete, the homegrown food product pathway can be a significant contributor to PFAS exposures (Brown et al. 2019; Stahl et al. 2011; Zhang et al. 2020).

8.3.4 Human Health Assessment of PFAS Associated with Other Exposure Pathways

Compared to drinking water and dietary exposures, the risks of PFAS in typical residential exposure scenarios for PFAS sites are likely to be much lower for other exposure pathways, such as the incidental ingestion of soils and dusts, dermal contact with soils and water, and inhalation of vapors (Table 8.1). These exposure pathways may be more relevant (compared to drinking water and diet) for some site-specific occupational exposure scenarios, as discussed in more detail below.

8.3.4.1 Incidental Soil Ingestion and Dust Inhalation

Incidental exposures to soil and dusts for PFAS can occur during a variety of soil disturbance activities, such as gardening and digging, hand-to-mouth activities that result in consumption of dusts or soils that enter living spaces, and intentional consumption of soils (pica eating disorder). As detailed in ITRC (2020), many US states and USEPA have calculated risk-based screening levels for these pathways (and many also include dermal exposures to soils) using well-established risk assessment models. For example, the residential exposure risk-based values for PFOS and PFOA in soil, as calculated by the USEPA RSL calculator, are 1.26 milligrams per kilogram (mg/kg) (USEPA 2020a; ITRC 2020). Most risk assessments for off-site residential areas will not focus primarily on this pathway, however, as elevated concentrations of PFAS in soil are primarily limited to PFAS release areas at industrial locations, airports, and military sites. Incidental soil and dust ingestion could be a potentially important pathway for areas that are converted from industrial or military sites to residential areas. The main transport route of concern for most sites is the transport of PFAS to residential areas via groundwater or surface water. The use of PFAS-impacted water for irrigation may result in transfer of PFAS to surface soils, but at relatively low concentrations. For example, in a study of residences within an area threatened by PFAS in groundwater, a survey of soils in gardens irrigated with groundwater revealed concentrations in soil several orders of magnitude below screening levels for ingestion of soil and dust (Scher et al. 2018).

In contrast, in cases in which industrial locations, airport, and military sites are being or have been converted for residential use, it is possible that risks via this exposure pathway could present a primary exposure pathway of concern. In some cases, incidental soil ingestion is likely to be the main exposure pathway for industrial and construction workers in areas where PFAS was released to soils, although risk-based screening levels would be higher than those for residential exposures since industrial exposures are generally estimated to be for a shorter duration. Concentrations of PFOS in soils measured at US Air Force facilities were noted to be as high as 9.7 mg/kg (Anderson et al. 2016), suggesting that soils in areas where PFAS were directly released may present risk potential to future residential land uses via incidental soil ingestion pathways.

8.3.4.2 Dermal Contact with Soils and Water

Dermal uptake of PFAS from water and soils is a potential route of exposure of concern considered for site-specific human health risk assessments, but is expected to be a relatively minor concern at most PFAS sites relative to other more direct exposure pathways due to the limited dermal uptake for PFAS (Stahl et al. 2011; Sunderland et al. 2019). A risk assessment modeling parameter based on PFOA (ATSDR 2018) has been used to evaluate dermal uptake of PFAS, and, as applied, generally indicates that dermal exposures are likely low in most cases. For example, risk-based screening levels accounting for the dermal uptake of PFOS from soil, as calculated by the USEPA (2020a) RSL calculator, for residential and industrial receptors were approximately two times higher (less conservative) than those for incidental soil ingestion. Bathing and recreational exposures such as swimming are likely to represent the highest long-term dermal exposures to PFAS in water; however, exposure has been assumed to be comparatively minimal or insignificant compared to direct exposures such as drinking or incidental ingestion. For example, in derivation of the LHA, USEPA (2016a, 2016b) did not include dermal uptake, and OEHHA (2019) noted that dermal uptake was relatively insignificant. Our own recreational exposure modeling (swimming scenario) indicated that dermal exposure of PFOS at an off-site surface water body being investigated for AFFF impacts was approximately 100 times less than that obtained via incidental ingestion of water. Thus, for typical residential and industrial/construction worker PFAS risk assessments involving dermal exposures to PFAS in groundwater, surface water, drinking water, soils, and sediments, the quantitative assessment of dermal exposures is not likely to trigger human health exposure concerns.

8.3.4.3 Vapor Inhalation

PFAAs such as PFOA and PFOS are not expected to volatilize from PFAS-impacted environmental media (USEPA 2016a, 2016b) such as soil and groundwater, which are the primary focus of most site-specific risk assessments. Thus, such assessments will not typically consider inhalation of volatile PFAS in air. PFAAs can be transported in air via particulate matter and water aerosols (Barton et al. 2006), although this exposure route is not true volatilization. Ingestion of PFAS associated with solid particles or aerosolized liquids can be accounted for via risk modeling of incidental soil and water ingestion, and specific approaches to address airborne particles and aerosols at most PFAS sites are not likely to be needed unless there are unique physical mechanisms that promote the formation of aerosols and particles.

In contrast to PFAAs, PFAS such as fluorotelomer alcohols (FTOHs) are considerably volatile and are among the more widely studied of the volatile PFAS. FTOHs are transient in the atmosphere with a lifetime of 20 days (Ellis et al. 2004). At most PFAS sites under evaluation, the current assumption is that volatile PFAS such as FTOHs historically released at the site will have transformed to stable, low-volatility PFAS such as PFAAs in soil or groundwater, or will they have diffused to the outdoor atmosphere. There is no evidence that FTOHs or other volatile PFAS are able to persist in groundwater or soils such that they present the type of risk concerns observed for chlorinated solvents for which indoor vapor intrusion pathways are significant.

Volatilization of PFAS from consumer products could be a potentially important source of ambient exposures (Stahl et al. 2011, Sunderland et al. 2019), although this exposure is expected to be minor relative to drinking water, diet, and incidental soil ingestion exposures that may be elevated due to site-specific PFAS releases. Additionally, direct exposures to atmospheres containing pure PFAS products, such as undiluted AFFF, could present inhalation exposures (Roth et al. 2020); however, these exposures are more specific to occupational industrial exposures and are generally not evaluated for typical off-site residents or industrial/construction workers that come into interact with environmental media.

8.4 ECOLOGICAL RISK ASSESSMENT

Given the current understanding of PFCA and PFSA releases, fate, exposure, and effects, the most important ecological PFAS risk issues at most sites are likely to be associated with risks to vertebrate wildlife associated with dietary intake via bioaccumulation of PFAS in aquatic and terrestrial food webs (Table 8.2). Avian and mammalian wildlife are relatively sensitive to PFAS, and bioaccumulation in terrestrial and aquatic food webs can result in exposures that are dominated by the more accumulative PFAS (Larson et al. 2018, Conder et al. 2020, Zodrow et al. 2020). Direct toxicity to aquatic life such as fish and invertebrates from exposure to sediment and surface water is also relevant at some sites, although for larger areas, concentrations in water associated with adverse effects to aquatic life are generally higher than those that could result in adverse effects to aquatic-dependent wildlife. Soil invertebrates and plants are generally less sensitive, with risk-based concentrations in soil being much higher than those associated with potential effects to terrestrial wildlife (Zodrow et al. 2020). The remainder of this section provides a detail of the state of ecological risk assessment approaches for aquatic life, vertebrate wildlife, and soil life.

8.4.1 Ecological Health Assessment of PFAS for Aquatic Life

Aquatic life (e.g., fish, pelagic life, benthic invertebrates, and aquatic plants) are exposed to PFAS through direct exposure to PFAS in surface water and aquatic

TABLE 8.2
An overview of ecological risk assessment approaches for PFAS at contaminated sites.

Receptor and Exposure Pathway(s)	Relative Importance as Risk-based Management Driver at Contaminated Sites	Screening Level Availability	Relative Maturity of Site Investigation Tools	Relative Maturity of Risk Assessment Tools
Aquatic Life (aquatic invertebrates, plants and fish): Direct exposures to surface water and sediment	**Moderate**: Prevalence in surface water and sediment, but organisms may be relatively less sensitive to PFAS	**Moderate**: Multiple sources of screening criteria for water, but not sediment	**Moderate**: Standard measurement and sampling methods available, except for sediment porewater	**Higher**: Standard aquatic life assessment approaches apply
Aquatic-dependent and terrestrial vertebrate wildlife: Exposures to sediment, water, soil, and diet	**Higher**: High bioaccumulation potential and sensitive effects in birds and mammals	**Lower**: Few sources of screening levels in sediment, soil, water, and diet items	**Moderate**: Standard measurement methods available, but optimal sampling methods and strategies needed	**Moderate**: Food web models and typical wildlife exposure tools apply
Soil Life (soil invertebrates and plants): Direct exposures to soil	**Lower**: Organisms relatively insensitive to PFAS	**Lower**: Few sources of screening levels in soil	**Higher**: Standard measurement and sampling methods available	**Moderate**: Standard terrestrial life assessment approaches apply, but are limited by paucity of effect concentrations

sediment. Due to their relatively high water solubility, PFAS have the potential to become widely distributed in aquatic environments. Acute PFAS studies result in a wide range of toxicities to aquatic organisms, and overall PFAS toxicity data shows moderate to low toxicity in aquatic life (ITRC 2020).

Ecological risk assessment approaches for PFAS for aquatic life follow standard chemical aquatic life risk assessment approaches (Table 8.2). The evaluation of potential risks for aquatic life with direct exposure to PFAS in media relies on the comparison of concentrations in external exposure media to protective, media-specific benchmarks. The majority of peer-reviewed literature and regulatory environmental quality benchmarks have been developed for PFOS and PFOA; however, other select PFAAs have been included in aquatic life evaluations thus far (ITRC 2020; Zodrow et al. 2020; Conder et al. 2020). USEPA National Recommended

Water Quality Criteria for Aquatic Life (NRWQC) (USEPA 2020b), the standard benchmarks for aquatic life, are currently not available for PFAS. Following extensive reviews of the peer-reviewed literature, Zodrow et al. (2020) used USEPA Great Lakes Initiative (USEPA 2012) methodology in developing acute and chronic screening levels for aquatic life for 23 PFAS. Qi et al. (2011), CRC CARE (2017), and Conder et al. (2020) used species sensitivity distributions (SSDs) to calculate chronic aquatic life screening levels for PFOA and PFOS. In addition to research groups, several US states and international agencies have used similar models to establish aquatic life criteria to protect aquatic organisms in their respective surface waters (ITRC 2020). Overall, the aquatic life screening levels developed by both these reviews indicate PFOA is several orders of magnitude less toxic to aquatic life than PFOS. However, care should be taken when using freshwater toxicity threshold values for evaluating marine systems, since threshold values for PFOS and PFOA in marine species were at least one to two orders of magnitude greater than those from freshwater (CRC CARE 2017; Conder et al. 2020; Zodrow et al. 2020).

In contrast to surface water aquatic life benchmarks, benchmark values for sediment are limited as aquatic toxicity studies are conducted with exposure to the chemical typically occurring in overlying water and/or sediment porewater and not sediment itself (McCarthy et al. 2017; Conder et al. 2020; Zodrow et al. 2020). Spiked sediment exposures studies for PFAS have been performed, primarily for the evaluation of bioaccumulation (Higgins et al. 2007) rather than the identification of adverse effects in benthic organisms. More work is needed to address the toxicity of PFAS in aquatic sediment.

8.4.2 Ecological Health Assessment of PFAS for Aquatic-dependent and Terrestrial Vertebrate Wildlife

Most areas at the point of PFAS releases (and many industrial areas where PFAS products were used) do not generally feature favorable habitats for terrestrial and aquatic-dependent wildlife. For example, AFFF training locations at military and civilian airports are usually located in a manicured area or an impermeable area (paved or cement) adjacent to an airfield. In most cases, these areas are highly disturbed, are not managed or meant to provide habitat for wildlife, and are not usually included in ecological risk assessments. However, these areas may require investigation and management when they serve as sources to downgradient areas that host ecological habitats. Off-site or downgradient areas are expected to pose the highest risk potential to wildlife, particularly those that feature aquatic habitat (Ahrens and Bundschuh 2014; Larson et al. 2018).

Wildlife receptors, specifically birds and mammals, are typically exposed to PFAS through uptake from dietary sources; however, insectivore or omnivore receptors are exposed via direct soil ingestion during foraging activities and consumption of diet items that have accumulated PFAS from soil, including invertebrates and plants (Table 8.2). For some species and sites, ingestion of surface water may also be relevant to evaluate, although for many chemicals, exposure via surface water ingestion is insignificant compared to exposure via food and incidental soil/sediment ingestion. Biomagnification is a key exposure pathway, and can result in higher concentrations in tissues of predators than those of its diet/prey (Gobas et al. 1999; Gobas, 1993), and studies have shown that there is a high bioaccumulative potential for PFAS in air breathing organisms, resulting in significant biomagnification in higher trophic level receptors of food webs (Kelly et al. 2009). For wildlife species where PFAS exposure occurs primarily via dietary uptake from biomagnification pathways, dietary intake modeling typical for ecological risk assessments is the recommended approach for an evaluation of potential risks.

Dietary intake modeling uses relevant exposure factors for each receptor group (terrestrial birds, terrestrial mammals, aquatic-dependent birds, and aquatic mammal) to determine a total daily intake (TDI) of PFAS via all potential exposure pathways. This approach requires concentrations of PFAS in dietary items, which can be obtained by measurement of PFAS in biota at sites, or the use of food web models to predict concentrations in biota using measured concentrations of PFAS in soil, sediment, or surface water. As noted above for the human health risk assessment of dietary items, the measurement of PFAS in biota is more challenging compared to other abiotic matrices such as drinking water. Food web models uses bioaccumulation metrics (e.g., bioaccumulation factors [BAFs], biomagnification factors [BMFs]) with measurements of PFAS in abiotic media to estimate concentrations in dietary items, including plants and benthic or pelagic invertebrates. For bioaccumulation metrics, the measurement of concentrations of chemicals in the whole body of animals (rather than organ-specific measurements) is preferred, as wildlife consumption of most animal prey usually occurs on a whole-body basis. However, BAFs can be very limited for some organisms, such as birds, while for others, including fish, the abundance in BAF data may lead to a large range of values to select from (Zodrow et al. 2020).

Once site-specific TDI values are calculated, they are compared to known toxicity reference values (TRVs) identified from toxicity data with exposure doses associated with a lack of adverse effects (termed no observed adverse effect level [NOAEL] values) or low adverse effects (termed lowest observed adverse effect level [LOAEL] values), per standard risk assessment practice (USEPA 1997). TRVs consider measures of exposure and effects that could adversely impact populations of wildlife (for example, chronic studies on reproduction, growth, and survival); however, available toxicological data for wildlife are limited compared to the range of surrogate organisms that are exposed to PFAS or within which PFAS have been detected, and, similar to other receptors, focuses on PFOS and PFOA. As with many other bioaccumulative and biomagnifying compounds, a common practice in ecological risk assessment (ERA) involves the use of available data from surrogate organisms. While this allows risk assessors to fill in some of the data gaps and uncertainties, caution should be taken and regulatory agencies or applicable stakeholders should be consulted during this process regarding selection of the surrogate organisms (for example, the closest taxonomic laboratory test species) and use of uncertainty factors.

While bird egg studies for PFAS are more abundant, oral dosing studies in birds, which are typically used for ERA, are less available. Due to the uncertainty as to whether naturally accumulated concentrations have the same adverse effect as concentrations administered via injection in ovo, caution should be taken when using these egg studies for TRVs. Using mammalian and avian chronic toxicity studies, the Environment and Climate Change Canada (ECCC) derived wildlife dietary TRVs.

For PFOS, the dietary TRV for mammals was 4.6 μg/kg wet weight (ww), while the avian dietary TRV was 8.2 μg/kg ww (Environment and Climate Change Canada (ECCC), 2018).

Despite the lack of data for many PFAS, risk assessment approaches for wildlife are possible. Recently, Conder et al. (2020), Gobas et al. (2020), and Zodrow et al. (2020) compiled bioaccumulation modeling parameters and approaches for terrestrial and aquatic food web modeling of a variety of commonly-detected PFAS at AFFF sites, and there are several sources of TRVs for the evaluation of predicted TDI values (Newsted et al. 2005; Conder et al. 2020; Gobas et al. 2020; Suski et al. 2020; Zodrow et al. 2020). These models can also be used to back-calculate screening levels for the media that make up the TDI. For example, following the Ecological Soil Screening Level (EcoSSL) methodology (USEPA 2005), Zodrow et al. (2020) calculated soil, sediment, and surface water risk-based screening levels (RBSLs) for terrestrial and aquatic-dependent birds and mammals based on food web models. Soil and surface water RBSLs for terrestrial wildlife and sediment and surface water RBSLs for aquatic wildlife were derived for the six individual PFAS for which both TRVs and BAFs were available. NOAEL-based RBSLs in soil range from 0.013 mg/kg for PFOS to 340 mg/kg for PFHxA; NOAEL-based RBSLs in sediment range from 1.4 μg/kg for PFOS to 370 mg/kg for PFBS; and NOAEL-based RBSLs in surface water had the greatest range of 75 ng/L for PFOS to 76 milligrams per liter(mg/L) for PFBS among aquatic receptors. These large ranges are likely due to factors such as incorporation of both mammal and bird species, limited toxicity data for birds and aquatic receptors, and limited bioaccumulation data for some PFAS.

States have developed wildlife screening values using various approaches. The Michigan Department of Community Health (MDCH 2015) derived a provisional PFOS surface water value for avian and mammalian wildlife. In California, the San Francisco Bay Regional Water Quality Control Board developed terrestrial habitat soil ecological screening levels based on values developed in Zodrow et al. (2020).

In general, the highest risk for PFAS are expected for smaller insectivore and omnivore receptors (e.g., shrews and other small rodents, small non-migratory birds), which tend to be lower in trophic level and spend more time foraging in small areas similar to or smaller in size than the impacted area. Compared to smaller, lower-trophic level organisms, larger mammalian and avian carnivores are expected to have lower exposures from site-specific PFAS sources, as they forage over larger areas that are relatively unimpacted, as compared to small organisms with small home ranges (Larson et al. 2018; Conder et al. 2020; Gobas et al. 2020; Suski et al. 2020; Zodrow et al. 2020).

8.4.3 Ecological Health Assessment of PFAS for Soil Life

Terrestrial ecosystems may also be contaminated by PFAS, resulting in exposures to receptors including terrestrial plants and soil invertebrates that are exposed directly to soil, as well as small terrestrial avian and mammalian insectivores or omnivores exposed directly to soil through incidental ingestion or from dietary items that have accumulated PFAS from soil. For plants, shorter-chain PFCAs generally accumulate in the leaves, resulting in the highest levels of PFAAs compared to other tissues

(Ghisi et al. 2019). Longer-chain PFCAs are accumulated by plants, albeit at lower levels, and these tend to accumulate in root tissues (Blaine et al. 2014).

Compared to aquatic invertebrates, there are relatively fewer studies on the effects of PFAS on terrestrial invertebrates and plants. Soil invertebrate toxicity studies are predominantly performed on the earthworm *Eisenia fetida* (Xu et al. 2013; He et al. 2016; Zhao et al. 2014; Rich et al. 2015) with results generally indicating toxicity occurring at parts per million (mg/kg) concentrations, which suggests a low to moderate toxicity of PFOS and PFOA (ITRC 2020). The majority of the terrestrial plant toxicity studies are typically performed on domestic crop plants (ITRC 2020). Additionally, the PFAS toxicity in terrestrial plants can be significantly influenced by organic matter in soil, where accumulation of PFOS and PFOA in plants and phytotoxicity may be reduced due to greater PFAS partitioning to the organic content (Zhao et al. 2011).

As with the evaluation of potential risks to aquatic life, the evaluation of potential risks for soil invertebrates and plants via direct exposure to PFAS in soil relies on the comparison of concentrations in external exposure media to protective, media-specific benchmarks. As part of the Strategic Environmental Research and Development Program effort, Conder et al. (2020) recommended no observed effect concentration (NOEC) values for soil invertebrates and plants based on individual chronic exposure studies. RBSLs for the protection of terrestrial plants and soil invertebrates for six PFAS were also calculated by Zodrow et al. (2020) using methods developed for calculation of EcoSSLs (USEPA 2005). While RBSLs are provided for PFAS other than PFOA and PFOS, these should be used with caution due to the limited data available. Internationally, there are several draft thresholds protective of direct toxicity of PFOS in soil. For example, the Canadian Council of Ministers of Environment (CCME 2018) developed several draft thresholds protective of direct toxicity of PFOS in soil. The threshold is based on an SSD of plant and invertebrate IC_{25} values (the concentration at which a 25% reduction in a non-lethal biological measurement, such as growth or reproduction, occurs).

Overall, while soil invertebrate and plant data are sparse and invertebrate data predominantly consist of earthworm data, screening levels for ecological risk assessments are available. Caution should be taken in applying toxicity values for earthworms and plants to site-specific terrestrial invertebrate and plant species as there are uncertainties due to the potential for interspecies differences for invertebrates and differences in phytotoxicity that may occur based on the soil organic carbon at the site. Compared to aquatic life and vertebrate toxicity, the soil invertebrate and plant toxicity may be of less concern as most effects occur in the parts per million range.

8.5 CONCLUSIONS

While detectable concentrations of PFAS are ubiquitous in the environment and may be elevated at thousands of sites worldwide, the application of human health and ecological risk assessment tools allow us to better understand the risk to human health and ecological receptors as well as make risk-based conclusions and management decisions on a site-specific basis. The development and application of these tools at PFAS-impacted sites is proceeding along a process similar to approaches used for other challenging contaminants such as dioxins and polychlorinated biphenyls. While regulatory agencies are beginning to develop advisory levels, to date these mostly focus on the human health drinking water exposure pathway. For ecological risk assessment, the approaches typically used have different levels of complexity, such that receptors most likely to have direct exposure to media risk are assessed by comparison to media-specific benchmarks, while wildlife that have multiple sources of exposure can be assessed using food web models that take all the routes of exposure, as well as food chain bioaccumulation into consideration. While these tools are not perfect, they are developed to result in a conservative approach to screening PFAS and calculating potential risks to receptors resulting in a reasonable approach to prioritizing risk and remediation.

During the risk assessment process, a number of cautions should be kept in mind. Most importantly, site-specific exceedances of benchmarks based on a lack of adverse effects (e.g., NOECs and NOAELs) or cancer risk levels do not necessarily imply the presence of adverse effects or unacceptable cancer levels. Risk assessment practices are generally conservative such that the risks of adverse effects are overestimated. Adverse risk outcomes may indicate the need for collection of additional data which can refine the assessment, more complete or site-specific evaluations, and further evaluation of the model approaches and assumptions. The potential uncertainty due to differences in sensitivity to PFAS needs to be acknowledged and discussed within the risk assessment process, especially for ERAs. While the detection of PFAS within organisms is clear evidence of exposure, whether or not this exposure results in toxicity is not clear; the availability of PFAS toxicological data beyond PFOS and PFOA is limited relative to the broad range of organisms within which PFAS have been detected (ITRC 2020). This may be more of an issue for terrestrial wildlife since chronic reproductive studies with dietary exposure are available for very few species and are typically only for PFOS and PFOA.

There are also a number of data gaps that, as they are filled, will help with the accuracy and acceptance of PFAS risk assessment. There are a number of additional PFAS that we know little about. The approach to dealing with the lack of information on the overwhelming number of PFAS is being debated; in the meantime, however, PFAS beyond PFOS and PFOA are beginning to be studied more, and this information will result in more accurate results for risk assessment. Another major challenge in effects assessment for PFAS, for both HHRAs and ERAs, is understanding the potential importance of mixtures. Considering the limited toxicity data available for just a few PFAS with respect to mammalian toxicity, direct toxicity to lower-trophic-level organisms or exposure to upper-trophic-level wildlife toxicity, the understanding of the relative toxicity, additivity, or synergistic effects of PFAS in mixtures is just beginning. For ERAs, some organisms, like reptiles and benthic invertebrates, do not currently have toxicity data. Benchmark or threshold concentrations for wildlife and aquatic organisms incorporates more uncertainty in their derivation due to the limited number of species for which data are available. As more species data becomes available, the accuracy of ERAs is likely to improve.

The understanding of site-specific human health and ecological risks at many PFAS sites is limited or completely unknown, presenting an immense challenge to decision-makers evaluating the results of site-specific PFAS measurements at hundreds to thousands of sites. PFAS still obey the laws of physics. Given this, we can continue to apply sound scientific principles of chemistry, biology, toxicology, and ecology that comprise the foundations of risk assessment, adding in judicious helpings of common sense, sound ethical practices, and reasonable regulatory policies. Despite the many data gaps and uncertainties highlighted in this chapter, risk assessment remains the most valid and powerful decision-making tool for understanding PFAS. The absence of a full and complete risk assessment knowledge base cannot be an excuse for inaction, nor should it be an excuse to abandon lessons learned from approximately 40 years of prior risk assessment practice involving a wide variety of diverse chemicals.

REFERENCES

AECOM Technical Services (AECOM). 2018a. Stage 2C Environmental Investigation - Ecological Risk Assessment – October 2018, *Army Aviation Centre Oakey (AACO), Oakey QLD*. 8 October. Available at https://www.defence.gov.au/Environment/PFAS/Docs/Oakey/Reports/201810OakeyERA.pdf (Accessed August 24, 2020).

AECOM Technical Services (AECOM). 2018b. *Ecological Risk Assessment, September 2018, RAAF Williamtown, Stage 2B Environmental Investigation*. 7 September. Available at: https://www.defence.gov.au/Environment/PFAS/docs/Williamtown/Reports/201810WilliamtownEcologicalRiskAssessmentReport.pdf (Accessed August 24, 2020).

Agency for Toxic Substances and Disease Registry (ATSDR). 2018. *Toxicological Profile for Perfluoroalkyls*. June. Available at: https://www.atsdr.cdc.gov/toxprofiles/tp200.pdf (Accessed August 24, 2020).

Ahrens, L. 2011. Polyfluoroalkyl compounds in the aquatic environment: a review of their occurrence and fate. *J. Environ. Monit.* (13):20–31.

Ahrens, L., and M. Bundschuh. 2014. Fate and effects of poly- and perfluoroalkyl substances in the aquatic environment: A review. *Environ. Toxicol. Chem.* 33(9):1921–1929.

Anderson, R., G. Long, R. Porter, and J. Anderson. 2016. Occurrence of select perfluoroalkyl substances at United States Air Force aqueous film-forming foam release sites other than fire-training areas: Field-validation of critical fate and transport properties. *Chemosphere* 150:678–685.

Barton, C.A., L.E. Butler, C.J. Zarzecki, J. Flaherty, and M. Kaiser. 2006. Characterizing perfluorooctanoate in ambient air near the fence line of a manufacturing facility: comparing 11,042 modeled and monitored values. *J. Air Waste Manage. Assoc.* 56:48–55.

Bertin, D.B., P. Ferrari, A. Labadie, J. Sapin, and O. Garric. 2014. Bioaccumulation of peruoroalkyl compounds in midge (*Chironomus riparius*) larvae exposed to sediment. *Environ. Pollut.* Elsevier, (189):27–34.

Bertin, D.P., B. Labadie, A. Ferrari, J. Sapin, O. Garric, H. Geffard, M. Budzinski, and Babut. 2016. Potential exposure routes and accumulation kinetics for poly- and perfluorinated alkyl compounds for a freshwater amphipod: Gammarus spp. (Crustacea). *Chemosphere* (155):380–387.

Blaine, A.C., C.D. Rich, L.S. Hundal, C. Lau, M.A. Mills, K.M. Harris, and C.P. Higgins. 2013. Uptake of Perfluoroalkyl acids into edible crops via land applied biosolids: field and greenhouse studies. *Environ. Sci. Technol.* (47):14062–14069.

Blaine, A.C., C.D. Rich, E.M. Sedlacko, K.C. Hyland, C. Stushnoff, E.R. Dickenson, and V.C. Higgins. 2014. Perfluoroalkyl acid uptake in lettuce (*Lactuca sativa*) and strawberry (Fragaria ananassa) irrigated with reclaimed water. *Environ. Sci. Technol.*. 48(24): 14361–14368.

Brown, J.B., J.M. Conder, J.A. Arblaster, and C. Higgins. 2019. An Approach for Developing Risk Based Screening Criteria for Consumption of Food Crops Impacted by Poly and

Perfluoroalkyl Substances. SETAC North America Focused Topic Meeting: Environmental Risk Assessment of PFAS. Durham, North Carolina.
Buck, R., J. Franklin, U. Berger, J. Conder, I. Cousins, P.D. Voogt, A. Jensen, K. Kannan, S. Mabury, and S. van Leeuwen. 2011. Perfluoroalkyl and polyfluoroalkyl substances in the environment: terminology, classifications, and origins. *Integr. Environ. Assess. Manag.* (7):513.
CCME. 2018. Canadian soil and groundwater quality guidelines for the protection of environmental and human health, *Perfluorooctane Sulfonate (PFOS) – Draft for Review. Canadian Council of Ministers of the Environment.* 30 July. Available at: https://www.watercanada.net/soil-and-groundwater-quality-guidelines-available-for-public-comment/ (Accessed August 24, 2020).
Chen, F., Z. Gong, and B.B. Kelly. 2016. Bioavailability and bioconcentration potential of perfluoroalkyl-phosphinic and -phosphonic acids in zebrafish (*Danio rerio*): Comparison to perfluorocarboxylates and perfluorosulfonates. *Sci. Total Environ.* (568):33–41.
Conder, J.M., R.A. Hoke, W. de Wolf, M.H. Russell, and R.C. Buck. 2008. Are PFCAs bioaccumulative? – A critical review and comparison with persistent lipophilic compounds. *Environ. Sci. Technol.* (42):995–1003.
Conder, J.M., J.A. Arblaster, E. Larson, J.B. Brown, and C. Higgins. 2020. Guidance for assessing the ecological risks of pfas to threatened and endangered species at aqueous film forming foam-impacted sites. *SERDP Contract Report ER18–1614.* Available at: https://www.serdp-estcp.org/Program-Areas/Environmental-Restoration/ER18–1614 (Accessed August 24, 2020).
Conder, J.M., and J.A. Arblaster. 2016. Development and use of wild game consumption rates in human health risk assessments. *Hum. Ecol. Risk. Assess.* 22:251–264.
CRC CARE. 2017. Assessment, management and remediation for PFOS and PFOA Part 3: ecological screening levels (CRC CARE Technical Report No. 38; p. 61). Retrieved from *Cooperative Research Centre (CRC) for Contamination Assessment and Remediation of the Environment website.* Available at: https://www.crccare.com/files/dmfile/CRCCARETechReport38Part3_AssessmentmanagementandremediationforPFOSandPFOA_ESLs2.pdf (Accessed August 24, 2020).
Dai, Z., X. Xia, J. Guo, and X. Jiang. 2013. Bioaccumulation and uptake routes of perfluoroalkyl acids in *Daphnia magna. Chemosphere* (90):1589–1596.
Environment and Climate Change Canada (ECCC). 2018. *Federal Environmental Quality Guidelines: Perfluorooctane Sulfonate (PFOS).* 30 July. Available at: https://www.canada.ca/en/environment-climate-change/services/evaluating-existing-substances/federal-environmental-quality-guidelines-perfluorooctane-sulfonate.html (Accessed August 24, 2020).
Ellis, D.A., J.W. Martin, A.O. De Silva, S.A. Mabury, M.D. Hurley, M.P. Sulbaek Andersen, and T.J. Wallington. 2004. Degradation of fluorotelomer alcohols: A likely atmospheric source of perfluorinated carboxylic acids. *Environ. Sci. Technol.* 38:3316–3321.
EnRisk. 2017. *DEHP-ID 183-D1 Human Health and Ecological Risk Assessment for PFAS: HMAS Albatross.* 10 November. Available at: https://www.defence.gov.au/Environment/PFAS/docs/Albatross/Reports/20171116HMASAlbatrossHHERAFullReport.pdf (Accessed August 24, 2020).
Fang, S., Y. Zhang, S. Zhao, L. Qiang, M. Chen, and L. Zhu. 2016. Bioaccumulation of perfluoroalkyl acids including the isomers of perfluorooctane sulfonate in carp (*Cyprinus carpio*) in a sediment/water microcosm. *Environ. Toxicol. Chem.* 35(10):3005–3013.
Field, J., J. Conder, R. Deeb, and C. Higgins. 2017. Frequently Asked Questions. per-and polyfluoroalkyl substances (pfass) associated with aqueous film forming use at United States military sites. *Developed for the Environmental Security Technology Certification Program (ESTCP).* August. Available at: https://clu-in.org/contaminantfocus/default.focus/sec/Per-_and_Polyfluoroalkyl_Substances_(PFASs)/cat/Occurrence/ (Accessed August 24, 2020).

Food Standards Australia New Zealand (FSANZ). 2017. *Perfluorinated Chemicals is Food*. April. Available at: https://www1.health.gov.au/internet/main/publishing.nsf/content/ohp-pfas-hbgv.htm (Accessed August 24, 2020).

Ghisi, R., T. Vamerali, and S. Manzetta. 2019. Accumulation of perfluorinated alkyl substances (PFAS) in agricultural plants: A review. *Environ. Res.* 169:326–341.

Gobas, F.A.P.C. 1993. A model for predicting the bioaccumulation of hydrophobic organic chemicals in aquatic food-webs: application to Lake Ontario. *Ecol. Model.* 69(1–2): 1–17.

Gobas, F.A.P.C., J.B. Wilcockson, R.W. Russell, and G.D. Haffner. 1999. Mechanism of biomagnification in fish under laboratory and field conditions. *Environ. Sci. Technol.* 33(1):133–141.

Gobas, F.A.P.C., B. Kelly, and J. Kim. 2020. A framework for assessing bioaccumulation and exposure risks of per- and polyfluoroalkyl substances in threatened and endangered species on aqueous film forming foam (AFFF)-impacted sites. *SERDP Contract Report ER18–1502*. Available at: https://www.serdp-estcp.org/Program-Areas/Environmental-Restoration/ER18-1502 (Accessed August 24, 2020).

Great Lakes Consortium (GLC). 2019. *Great Lakes Consortium for Fish Consumption Advisories, Best Practice for Perfluorooctane Sulfonate (PFOS) Guidelines*. November. Available at: https://www.health.state.mn.us/communities/environment/fish/docs/consortium/bestpracticepfos.pdf (Accessed August 24, 2020).

Guelfo, J.L., and C.P. Higgins. 2013. Subsurface transport potential of perfluoroalkyl acids at aqueous film-forming foam (AFFF)-impacted sites. *Environ. Sci. Technol.* (47):4164–4171.

Guelfo, J.L., T. Marlow, D.M. Klein, D.A. Savitz, S. Frickel, M. Crimi, and E.M. Suuberg. 2018. Evaluation and management strategies for per- and polyfluoroalkyl substances (PFASs) in drinking water aquifers: Perspectives from impacted U.S. northeast communities. *Environ. Health Perspect.* 126:1–13.

He, W., M. Megharaj, and R. Naidu. 2016. Toxicity of perfluorooctanoic acid towards earthworm and enzymatic activities in soil. *Environ. Monit. Assess.* 188(7).

Hekster, F.M., R.W.P.M. Laane, and P. de Voogt. 2003. Environmental and toxicity effects of perfluoroalkylated substances. *Rev. Environ. Contam. Toxicol.* (179):99–121.

Higgins, C.P., and R.G. Luthy. 2006. Sorption of perfluorinated surfactants on sediments. *Environ. Sci. Technol.* (40):7251–7256.

Higgins, C.P., Pamela B. Mcleod, L.A. Macmanus-Spencer, and Luthy. 2007. Bioaccumulation of perfluorochemicals in sediments by the aquatic oligochaete *Lumbriculus variegatus*. *Environ. Sci. Technol.* (41):4600–4606.

Houtz, E.F., C.P. Higgins, J.A. Field, and D.L. Sedlak. 2013. Persistence of perfluoroalkyl acid precursors in AFFF-impacted groundwater and soil. *Environ. Sci. Technol.* (47):8187–8195.

Hu, X.C., D.Q. Andrews, A.B. Lindstrom, T.A. Bruton, L.A. Schaider, P. Grandjean, R. Lohmann, C.C. Carignan, A. Blum, S.A. Balan, C.P. Higgins, and E.M. Sunderland. 2016. Detection of poly-and perfluoroalkyl substances (PFAS) in United States drinking water linked to industrial site, military fire training areas, and wastewater treatment plants. *Environ. Sci. Technol. Lett.* (3):344–350.

ITRC. 2020. *PFAS Technical and Regulatory Guidance Document and Fact Sheets PFAS-1*. Washington, D.C.: Interstate Technology & Regulatory Council, PFAS Team. Available at: https://pfas-1.itrcweb.org (Accessed August 24, 2020).

Kelly, B.C., M.G. Ikonomou, J.D. Blair, B. Surridge, D. Hoover, R. Grace, F.A.P.C. Gobas. 2009. Perfluoroalkyl contaminants in an Arctic marine food web: Trophic magnification and wildlife exposure. *Environ. Sci. Technol.* 43:4037–4043.

Larson, E.S., J.M. Conder, and J.A. Arblaster. 2018. Modeling avian exposures to perfluoroalkyl substances in aquatic habitats impacted by historical aqueous film forming foam releases. *Chemosphere* 201:335–341.

Lau, C. 2012. Perfluorinated Compounds. In A. Luch (Ed.), *Molecular, Clinical and Environmental Toxicology: Volume 3: Environmental Toxicology* (pp. 47–86).

Maine PFAS Task Force. 2020. *Managing PFAS in Maine, Final Report from the Maine PFAS Task Force*. January. Available at: https://www.maine.gov/pfastaskforce/materials/report/PFAS-Task-Force-Report-FINAL-Jan2020.pdf (Accessed August 24, 2020).

Martin, J.W., S.A. Mabury, K.R. Solomon, and D.C. Muir. 2003a. Dietary accumulation of perfluorinated acids in juvenile rainbow trout (*Oncorhynchus mykiss*). *Environ. Toxicol. Chem.* 22:189–195.

Martin, J.W., S.A. Mabury, K.R. Solomon, and D.C. Muir. 2003b. Bioconcentration and tissue distribution of perfluorinated acids in rainbow trout (*Oncorhynchus mykiss*). *Environ. Toxicol. Chem.* 22:196–204.

Massachusetts Department of Environmental Protection (MADEP). 2019. *Fact Sheet: Interim Guidance on Sampling and Analysis for PFAS at Disposal Sites Regulated under the Massachusetts Contingency Plan*. 19 June. Available at: https://www.mass.gov/doc/interim-guidance-on-sampling-and-analysis-for-pfas-at-disposal-sites-regulated-under-the/download (Accessed August 24, 2020).

McCarthy, C., W. Kappleman, and W. DiGuiseppi. 2017. Ecological considerations of per- and polyfluoroalkyl substances (PFAS). *Curr. Pollut. Rep.* 3:289–301.

Michigan Department of Community Health (MDCH). 2015. USEPA great lakes restoration initiative (GLRI) project. Final Report. *Michigan Department of Community Health*. 30 July. Available at: https://www.michigan.gov/documents/mdch/MDCH_GL-00E01122-0_Final_Report_493494_7.pdf (Accessed August 24, 2020).

Michigan Department of Environmental Quality (MDEQ). 2018. *General PFAS Sampling Guidance*. 16 October. Available at: https://www.michigan.gov/documents/pfasresponse/General_PFAS_Sampling_Guidance_634597_7.pdf (Accessed August 24, 2020).

Michigan PFAS Action Response Team (MPART). 2020. *Consumption guidelines for fish with elevated PFOS levels*. Available at: https://www.michigan.gov/pfasresponse/0,9038,7–365-86512_88987_88989–481,104--,00.html (Accessed August 24, 2020).

Minnesota Department of Natural Resources. 2020. *Informational Packet DNR Sampling Shows Elevated Levels of PFAS Contamination in Fish and Surface Water in Starkweather Creek and Lake Monona*. January. Available at: https://dnr.wi.gov/topic/Contaminants/documents/pfas/Starkweather20200115.pdf (Accessed August 24, 2020).

Minnesota Pollution Control Agency (MPCA). 2013. *Aquatic Life Criteria and Water Quality Standards*. Available at: https://www.pca.state.mn.us/water/water-quality-standards (Accessed August 24, 2020).

Muller, C.E., A.O.D. Silva, J. Small, M. Williamson, X. Wang, A. Morris, S. Katz, M. Gamberg, and D.C.G. Muir. 2011. Biomagnification of perfluorinated compounds in a remote terrestrial food chain: Lichen-caribou-wolf. *Environ. Sci. Technol.* (45): 8665–8673.

New Jersey Department of Environmental Protection (NJDEP). 2019. *Investigation of Levels of Perfluorinated Compounds in New Jersey Fish, Surface Water, and Sediment*. June 2018, updated April 2019. Available at: https://www.nj.gov/dep/dsr/publications/Investigation%20of%20Levels%20of%20Perfluorinated%20Compounds%20in%20New%20Jersey%20Fish,%20Surface%20Water,%20and%20Sediment.pdf (Accessed August 24, 2020).

Newsted, J.L., P.D. Jones, K. Coady, and J.P. Giesy. 2005. Avian toxicity reference values for perfluorooctane sulfonate. *Environ. Sci. Technol.* 39:9357–9362.

New York State Department of Health. 2020. *Hudson Valley Region Fish Advisories*. Available at: https://www.health.ny.gov/environmental/outdoors/fish/health_advisories/regional/hudson_valley_and_capital_district.htm (Accessed August 24, 2020).

Office of Environmental Health Hazard Assessment (OEHHA). 2019. *Notification Level Recommendations, Perfluorooctanoic Acid and Perfluorooctane Sulfonate in Drinking*

Water. August. Available at: https://oehha.ca.gov/media/downloads/water/chemicals/nl/final-pfoa-pfosnl082119.pdf (Accessed August 24, 2020).

Post, G.B., P.D. Cohn, and K.R. Cooper. 2012. Perfluorooctanoic acid (PFOA), an emerging drinking water contaminant: A critical review of recent literature. *Environ. Res.* 116:93–117.

Prosser, R.S., K. Mahon, P.K. Sibley, D. Poirier, and T. Watson-Leung. 2016. Bioaccumulation of perfluorinated carboxylates and sulfonates and polychlorinated biphenyls in laboratory-cultured Hexagenia spp., *Lumbriculus variegatus* and *Pimephales promelas* from field-collected sediments. *Sci. Total Environ.* (543):715–726.

Qi, P., Y. Wang, J. Mu, and J. Wang. 2011. Aquatic predicted no-effect-concentration derivation for perfluorooctane sulfonic acid. *Environ. Toxicol. Chem.* (30):836–842.

Rankin, K., S.A. Mabury, T.M. Jenkins, and J.W. Washington. 2016. A North American and global survey of perfluoroalkyl substances in surface soils: Distribution patterns and mode of occurrence. *Chemosphere* (161):333–341.

Rich, C.D., A.C. Blaine, L. Hundal, and C. Higgins. 2015. Bioaccumulation of perfluoroalkyl acids by earthworms (*Eisenia fetida*) exposed to contaminated *Soils. Environ. Sd. Technol.* (49):881–888.

Rodowa, A.E., E. Christie, J. Sedlak, G.F. Peaslee, D. Bogdan, B. DiGuiseppi, and J.A. Field. 2020. Field sampling materials unlikely source of contamination for perfluoroalkyl and polyfluoroalkyl substances in field samples. *Environ. Sci. Technol. Lett.* (7):156–163.

Roth, J., I. Abusallout, T. Hill, C. Holton, U. Thapa, and D. Hanigan. 2020. Release of volatile per- and polyfluoroalkyl substances from aqueous film-forming foam. *Environ. Sci. Technol. Lett.* (7):164–170.

Scher, D.P., J.E. Kelly, C.A. Huset, K.M. Barry, R.W. Hoffbeck, V.L. Yingling, and R.B. Messing. 2018. Occurrence of perfluoroalkyl substances (PFAS) in garden produce at homes with a history of PFAS-contaminated drinking water. *Chemosphere* (196):548–555.

Sepulvado, J.G., A.C. Blaine, L.S. Hundal, and C.P. Higgins. 2011. Occurrence and fate of perfluorochemicals in soil following the land application of municipal biosolids. *Environ. Sci. Technol.* (45):8106–8112.

Stahl, T., D. Mattern, and H. Brunn. 2011. Toxicology of perfluorinated compounds. *Environ. Sci. Eur.* (23):38.

Stahl, L.L., B.D. Snyder, A.R. Olsen, T.M. Kincaid, J.B. Wathen, and H.B. McCarty. 2014. Perfluorinated compounds in fish from United States urban rivers and the Great Lakes. *Sci. Total Environ.* (499):185–195.

Strynar, M.J., A.B. Lindstrom, S.F. Nakayama, P.P. Egeghy, and L.J. Helfant. 2012. Pilot scale application of a method for the analysis of perfluorinated compounds in surface soils. *Chemosphere* (86):252–257.

Sunderland, E.M., X.C. Hu, C. Dassuncao, A.K. Tokranov, C.C. Wagner, and J.G. Allen. 2019. A review of the pathways of human exposure to poly- and perfluoroalkyl substances (PFASs) and present understanding of health effects. *J. Expo. Sci. Environ. Epidemiol.* (29):131–147.

Suski, J.G., C.J. Salice, E. Yan, and M. Dhruv. 2020. Investigating potential risk to threatened and endangered species from per- and polyfluoroalkyl substances (PFAS) on department of defense (DoD) sites. *SERDP Contract Report ER18–1626*. Available at: https://www.serdp-estcp.org/Program-Areas/Environmental-Restoration/ER18-1626 (Accessed August 24, 2020).

ToxConsult. 2016. Toxicity profiles for the perfluorinated compounds PFOS, PFOA, 6:2 FtS and 8:2 FtS. Prepared for AECOM Australia Pty Ltd. ToxConsult document ToxCR011115-RTd5. 4 July.

USEPA. 1989. Risk assessment guidance for superfund (rags) part a: Human health evaluation manual. *EPA/540/1–89/002*. December. Available at: https://www.epa.gov/risk/risk-assessment-guidance-superfund-rags-part (Accessed August 24, 2020).

USEPA. 1997. Ecological risk assessment guidance for superfund: Process for designing and conducting ecological risk assessments. interim final. *Office of Solid Waste and Emergency Response, Edison, NJ. EPA 540-R-97-OCS.EPA*. Available at: https://nepis.epa.gov/Exe/ZyPDF.cgi/9100VKNT.PDF?Dockey=9100VKNT.PDF (Accessed August 24, 2020).

USEPA. 2000a. Methodology for deriving ambient water quality criteria for the protection of human health. *EPA-822-B-00-004*. Available at: https://www.epa.gov/sites/production/files/2018-10/documents/methodology-wqc-protection-hh-2000.pdf (Accessed August 24, 2020).

USEPA. 2000b. Guidance for assessing chemical contaminant data for use in fish advisories. Volume 2 risk assessment and fish consumption limits third edition. *EPA 823-B-00-008* November. Available at: https://www.epa.gov/fish-tech/guidance-assessing-chemical-contaminant-data-use-fish-advisories-documents (Accessed August 24, 2020).

USEPA. 2005. Guidance for developing ecological soil screening levels. Office of solid waste and emergency response. November 2003; Revised February 2005. OSWER Directive 9285.7–55. Available at: https://www.epa.gov/sites/production/files/2015-09/documents/ecossl_guidance_chapters.pdf (Accessed August 24, 2020).

USEPA. 2011. Human health exposure factors handbook. 2011 Edition. *EPA/600/R-090/052F*. September. Available at: https://cfpub.epa.gov/si/si_public_record_report.cfm?Lab=NCEA&dirEntryId=236252 (Accessed August 24, 2020).

USEPA. 2012. Water quality guidance for the great lakes system. *Part 132*. Available at: https://www.govinfo.gov/app/details/CFR-2012-title40-vol23/CFR-2012-title40-vol23-part132 (Accessed June 30, 2020).

USEPA. 2016a. Drinking water health advisory for perfluorooctane sulfonate (PFOS). United States Environmental Protection Agency. Office of Water. EPA 822-R-16-004. May. Available at: https://www.epa.gov/sites/production/files/2016-05/documents/pfos_health_advisory_final-plain.pdf (Accessed August 24, 2020).

USEPA. 2016b. Drinking water health advisory for perfluorooctanoic acid (PFOA). United States Environmental Protection Agency. Office of Water. EPA 822-R-16-005. May. Available at: https://www.epa.gov/sites/production/files/2016-05/documents/pfoa_health_advisory_final_508.pdf (Accessed August 24, 2020).

USEPA. 2016c. Health effects support document for perfluorooctane sulfonate (PFOS). United States Environmental Protection Agency. Office of Water. EPA 822-R-16-002. May. Available at: https://www.epa.gov/sites/production/files/2016-05/documents/pfos_hesd_final_508.pdf (Accessed August 24, 2020).

USEPA. 2016d. Health effects support document for perfluorooctanoic acid (PFOA), United States Environmental Protection Agency. Office of Water. EPA 822-R-16-003. May. Available at: https://www.epa.gov/sites/production/files/2016-05/documents/pfoa_hesd_final-plain.pdf (Accessed August 24, 2020).

USEPA. 2019. Interim *Recommendations to Address Groundwater Contaminated with Perfluorooctanoic Acid and Perfluorooctane Sulfonate*. December 19. Available at: https://www.epa.gov/sites/production/files/2019–12/documents/epas_interim_recomendations_for_addressing_groundwater_contaminated_with_pfoa_and_pfos_dec_2019.pdf (Accessed August 24, 2020).

USEPA. 2020a. *Regional Screening Levels for Chemical Contaminants at Superfund Sites*. Available at: https://www.epa.gov/risk/regional-screening-levels-rsls-generic-tables (Accessed August 24, 2020).

USEPA. 2020b. *National Recommended Water Quality Criteria for Aquatic Life*. July 30. Available at: https://www.epa.gov/wqc/national-recommended-water-quality-criteria-aquatic-life-criteria-table (Accessed August 24, 2020).

van Asselt, E., J. Kowalczyk, J. van Eijkeren, M. Zeilmaker, S. Ehlers, P. Furst, M. Lahrssen-Wiederhold, and H. van der Fels-Klerx. 2013. Transfer of perfluorooctane sulfonic acid (PFOS) from contaminated feed to dairy milk. *Food Chem.* (141):1489–1495.

Xu, D., C. Li, Y. Wen, and W. Liu. 2013. Antioxidant defense system responses and DNA damage of earthworms exposed to Perfluorooctane sulfonate (PFOS). *Environ. Pollut.* (174):121–127. https://doi.org/10.1016/j.envpol.2012.10.030

Ye, X., M.J. Strynar, S.F. Nakayama, J. Varns, L. Helfant, J. Lazorchak, A.B. Lindstrom. 2008. Perfluorinated compounds in whole fish homogenates from the Ohio, Missouri, and Upper Mississippi Rivers, USA. *Environ. Pollut.* (156):1227–1232.

Zafeiraki, E., D. Costopoulou, I. Vassiliadou, L. Leondiadis, E. Dassenakis, R.L. Hoogenboom, and S.P. van Leeuwen. 2016. Perfluoroalkylated substances (PFASs) in home and commercially produced chicken eggs from the Netherlands and Greece. *Chemosphere* (144):2106–2112.

Zhang, M., P. Wang, Y. Lu, X. Lu, A. Zhang, Z. Liu, Y. Zhang, K. Khan, and S. Sarvajayakesavalu. 2020. Bioaccumulation and human exposure of perfluoroalkyl acids (PFAAs) in vegetables from the largest vegetable production base of China. *Environ. Int.* (135): 105347.

Zhao, H., C. Chen, X. Zhang, and J. Chen. 2011. Phytotoxicity of PFOS and PFOA to *Brassica chinensis* in different Chinese soils. *Ecotoxicol. Environ. Saf.* 74(5):1343–1347.

Zhao, S., S. Fang, L. Zhu, L. Liu, Z. Liu, and Y. Zhang. 2014. Mutual impacts of wheat (*Triticum aestivum L.*) and earthworms (*Eisenia fetida*) on the bioavailability of perfluoroalkyl substances (PFASs) in soil. *Environ. Pollut.* (184):495–501.

Zodrow, J., M. Frenchmeyer, K. Dally, E. Osborn, P. Anderson, and C. Divine. 2020. Per- and polyfluoroalkyl substances ecological risk-based screening levels. *SERDP Contract Report ER18–1653.* Available at: https://www.serdp-estcp.org/Program-Areas/Environmental-Restoration/ER18-1653 (Accessed August 24, 2020).

Section III

Remediation

9 Advances in Remediation of PFAS-impacted Waters

John Anderson
Arcadis

Pingping Meng
North Carolina State University, USA

Tim Sidnell
University of Surrey, United Kingdom

Ian Ross
Tetra Tech, United Kingdom

CONTENTS

9.1 Introduction 190
9.2 Adsorption 191
9.2.1 Activated Carbon 191
9.2.1.1 Granular Activated Carbon 192
9.2.1.2 Powdered Activated Carbon 193
9.2.1.3 Super-fine Powdered Activated Carbon 193
9.2.2 Ion Exchange 193
9.2.3 Functionalized Adsorbents 194
9.3 Separation 194
9.3.1 Membrane Treatment 195
9.3.1.1 Nanofiltration (NF) 196
9.3.1.2 Reverse Osmosis 196
9.3.2 Foam Fractionation 197
9.4 Destruction 198
9.4.1 Thermal Destruction 198

9.4.1.1 Incineration 199
9.4.1.2 Cement Kilns 199
9.4.2 Sonolysis 200
9.4.3 Electrochemical Treatment 202
References 202

9.1 INTRODUCTION

The unique physical and chemical properties that make PFAS valuable commercial chemicals also impart resistance to many conventional water and wastewater treatment technologies used for remediation. There are, however, PFAS-relevant liquid treatment technologies available and under development. This chapter identifies and evaluates PFAS-relevant treatment technologies that are currently used in the industry and/or are under development for future use to meet treatment objectives for various liquid waste streams. PFAS treatment is the primary focus in this chapter; however, other characteristic constituents, such as total organic carbon, total suspended solids, and metal ion concentrations, may influence the efficacy of certain treatment technologies for addressing PFAS when in solution.

Treatment technologies for PFAS are divided into those using three potential mechanisms for removal: adsorption, separation, and destruction. A visual representation of PFAS-relevant liquid treatment technologies based on their current stage of development and effectiveness is presented in Figure 9.1.

PFAS treatment remains a young and dynamic field with advances and innovation happening frequently. In Figure 9.1 the placement of each technology relative to the vertical axis is based on its maturity as a PFAS treatment technology, not its maturity in the water and wastewater treatment industry. Similarly, placement along the horizontal axis is based on a technology's understood effectiveness at PFAS-specific treatment. For example, activated carbon, most commonly used in the granular form (granular activated carbon [GAC]), exploits an adsorptive mechanism to treat

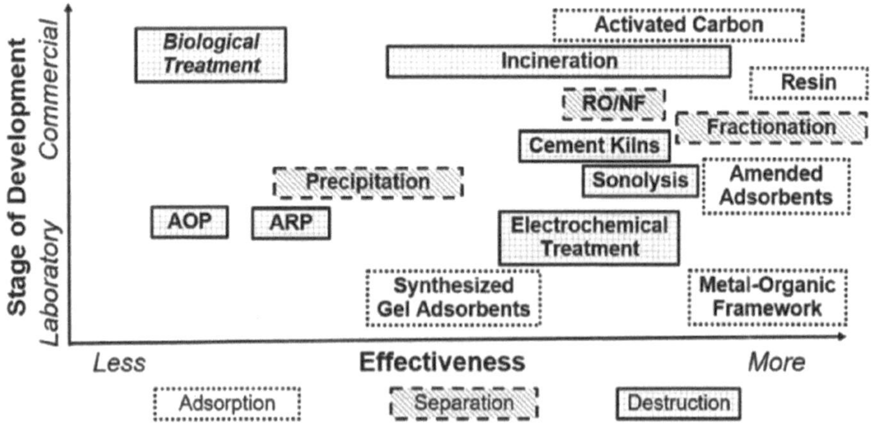

FIGURE 9.1 Treatment Technologies for PFAS in Liquids (Adapted from Ross et al. 2018).

PFAS-impacted liquids. It is a mature technology and generally feasible to implement, placing it in the upper-right quadrant of Figure 9.1. Conversely, advanced oxidation processes (AOP) are a mature technology for the water and wastewater treatment industry, but AOP has not developed significantly beyond the laboratory scale for PFAS applications (Bruton and Sedlak 2018).

PFAS-impacted waste streams are typically complex with respect to their chemistry, as several PFAS are usually present simultaneously and may require several different treatment technologies to achieve relevant discharge requirements. The treatment technologies discussed herein can be challenged by non-PFAS constituents that may be present in the waste stream, which may be co-contaminants or naturally occurring organics (such as humic or fulvic acids) or inorganics (such as sulfate or nitrate). PFAS treatment systems often employ multiple treatment mechanisms in series to maximize the overall effectiveness, commonly referred to as a treatment train.

9.2 ADSORPTION

Adsorptive technologies remove PFAS by promoting electrostatic and/or hydrophobic adsorption to a relatively inert medium or matrix. Intuitively, these technologies do not destroy PFAS; instead, they transfer PFAS associated with liquids onto solid-phase media, simultaneously concentrating PFAS and reducing the waste volume to be destroyed. Adsorptive technologies may rely on the properties of the matrix employed or matrices can be modified by adding functional groups (such as quaternary amines) to promote PFAS adsorption.

9.2.1 Activated Carbon

Activated carbon is a highly porous media made from carbon-rich raw materials (e.g., coal, coconut shell, wood, or bamboo) that can be used for treatment of water and wastewater to remove PFAS and some non-target constituents. Its high specific surface area allows for effective removal of contaminants from water primarily through adsorption mechanisms based on hydrophobic interaction.

Activated carbon can be utilized in several forms, listed in order of decreasing particle size: GAC, powdered activated carbon (PAC), and sub-micron powdered activated carbon (SPAC). GAC is a primary technology for adsorption of PFAS with numerous examples of functioning applications. Both PAC and SPAC have been studied via bench- and pilot-scale testing, but neither technology has been broadly developed for commercial-scale applications to address PFAS. The following subsections will further characterize each type of activated carbon.

Activated carbon treatment, while ubiquitous across the PFAS industry, does have some negative characteristics that can preclude it from use. These include rapid breakthrough for short-chain PFAS and increased operational complexity related to frequent media changeouts as compared to other adsorbent media. Short- and ultra-short chain PFAS are more hydrophilic than long-chain PFAS and are not amphiphiles, making them less adsorptive to GAC (Ateia et al. 2019). Multiple studies indicate reduced capacity to treat short-chain PFAS (Appleman et al. 2014;

Xiao et al. 2017b; Westreich et al. 2018). Faster short-chain breakthrough has been attributed to lower hydrophobic affinity based on perfluoroalkyl chain length (Dickenson and Higgins 2016).

PFAS adsorption onto activated carbon in the presence of non-target constituents, such as dissolved organic matter (DOM), is dependent on the characteristics of the DOM and of the target PFAS. Evidence shows that DOM can decrease adsorption of PFAS on activated carbon (Pramanik et al. 2015; Appleman et al. 2013), increase adsorption on activated carbon (Kothawala et al. 2017), or show minimal effect for adsorption of PFAS on activated carbon (Siriwardena et al. 2019). Presence of DOM and other non-target constituents may reduce available adsorption sites by competing for adsorption sites and potentially hindering access to pores (Appleman 2013; Pramanik et al. 2015).

9.2.1.1 Granular Activated Carbon

Granular activated carbon (GAC) media are frequently used as interim response measures to achieve immediate compliance with regulatory standards and represent the state of the practice for managing PFAS impacts (Horst et al. 2018). There is a broad array of available GAC materials, including those materials that are intended for physical separation, general adsorption, and those that are optimized for PFAS removal. The most common types of GAC used for PFAS adsorption are coal-based (bituminous and lignite) and coconut-based carbons (Westreich et al. 2018). The inherent structure and the process used to activate the carbon influence treatment performance. As an example, enhanced coconut shell GAC has a reduced microporosity and increased mesoporosity to increase PFAS removal efficiency. Research has demonstrated that a higher mesoporosity (i.e., pores with widths 20-500 Å) was observed to be more favorable for PFAS adsorption to GAC, which was explained as counteracting the expected non-target constituent competition (Bell et al. 2019).

GAC has been proven most effective at managing perfluorooctanoic acid (PFOA) and perfluorooctanesulfonic acid (PFOS) at influent concentrations, ranging into the parts per billion (Du et al. 2014), and can produce concentrations into single digit parts per trillion in the treated water. GAC is widely used for organics removal throughout the water and wastewater treatment industry and is a fully demonstrated and reliable technology (ITRC 2018). Rapid small-scale column testing (RSSCT) can be used to provide proof-of-concept testing and comparative and scalable data for developing cost–benefit analyses. RSSCT is scaled such that contaminant breakthrough can be assessed in as little as 1 to 15% of runtime, as compared to an applicable pilot-scale system (Crittenden et al. 1991, Westreich et al. 2018). Depending on the application, data collected via RSSCT can be used in lieu of pilot testing for design of full-scale GAC adsorption systems.

As a primary PFAS removal technology, GAC is most frequently deployed within fixed-bed vessels in a flow-through configuration. Typical vessel design arrangements include at least two fixed-bed vessels in series, which allows monitoring of media beds for breakthrough. When targeting a given effluent PFAS concentration in wastewater, higher influent concentrations of PFAS will result in increased operations and maintenance costs due to more frequent media reactivation or disposal. Spent GAC is reactivated or incinerated depending on influent concentrations. GAC systems require minimal operator input due to typically simple system design.

9.2.1.2 Powdered Activated Carbon

Powdered activated carbon (PAC) is a more finely ground carbon than GAC, with particles that pass the #80 mesh sieve, or less than 0.177 millimeter (ASTM International 2019). It is well established that PAC requires less time for adsorption to reach equilibrium owing to its smaller grain size.

PAC is not commonly used in a fixed-bed vessel configuration due to the very high head loss across the bed that would occur due to its considerably smaller size (than GAC). Consequently, the smaller grain size also leads to propensity for clogging. PAC is typically added directly to process units where the PAC remains fluidized, avoiding flow-through kinetics complications (e.g., during coagulation).

9.2.1.3 Super-fine Powdered Activated Carbon

Super-fine powdered activated carbon (SPAC) is a more finely ground carbon than either GAC or PAC, with particles of diameters less than 1 micron. SPAC has some of the same application limitations as PAC, including potentially high pump back-pressure, and has the added challenge of maintaining the sub-micron media within a vessel during flow-through conditions. Murray et al. (2019) tested SPAC coupled with ceramic microfiltration (CMF) on contaminated groundwater and diluted fire-fighting containment tank water and compared it with GAC. SPAC was dosed into a mixed tank to provide sufficient contact time. SPAC coupled with CMF had three times more perfluorobutanesulfonic acid (PFBS) loading than GAC in the diluted firefighting containment tank water, and the loading rate increased nearly eightfold in the more contaminated groundwater. The study reported minimal CMF fouling by SPAC, but identified that SPAC would need to be "continually drained and replenished" to maintain expected adsorption loading rates.

9.2.2 Ion Exchange

Ion exchange (IX) resins can be manufactured to have positively charged (i.e., cationic), or negatively charged (i.e., anionic) functional groups depending on the charge characteristics of the constituents targeted. Most PFAS of current regulatory interest are anionic under environmental pH conditions (i.e., circumneutral), and zwitterionic and cationic PFAS tend to be less mobile in aquifers so may not travel to locations in need of treatment, such as drinking water supply wells. Therefore, anion exchange (AIX) resins are the most suitable for removing them (Deng et al. 2010). AIX resins contain cationic functional groups bound by a polyacrylic or polystyrene bead and exchange inert anionic counterions with anionic contaminant molecules. PFAS interactions with AIX resins are not limited to ion exchange. Additionally, PFAS can adsorb to AIX resin via a combination of mechanisms including electrostatic or hydrophobic interactions and aggregation.

Functional group characteristics for both resin (greater hydrophobicity increases hydrophobic attraction) and PFAS (presence of a negatively charged functional group increases electrostatic interaction) drive the effectiveness of resins for short-chain PFAS adsorption (Zaggia et al. 2016, McCleaf et al. 2017). Research into management of ultra-short PFAS is still in the exploratory phase with discovery of these PFAS only being reported recently.

IX resins are commonly used in water treatment to remove inorganic anions, such as sulfate, chromate, nitrate, chloride, radionuclides, and perchlorate. Many anions are present within municipal waste streams and can be present at concentrations that are orders of magnitude greater than PFAS, resulting in direct competition with PFAS for adsorption sites (Ross et al. 2018) and reducing sorptive capacity for PFAS (Ateia et al. 2019, Zeng et al. 2020). High concentrations of total dissolved solids and competing anions exert a considerable ionic strength influence, which may suppress electrostatic adsorption of PFAS onto the resins (Maimaiti et al. 2018).

AIX resins are most frequently used in water treatment within fixed-bed vessels in a flow-through configuration. Typical vessel design arrangements include at least two fixed-bed vessels in series, which allows monitoring of media beds for breakthrough. AIX resins can be deployed as single-use and disposable or multi-use and regenerable. At lower concentrations of PFAS (i.e., concentrations in the hundreds of micrograms per liter), AIX resins are typically installed as single-use and disposable units because there is less cost in disposing of single-use resin than managing regenerative solutions that may involve additional hazardous material management considerations. At high concentrations, regenerable resins may be utilized when the cost of managing regenerative solutions approximates the cost of single-use resin disposal. IX resins, generally, are a well-developed and highly implementable process, and AIX resins for PFAS have been developed and are considered a fully demonstrated technology.

9.2.3 Functionalized Adsorbents

Recent research efforts have expanded to alternative adsorptive materials, including inert media functionalized to adsorb PFAS from an aqueous media. Examples of these specialized media for aqueous use include, among others, organically modified silica (Horst et al. 2018; Stebel et al. 2019), crosslinked β-cyclodextrin (Alsbaiee et al. 2016; Xiao et al. 2017a), and modified cellulose (Ateia et al. 2018). Additional amended adsorbents are under development for PFAS-impacted solids and are not detailed in this chapter.

Development of novel adsorbents remains at the laboratory scale of development at the time of this writing. Companies like ABS Materials, Inc., the manufacturer of Osorb®, an organically modified silica, Cyclopure, the manufacturer of an amended crosslinked β-cyclodextrin (see Chapter 14 by Ling et al.), and Puraffinity, the manufacturer of a modified cellulose adsorbent, are in the process of developing a commercial product using these materials.

9.3 SEPARATION

Separation technologies remove PFAS from aqueous matrices by either absorbing the PFAS into a material, in the case of precipitation or fractionation, or physically excluding PFAS based on molecular size, such as with membrane exclusion technology. Furthermore, these technologies do not destroy PFAS. Instead, they convert lower-concentration high-volume waste streams into higher-concentration lower-volume waste streams, aiming at reducing waste disposal costs by reducing waste volume to be managed by a further treatment stage. Generally, these technologies are commercially developed due to their use in other industries, including water and wastewater treatment, but are less mature for managing PFAS-impacted wastes.

9.3.1 Membrane Treatment

Membrane filtration is a physical exclusion process that rejects aqueous phase constituents via size exclusion and/or diffusive mechanisms. Microfiltration (MF) and ultrafiltration (UF) are classified as membrane filtration processes, operating at low pressures (15 to 60 pounds per square inch (psi) for MF, 30 to 100 psi for UF) while providing reliable removal of macromolecules, colloids, particles, turbidity, and microorganisms. MF/UF are made of organic and inorganic microporous materials that allow passage of water while retaining particles and macromolecules. The size cut-offs for MF and UF vary between sources, but commonly MF membranes are considered to have nominal pore sizes of 0.1 to 10 μm, and 0.01 to 0.1 μm for UF.

Nanofiltration (NF) and reverse osmosis (RO) are technologies using semipermeable polymeric materials used widely for removal of salts and/or dissolved organics from water (Gude 2016). NF and RO are most commonly thin-film composite materials consisting of a nonwoven polymeric fabric, a porous polysulfone support layer, and a polyamide barrier film (Cadotte and Petersen 1981). A hydrostatic pressure is applied across the membrane to counteract the thermodynamic driving force of osmosis and pushes water through the membrane.

The key distinguishing traits for an NF membrane are its operating pressure and the compounds that the membrane rejects. However, the functionality of an NF membrane is not different from a RO membrane. For this reason, some refer to NF as 'loose RO' (Bellona 2014). Operating pressures for NF are often around 80 to 100 psi, but can be as high as 150 psi. Operating pressure for RO systems depends upon the total dissolved solids (TDS) content of the water being treated and can range from 200 to 400 psi for waters containing 100 to 1,000 milligrams per liter (mg/L) of TDS, or as high as 800 to 1,200 psi for desalination membranes, where TDS is 5,000 to 50,000 mg/L.

While NF and RO may seem very similar to MF and UF because they all clean water by pushing water through the membrane, the underlying mechanisms by which each operate are different. MF and UF work as filtration processes operating on the principle of physical separation via size exclusion. Constituents in the water that are bigger than the MF/UF pores will be rejected and will not pass through, while constituents smaller than the pores may pass through. For NF and RO, solutes are rejected by the semipermeable barrier film (membrane) based not only upon their size (similar to MF/UF), but also their ability to sorb into, diffuse through, and desorb out of the membrane.

In the process of creating a clean permeate stream, NF and RO also produce a high-strength reject stream. For silica and/or compounds with high hardness levels, NF/RO may concentrate these species to a point above their respective solubility limits, leading to the formation of scale on the membrane. As such, it is important to pay careful attention to water chemistry when considering NF/RO to understand the potential for scaling. In some cases, it may be necessary to pre-treat water with anti-scalants to remove scale-forming compounds prior to NF/RO treatment.

9.3.1.1 Nanofiltration (NF)

Nanofiltration membranes are often marketed based on a molecular weight cut off (MWCO), typically ranging from 200 to several thousand Daltons (Da). Appleman et al. (2013) demonstrated greater than 94% removal for PFCA and PFSA of chain lengths, varying from C4 to C9 for both deionized and artificially generated groundwater matrices using a NF membrane with a MWCO of 270 Da. Subsequent testing by Franke et al. (2019) indicated that PFAS with molecular weights below 270 Da were concentrated, indicating a likelihood that size exclusion was not the only mechanism for removal, and suggesting that electrostatic repulsion and diffusion into the membrane polymer could be responsible for short-chain PFAS removal. Much of the work presented was performed with the same commercial NF membrane. Additional research indicates that fabricated NF membranes designed for PFAS selectivity may provide effective capacity while increasing water recovery rates (Boo et al. 2018). However, while NF is a fully developed technology for drinking and wastewater treatment, much of the characterized PFAS-related work for NF remains on the bench-scale with implications of these developments unclear for field-scale application. NF is considered a highly implementable technology commercially and is sometimes combined with reverse osmosis (RO) as part of a more complex treatment train.

9.3.1.2 Reverse Osmosis

Reverse osmosis (RO) technology is frequently used in water treatment for seawater desalination and is considered well-developed for this application (Yang et al. 2019). RO is also widely applied as part of industrial wastewater recycle systems, using pretreated industrial wastewater to produce high-quality permeate for reuse. RO technology for PFAS exclusion has been studied since at least 2006 when it was reported to reject greater than 99% of PFOS at a significant cost to permeate flux (Tang et al., 2006). Although some PFAS have molecular weights below the effective MWCO rating for NF, even short-chain PFAS are effectively rejected by RO membranes. Migration of compounds through RO is not only a function of molecular weight, but also is strongly influenced by charge and the ability of the compound to sorb into and diffuse through the barrier film. One study reported that RO has lower MWCO ratings, down to approximately 100 Da (Thompson, et al. 2011); however, in practice, some compounds with molecular weights below this value are rejected (Catarino et al. 2007). As a point of reference, perfluorobutanoic acid has an approximate mass of 214 Da and PFBS has an approximate mass of 300 Da. The ability of RO to reject very small ions and molecules make it effective at PFAS removal and other non-target constituent removal, including metal ions and aqueous salts. RO membranes are able to effectively reject monovalent ions (as well as ions of higher charge). In comparison, NF is able to effectively reject divalent ions (and ions of higher charge), but are not effective at rejecting monovalent ions. Similarly, RO is able to reject smaller organics than NF. Because of these rejection capabilities, more material is rejected by RO than NF, meaning that in a PFAS treatment application the reject stream is higher in both PFAS and non-target constituent concentrations.

As noted earlier, markedly higher pressures are required for RO than for NF systems resulting in higher energy consumption for RO than NF systems. A variety of

RO membrane types are commercially available, spanning a range of rejection efficiencies (i.e., 'membrane tightness').

9.3.2 Foam Fractionation

Fractionation is a separation technology using nano- to micron-sized gas bubbles, such as ozone or air, to treat water. Due to the amphiphilic properties exhibited by many PFAS, fractionation could be a promising approach for removing those that are surfactants. Amphiphilic PFAS can be removed from an aqueous solution as foams or aerosols formed during aeration. Gaseous bubbles are added or formed at the bottom of the reaction vessels and rise to the top. Bubble interaction with PFAS generate aerosols or foams. By removing the continuously generated aerosols or foams, amphiphilic PFAS in the aqueous phase are removed and concentrated in a small volume of concentrated foam.

The partitioning of amphiphilic PFAS from the liquid phase to gaseous bubbles is drawing increasing attention as a tool to enable their removal from aqueous waste streams to form a concentrate (Horst et al. 2018). Significantly higher PFAS concentrations have been detected in the surface microlayer of coastal waters in China (Ju et al., 2008), indicating PFAS tend to accumulate at the air–water interface. The preference for amphiphilic PFAS to accumulate at the air–water interface has also been investigated to develop efficient PFAS remediation strategies. Air bubbles were proven to assist in the adsorption of PFOS on some adsorbents (Meng et al. 2014, 2017), and the removal of amphipillic PFAS by activated carbon was improved with the assistance of aeration during adsorption (Meng et al. 2019).

Based on the preference of amphiphilic PFAS to partition to gaseous bubbles and the applicability of aeration during water treatment, aeration alone has been considered for PFAS remediation under different circumstances. For example, aeration has been used to treat wastewater from the electroplating industry (Ebersbach et al. 2016). During aeration, amphiphilic PFAS were highly concentrated at the surface of air bubbles as they rose from the bottom of a vessel to the surface. When the air bubbles finally reached the surface of the wastewater and burst, PFAS-enriched aerosols were formed and transmitted to the atmospheric phase. By capturing the condensed aerosols, PFAS in the wastewater was selectively removed without matrix interference, with 99.8% of the initial amount recovered in the collected aerosols. However, the method of Ebersbach et al. was concentration dependent. When the concentration of PFAS in the wastewater is higher than a threshold value (5 mg/L for 6:2 FTSA), a layer of foam is formed at the surface of the wastewater which inhibits continuous aerosol formation. However, the foam can be removed and contains concentrated PFAS. Nevertheless, as a result of excessive foaming over a concentration of approximately 20 mg/L of total surfactant, this method may not be applicable for treating aqueous film-forming foam (AFFF)-contaminated wastewater, which usually has significantly higher PFAS concentrations. Foam fractionation has also been used for separating PFOS and PFOA from diluted water solutions with high concentrations of metallic activators (Lee et al. 2017). Lee et al. found their fractionation method was very sensitive to the pH value of the wastewater, as high concentrations of metallic activators are involved. Fractionation would be adaptable to a wider range of pH if there are no chemicals involved. A previous study (Meng et al. 2019) showed that by simply aerating the AFFF solution and removing the formed foams on the surface for one hour, over 99% of the PFOS was removed from the AFFF-impacted wastewater. Fractionation could be an efficient and cost-effective treatment for AFFF-impacted water when not inhibited by the other components in the matrix.

Among the few pilot- or full-scale fractionation studies, a fractionation field study using ozone and a polishing adsorptive media achieved greater than 99.99% total PFAS removal from a mixed waste of brackish surface water and untreated wastewater from a sewer (Storch et al. 2018). Fractionation with ozone gas creates a further challenge to treatment, as ozone converts amphiphilic precursors into shorter-chain perfluoroalkyl acids (PFAAs) which are not amphiphiles and cannot be extracted efficiently. There is potential to create ultrashort PFAAs in addition to short-chain PFAAs as a result of fluorotelomer chain shortening during the oxidation process. This means that ozone use can form PFAAs, which may then require a further treatment stage to remove the generated shorter-chain PFAS. When using ozone, common co-contaminants such as petroleum hydrocarbons could also be oxidized during the separation process if residence time in fractionating vessels is adequate. PFAS are not destroyed during the fractionation process (even with ozone) due to the high chemical stability of the perfluoroalkyl group. Therefore, fractionation can only transfer the PFAS from a large volume of low PFAS-density wastewater to a much smaller volume of concentrated PFAS concentrate (typically less than 5% of influent water volume), which requires additional PFAS removal via processes such as ion exchange in highly concentrated doses.

Overall, fractionation may be a cost-effective application that does not require additional pre-treatment and is compatible with multiple flow rates. This is a promising technology with further pilot- and full-scale applications progressing.

9.4 DESTRUCTION

The ultimate goal of treating water containing PFAS is destruction of the PFAS. The unique chemistry that has made PFAS so valuable as a commodity in the fire response and commercial spaces makes this class of compounds difficult or costly to destroy in water. There are two reported primary pathways for PFAS destruction: molecular decomposition via thermal destruction (i.e., pyrolysis) and sequential defluorination. Individual PFAS react to destructive mechanisms differently based on factors such as chain length and functional group. Partial defluorination of a PFAS is still considered destruction, but may generate unwanted by-products. As such, for the purposes of this evaluation, destruction is defined as mineralization, which is the complete defluorination of PFAS into fluoride ions and carbon dioxide (Horst et al. 2020). This section reviews destructive technologies that have been studied as viable options for PFAS treatment.

9.4.1 Thermal Destruction

Functionally, thermal destruction is achieved either by thermal oxidation or pyrolysis. With pyrolysis, PFAS molecules break apart into smaller pieces, eventually forming fluoride ions or hydrogen fluoride gas and a nonfluorinated carbonaceous species (i.e., carbon dioxide). Primarily, thermal destruction of PFAS in the United States is achieved by incineration, although other technologies such as cement kilns and infrared can achieve a destructive result. Because PFAS represents a large class of compounds that have varying structures, the required conditions for mineralization vary and must be understood for the specific PFAS compounds present in a given wastewater. Molecular structure variations such as branched vs. linear composition, chain length, and carbon substitution affect the required temperature to achieve complete mineralization. The following subsections explain the capabilities of each of the available thermal destruction technologies.

9.4.1.1 Incineration

Incineration has historically been the accepted method for PFAS destruction. For unimolecular decomposition, fluorinated organic compounds require temperatures above 1000 °C (degrees Celsius) to achieve 99.99% destruction in residence time of one second (USEPA 2019). A review of destructive technologies described that PFOA gasification begins at 200 °C and upwards of 600 °C for PFOS (Horst et al. 2020). Complete mineralization of PFOS adsorbed to GAC is reported at 1000 °C (Watanabe et al. 2016). However, as Horst et al. (2020) discuss, reported evidence in the literature is conflicting on required temperatures for complete PFAS defluorination. There are unresolved questions on the fate of PFAS present in liquid waste streams subject to incineration. When incinerating liquids, temperatures required for PFAS thermal destruction result in steam discharge from the stack, which has been observed during long-term flue gas sampling to include PFOS and PFOA. Post-combustion data indicates residuals of PFOS were observed in laboratory incineration trials and in municipal solid waste ash. Further, laboratory-scale incineration testing indicated complete destruction of PFOS and two associated perfluoroalkyl sufonamides with good sulfur recoveries, but reported evidence of organofluorine intermediates in the reactor effluent, including carbon tetrafluoride (Yamada and Taylor 2003). Due to ongoing concerns of incomplete PFAS combustion in liquid matrices during incineration, attitudes regarding the risk of incinerating PFAS-impacted liquids are changing. The US Environmental Protection Agency held a PFAS thermal destruction workshop in Spring of 2020 and published an interim guidance document on PFAS destruction and disposal. This document identifies the unknowns regarding PFAS stack discharge due to a lack of "standardized validated methodology" for quantifying PFAS in stack emissions. As a result, the document presents thermal destruction as having "higher levels of uncertainties regarding their capacity to manage the migration of PFAS into the environment" (USEPA 2020).

9.4.1.2 Cement Kilns

There is significant interest in the use of cement kilns for co-processing of hazardous wastes. The United Nations Environmental Programme has co-authored several technical guideline documents relating to the co-processing of hazardous wastes, including persistent organic pollutants in cement kilns (Stockholm Convention 2007; Basel Convention 2011). The cement manufacturing process generally consists of two thermal processes: the decomposition of calcium carbonate in the precalciner at

900 °C, or calcination, followed by the formation of calcium silicates, aluminates, and ferrites in the main burner at between 1400 °C and 1500 °C, or clinkering (Stockholm Convention on Persistent Organic Pollutants, 2007). Residence times during the clinkering process are typically between six and fifteen seconds (Ueno et al. 1997, Basel Convention on the Control of Transboundary Movements of Hazardous Wastes and their Disposal, 2011). When compared to incineration (typically two seconds in the burn chamber), the extended residence time of the cement kiln provides additional capability for PFAS destruction. Unrelated work focused on the role of calcium in the mineralization of PFOS suggests that the presence of calcium during thermal destruction can lead to decomposition at lower temperatures (Wang et al. 2011; Wang, 2013).

While published literature on the destruction of PFAS in cement kilns is lacking, at least one cement kiln in Australia has successfully treated foam concentrate as well as other PFAS wastes. There is no evidence of this practice being used in the United States or Europe at this time.

9.4.2 Sonolysis

Sonolysis uses sound waves to facilitate cavitation in water. Bubbles form and subsequently collapse, releasing up to 5,000 degrees Kelvin of heat energy. PFAS congregate at the bubble surface, and, upon collapse of the bubble, are destroyed by the released energy (Ross et al. 2018). Studies have shown complete mineralization of PFOA and PFOS into CO, CO_2, F^- and SO_4^{2-} using sonolysis and have identified theoretical chemical pathways of destruction (Vecitis et al. 2008, 2010; Rodriguez-Freire et al. 2015; Wood et al. 2020). Small quantities (typically less than 1% of the initial PFAS mass) of shortened PFAS can be generated during sonolysis, but may be degraded with continued treatment following a sequential chain-shortening process (Moriwaki et al. 2005; Wood et al. 2020). When compared to other potential destruction technologies, complete PFAS mineralization is more common during sonolysis, especially for PFOS. For example, plasma treatment of PFAS can leave as much as 97.5% of fluorine remaining as fluorinated organics (Singh et al. 2019; Stratton et al. 2017). Chapter 12 by Singh et al. provides an in-depth discussion on this technology. Other PFAS structures have also been investigated and their structural effects determined. For example, it was shown that the increasing the length of the perfluorinated tail increases reaction rates due to increased hydrophobicity (and hence partitioning to the bubble interface) and enhanced anchorage at the interface to mitigate repulsion between sorbed PFAS molecules (which increases bubble surface concentrations) (Campbell et al. 2009). Note that the reaction rate of PFAS sonolysis is first-order at low concentrations since increased concentrations leads to increased bubble-uptake and thus a greater pseudo-first-order rate constant. However, at high concentrations the bubble surfaces become saturated and the reaction becomes zero order (Moriwaki et al. 2005; Vecitis et al. 2008, Wood et al. 2020).

Additional work varying applied frequencies and solution amendments has shown promising increases in PFAS removal and defluorination (Campbell et al. 2009; Gole et al. 2018a; Lei et al. 2020; Wood et al. 2020). Ultrasonic frequencies of around 350 kilohertz (kHz) appear to be optimal for treating of amphiphilic PFAS (Campbell et al. 2009; Wood et al. 2020). However, shorter-chain PFAS are better treated at around 600 kHz (Campbell et al. 2009;). Since the short chains less readily partition to the bubble interface, higher frequencies cause cavitating bubbles to oscillate at a higher frequency, thus moving the wall of the expanding and contracting bubble into the path of the PFAS faster than the PFAS can diffuse toward the bubble (Campbell et al. 2009, 2015). Other parametric effects have also been studied, such as pH (Cheng et al. 2008; Gole et al. 2018a), ultrasonic power (Campbell et al. 2015), bubble gas choice (Moriwaki et al. 2005; Hori et al. 2012; Lee et al. 2016), and the effects of co-contaminants, both organic (Cheng et al. 2008) and inorganic (Cheng et al. 2010).

It is important that these parametric, structural, and matrix effects are understood for simple systems since sonochemical reactions are extremely complicated and the modeling or measurement of different reaction condition effects in isolation is challenging. However, with much work on these parametric effects, several works have begun testing on more convoluted matrices, including; landfill leachate (Cheng et al. 2008, 2010), PFAS containing aqueous firefighting foams (AFFFs) (Vecitis et al. 2010; Rodriguez-Friere et al. 2016; Gole et al. 2018b), and artificially contaminated soil slurries (Lei et al. 2020).

While sonolysis shows significant promise for PFAS destruction, the technology has only advanced beyond laboratory scale in two select works. It is also noted that sonolysis has been the subject of study for treating other environmental constituents, working well at bench-scale but lacking commercialization (Lifka et al. 2003; Gonzalez-Garcia et al. 2010a). To date, most studies have utilized reactors of several hundred milliliters; however, two studies have assessed the capabilities of a large ultrasonic reactor (91 liters) to degrade PFOS (Gole et al. 2018a) and diluted AFFFs (Gole et al. 2018b). Sonolysis has also been combined with several other destructive technologies which enhance treatment, including photocatalytic decomposition (Panchangam et al. 2009; Sekiguchi et al. 2017), persulfate oxidation (Hori et al. 2012; Hao et al. 2014) carbonate radical oxidation (Thi et al. 2014), and periodate oxidation (Lee et al. 2016), among others.

A reported disadvantage of PFAS sonolysis might include moderate energy consumption when compared to treatments such as plasma (Wood et al. 2020); however, plasma has been demonstrated in some cases to be unable to fully mineralize PFAS (Singh et al. 2019; Stratton et al. 2017). Sonolysis is also more cost-effective at higher concentrations (Vecitis et al. 2008), making it potentially suitable for direct treatment of certain wastes such as firefighting foams, but not for destruction of trace-level contamination found in most waterways. It has been shown that competitive adsorption to the bubble interface of co-organics can reduce PFAS treatment efficiency (Cheng et al. 2008), and some inorganics found in soils can both augment and diminish reaction rates in agreement with Hofmeister series effects (Cheng et al. 2010). At the industrial scale, sonolysis would likely benefit from a pre-treatment process which both removes interfering organics and concentrates PFAS to a level which permits zero order reaction rates. As an example, foam fractionation combined with GAC or membrane-based separation may be a suitable pretreatment train ahead of sonolysis.

9.4.3 Electrochemical Treatment

Electrochemical treatment uses reduction–oxidation reactions as part of an electrolytic cell to directly oxidize constituents or generate hydroxyl radicals at the anode (Martinez-Huitle and Ferro 2006). Some potential mechanisms for destruction in the electrolytic cell include oxidation via a direct electron transfer from the anode and reduction via interactions with hydroxyl radicals generated from the electron flow from anode to cathode (Chaplin 2014).

Research into PFOA and PFOS degradation has focused on mixed metal oxides (including lead oxides, titanium/tin oxides, and titanium oxides) and boron-doped diamond (BDD) anodes (Yang et al. 2015; Schaefer et al. 2017; Fang et al. 2019). The titanium and tin oxide anode achieved 99% removal of PFOA in 30 minutes in clean water, but partial oxidation products were formed (Yang et al. 2015). Use of lead oxide from a lead-acid battery reportedly reduced PFOA concentrations by greater than 99% with nearly 60% defluorination (Fang et al. 2019). Fluoride formation was monitored during a titanium oxide–anodic oxidation experiment, and approximately 15% fluoride generation was reported after 30 minutes and 71.5% at 2.5 hours of reaction time (Huang 2019), indicating that electrochemical oxidation can generate nearly complete destruction but that reaction times to achieve it may not yet be feasible for field application. Several studies have demonstrated PFOA and PFOS destruction and the subsequent generation of shorter-chain PFAS compounds using a BDD electrode (Trautmann et al. 2015; Schaefer et al. 2017). Electrochemical treatment using BDD, lead oxide, and platinum anodes has shown evidence of perchlorate generation as a by-product (Chaplin 2014; Trautmann et al. 2015), which would require additional treatment in a real-world application. The major disadvantage of electrochemical treatment is the creation of oxyanions from common inorganics such as chloride and bromide, which can be stoichiometrically converted into perchlorate and bromate (Gomez-Ruiz et al. 2017). These are contaminants which usually require treatment prior to their discharge. Additional material on electrochemical treatment, advanced oxidation, and advanced reduction processes is provided by Lester in Chapter 13.

REFERENCES

Alsbaiee, Alaaeddin, et al. 2016. "Rapid removal of organic micropollutants from water by a porous β-cyclodextrin polymer." *Nature* 529.7585: 190–194.

Appleman, Timothy D., et al. 2013. "Nanofiltration and granular activated carbon treatment of perfluoroalkyl acids." *Journal of hazardous materials* 260: 740–746.

Appleman, Timothy D., et al. 2014. "Treatment of poly-and perfluoroalkyl substances in US full-scale water treatment systems." *Water research* 51: 246–255.

ASTM International 2019. "ASTM D5158–98: Standard Test Method for Determination of Particle Size of Powdered Activated Carbon by Air-Jet Sieving", *ASTM International*, West Conshohocken, PA. Available at: https://www.astm.org/Standards/D5158.htm (accessed November 15, 2020).

Ateia, Mohamed, et al. 2018. "Rapid removal of poly-and perfluorinated alkyl substances by poly (ethylenimine)-functionalized cellulose microcrystals at environmentally relevant conditions." *Environmental Science & Technology Letters* 5.12: 764–769.

Ateia, Mohamed, et al. 2019. "The overlooked short-and ultrashort-chain poly-and perfluorinated substances: A review." *Chemosphere* 220: 866–882.

Barpaga, Dushyant, et al. 2019. "Probing the Sorption of Perfluorooctanesulfonate Using Mesoporous Metal–Organic Frameworks from Aqueous Solutions." *Inorganic Chemistry* 58.13: 8339–8346.

Barzen-Hanson, Krista A., and Jennifer A. Field 2015. "Discovery and implications of C2 and C3 perfluoroalkyl sulfonates in aqueous film-forming foams and groundwater." *Environmental Science & Technology Letters* 2.4: 95–99.

Basel Convention on the Control of Transboundary Movements of Hazardous Wastes and their Disposal 2011. "*Technical guidelines on the environmentally sound co-processing of hazardous wastes in cement kilns*" (Colombia: United Nations Environmental Programme) Available at: http://www.basel.int/Portals/4/Basel%20Convention/docs/pub/techguid/cement/tg-cement-e.pdf (accessed November 15, 2020).

Bell, Caitlin H., et al., eds 2019. Emerging contaminants handbook. CRC Press. Chapter 3 Per- and Polyfluoroalkyl Substances.

Bellona, Christopher 2014. "Nanofiltration–theory and application." In J. Kucera (Ed.) *Desalination: Water from Water*": 205–253.

Boo, Chanhee, et al. 2018. "High Performance Nanofiltration Membrane for Effective Removal of Perfluoroalkyl Substances at High Water Recovery." *Environmental Science and Technology* 52: 7279–7288.

Bruton, Thomas A., and David L. Sedlak 2018. "Treatment of perfluoroalkyl acids by heat-activated persulfate under conditions representative of in situ chemical oxidation." *Chemosphere* 206: 457–464.

Buck, Robert C., et al. 2011. "Perfluoroalkyl and polyfluoroalkyl substances in the environment: terminology, classification, and origins." *Integrated environmental assessment and management* 7.4: 513–541.

Burns, Darcy C., et al. 2008. "Experimental p K a determination for perfluorooctanoic acid (PFOA) and the potential impact of p K a concentration dependence on laboratory-measured partitioning phenomena and environmental modeling." *Environmental science & technology* 42.24: 9283–9288.

Cadotte, John E., & Robert J. Petersen 1981. "Thin-film composite reverse-osmosis membranes: Origin, development, and recent advances". In A. F. Turbak (Ed.), *Synthetic membranes: ACS symposium series* (Vol. 153). American Chemical Society.

Campbell, Tammy Y., et al. 2009. "Perfluorinated Surfactant Chain-Length Effects on Sonochemical Kinetics." *Journal of Physical Chemistry* 113, 9834–9842

Campbell, Tammy Y., et al. 2015. "Sonochemical degradation of perfluorinated surfactants: Power and multiple frequency effects" *Separation and Purification Technology* 156, 1019–1027.

Catarino, Margarida, et al. 2007. "Alcohol removal from beer by reverse osmosis." *Separation Science and Technology* 42.13: 3011–3027.

Chaplin, Brian P 2014. "Critical review of electrochemical advanced oxidation processes for water treatment applications." *Environmental Science: Processes & Impacts* 16.6: 1182–1203.

Cheng, Jie., et al. 2008. "Sonochemical Degradation of Perfluorooctane Sulfonate (PFOS) and Perfluorooctanoate (PFOA) in Landfill Groundwater: Environmental Matrix Effects." *Environmental Science and Technology* 42: 8057–8063.

Cheng, Hie., et al. 2010. "Sonochemical Degradation of Perfluorooctane Sulfonate (PFOS) and Perfluorooctanoate (PFOA) in Groundwater: Kinetic Effects of Matrix Inorganics" *Environmental Science and Technology* 44: 445–450.

Cheng, Yu H., et al. 2020. "Metal–Organic Framework-Based Microfluidic Impedance Sensor Platform for Ultrasensitive Detection of Perfluorooctanesulfonate." *ACS Applied Materials & Interfaces* 12.9: 10503–10514.

Crittenden, John C., et al. 1991. "Predicting GAC performance with rapid small-scale column tests." *Journal-American Water Works Association* 83.1: 77–87.

Deng, Shubo, et al. 2010. "Removal of perfluorooctane sulfonate from wastewater by anion exchange resins: effects of resin properties and solution chemistry." *Water Research* 44.18: 5188–5195.

Deng, Yang, and Renzun Zhao 2015. "Advanced Oxidation Processes (AOPs) in Wastewater Treatment." *Curr Pollution Rep* 1: 167–176.

Dickenson, E. R. V., and C. Higgins 2016. "Treatment mitigation strategies for poly-and perfluoroalkyl substances." Water Research Foundation Web Report 4322.

Dickson, Michael D. 2013. "Method for treating industrial waste". Australia patent application 2,012,289,835.

Dickson, Michael D. 2014. United States US 20140190896A1 Patent Application Publication.

Du, Ziwen, et al. 2014. "Adsorption behavior and mechanism of perfluorinated compounds on various adsorbents—A review." *Journal of hazardous materials* 274: 443–454.

Ebersbach, Ina, et al. 2016. "An alternative treatment method for fluorosurfactant-containing wastewater by aerosol-mediated separation." *Water Research* 101: 333–340.

Eriksson, Ulrika, Peter Haglund, and Anna Kärrman 2017. "Contribution of precursor compounds to the release of per-and polyfluoroalkyl substances (PFAS) from wastewater treatment plants (WWTPs)." *Journal of Environmental Sciences* 61: 80–90.

Fang, Cheng, et al. 2019. "Removal of PFAS from aqueous solution using PbO2 from lead-acid battery." *Chemosphere* 219: 36–44.

Field, Jennifer A., and Jimmy Seow 2017. "Properties, occurrence, and fate of fluorotelomer sulfonates." *Critical Reviews in Environmental Science and Technology* 47.8: 643–691.

Franke, Vera, et al. 2019. "Efficient removal of per-and polyfluoroalkyl substances (PFAS) in drinking water treatment: nanofiltration combined with active carbon or anion exchange." *Environmental Science: Water Research & Technology* 5.11: 1836–1843.

Gole, Vitthal L., et al. 2018a. "Treatment of perfluorooctane sulfonic acid (PFOS) using a large-scale sonochemical reactor" *Separation and Purification Technology* 194, 104–110.

Gole, Vitthal L., et al. 2018b. "Sono-chemical treatment of per- and poly-fluoroalkyl compounds in aqueous film-forming foams by use of a large-scale multi-transducer dual-frequency based acoustic reactor" *Ultrasonics - Sonochemistry* 45: 213–222.

Gomez-Ruiz, Beatriz, et al. 2017. "Efficient electrochemical degradation of poly-and perfluoroalkyl substances (PFASs) from the effluents of an industrial wastewater treatment plant." *Chemical Engineering Journal* 322: 196–204.

Gude, Veera Gnaneswar. 2016. "Desalination and sustainability–an appraisal and current perspective." *Water Research* 89: 87–106.

González-García, José, et al. 2010. "Sonochemical treatment of water polluted by chlorinated organocompounds. A review." *Water* 2.1: 28–74.

Hao, Feifei, et al., 2014. "Intensification of sonochemical degradation of ammonium perfluorooctanoate by persulfate oxidant" *Ultrasonics Sonochemistry* 21: 554–558.

Health Canada 2019. "*Water Talk – Perfluoroalkylated substances in drinking water*" Modified 2019-06-04 Available at: https://www.canada.ca/en/services/health/publications/healthy-living/water-talk-drinking-water-screening-values-perfluoroalkylated-substances.html (accessed November 15, 2020).

Hori, Hisao, et al. 2005. "Efficient decomposition of environmentally persistent perfluorocarboxylic acids by use of persulfate as a photochemical oxidant." *Environmental Science & Technology* 39.7: 2383–2388.

Hori, Hisao, et al. 2012. "Efficient decomposition of perfluoroether carboxylic acids in water with a combination of persulfate oxidant and ultrasonic irradiation." *Journal of Fluorine Chemistry* 141: 5–10.

Horst, John, et al. 2018. "Water Treatment Technologies for PFAS: The Next Generation." *Groundwater Monitoring & Remediation* 38.2: 13–23.

Horst, John, et al. 2020. "Understanding and Managing the Potential By-Products of PFAS Destruction." *Groundwater Monitoring & Remediation* 40. 2.

Houtz, Erika F., and David L. Sedlak 2012. "Oxidative conversion as a means of detecting precursors to perfluoroalkyl acids in urban runoff." *Environmental Science & Technology* 46.17: 9342–9349.

Houtz, Erika F., et al. 2013. "Persistence of perfluoroalkyl acid precursors in AFFF-impacted groundwater and soil." *Environmental Science & Technology* 47.15: 8187–8195.

Houtz, Erika F., et al. 2016. "Poly-and perfluoroalkyl substances in wastewater: Significance of unknown precursors, manufacturing shifts, and likely AFFF impacts." *Water Research* 95: 142–149.

Huang, Po-Jung, et al. 2018. "Reusable functionalized hydrogel sorbents for removing long- and short-chain perfluoroalkyl acids (PFAAs) and GenX from aqueous solution." *ACS Omega* 3.12: 17447–17455.

Huang, Qingguo 2019. "Final Report: A Novel Reactive Electrochemical Membrane System for Treatment of Mixed Contaminants". *SERDP Project ER-2717*. Available at: https://www.serdp-estcp.org/Program-Areas/Environmental-Restoration/Contaminated-Groundwater/ER-2717/ER-2717-FR (accessed November 15, 2020).

IAC 2019. *Illinois Administrative Code. Title 35, Part 620*, 19 December 2019. DRAFT. Available at: https://www2.illinois.gov/epa/public-notices/Documents/General%20Notices/Part%20620.pdf (accessed November 15, 2020).

Interstate Technology Regulatory Council (ITRC) 2018. "*Remediation Technologies and Methods for Per- and Polyfluoroalkyl Substances (PFAS)*". Available at: https://pfas-1.itrcweb.org/fact_sheets_page/pfas_fact_sheet_remediation_3_15_18.pdf#:~:text=Remediation%20Technologies%20and%20Methods%20for%20Per-%20and%20Polyfluoroalkyl,flow-through%20vessels%20generally%20operated%20in%20series%20%28lead-lag%20configuration%29 (accessed November 15, 2020).

Ju, Xiaodong, et al. 2008. "Perfluorinated surfactants in surface, subsurface water and microlayer from Dalian coastal waters in China." *Environmental Science & Technology* 42: 3538–3542.

Karim, M. A., and J. L. Mark 2017. "A preliminary comparative analysis of MBR and CAS wastewater treatment systems." *Int J Water Wastewater Treat* 3.2.

Kothawala, Dolly N., et al. 2017. "Influence of dissolved organic matter concentration and composition on the removal efficiency of perfluoroalkyl substances (PFAS) during drinking water treatment." *Water Research* 121: 320–328.

Kumarasamy, Elango, et al. 2019. "Ionic Fluorogels for Remediation of Per-and Polyfluorinated Alkyl Substances from Water." *ACS Central Science* 6: 487–492.

Lee, Yu-Chi, et al. 2016. "Efficient sonochemical degradation of perfluorooctanoic acid using periodate" *Ultrasonics Sonochemistry* 31: 499–505.

Lee, Yu-Chi, et al. 2017. "Recovery of perfluorooctane sulfonate (PFOS) and perfluorooctanoate (PFOA) from dilute water solution by foam flotation." *Separation and Purification Technology* 173: 280–285.

Lei, Yong-Jia, et al. 2020. "Synergistic degradation of PFAS in water and soil by dual-frequency ultrasonic activated persulfate." *Chemical Engineering Journal* 388: 124215.

Li, Yuxiang, et al. 2017. "A mesoporous cationic thorium-organic framework that rapidly traps anionic persistent organic pollutants." *Nature communications* 8.1: 1–11.

Lifka, Jan, Bernd Ondruschka, and Jörg Hofmann 2003. "The use of ultrasound for the degradation of pollutants in water: aquasonolysis–a review." *Engineering in life sciences* 3.6: 253–262.

Lin, Hui, et al. 2015. "Efficient sorption and removal of perfluoroalkyl acids (PFAAs) from aqueous solution by metal hydroxides generated in situ by electrocoagulation." *Environmental Science & Technology* 49.17: 10562–10569.

Liu, Yang, et al. 2018. "Removal of perfluorooctanoic acid in simulated and natural waters with different electrode materials by electrocoagulation." *Chemosphere* 201: 303–309.

Maimaiti, Ayiguli, et al. 2018. "Competitive adsorption of perfluoroalkyl substances on anion exchange resins in simulated AFFF-impacted groundwater." *Chemical Engineering Journal* 348: 494–502.

Martinez-Huitle, Carlos A., and Sergio Ferro 2006. "Electrochemical oxidation of organic pollutants for the wastewater treatment: direct and indirect processes." *Chemical Society Reviews* 35.12: 1324–1340.

Merino, Nancy, et al. 2016. "Degradation and removal methods for perfluoroalkyl and polyfluoroalkyl substances in water." *Environmental Engineering Science* 33.9: 615–649.

McCleaf, Philip, et al. 2017. "Removal efficiency of multiple poly-and perfluoroalkyl substances (PFAS) in drinking water using granular activated carbon (GAC) and anion exchange (AE) column tests." *Water Research* 120: 77–87.

Meng, Pingping, et al. 2014. "Role of air bubbles overlooked in the adsorption of perfluorooctanesulfonate on hydrophobic carbonaceous adsorbents." *Environmental science & technology* 48: 13785–13792.

Meng, Pingping, et al. 2017. "Superhigh adsorption of perfluorooctane sulfonate on aminated polyacrylonitrile fibers with the assistance of air bubbles." *Chemical Engineering Journal* 315: 108–114.

Meng, Pingping, et al. 2020. "Role of the air-water interface in removing perfluoroalkyl acids from drinking water by activated carbon treatment." *Journal of Hazardous Materials*, 386.

Mollah, Mohammad Y. A., et al. 2004. "Fundamentals, present and future perspectives of electrocoagulation." *Journal of hazardous materials* 114.1–3: 199–210.

Moriwaki, Hiroshi, et al. 2005. "Sonochemical Decomposition of Perfluorooctane Sulfonate and Perfluorooctanoic Acid." *Environmental Science and Technology* 39: 3388–3392.

Panchangam, Sri Chandana, et al. 2009. "Sonication-assisted photocatalytic decomposition of perfluorooctanoic acid" *Chemosphere* 75: 654–660.

Park, Hyunwoong, et al. 2011. "Reductive degradation of perfluoroalkyl compounds with aquated electrons generated from iodide photolysis at 254 nm." *Photochemical & Photobiological Sciences* 10.12: 1945–1953.

Phan Thi, Lan-Anh, et al. 2014. "Enhancing decomposition rate of perfluorooctanoic acid by carbonate radical assisted sonochemical treatment." *Ultrasonics Sonochemistry* 21: 1875–1880.

Pramanik, Biplob Kumar, Sagor Kumar Pramanik, and Fatihah Suja 2015. "A comparative study of coagulation, granular-and powdered-activated carbon for the removal of perfluorooctane sulfonate and perfluorooctanoate in drinking water treatment." *Environmental technology* 36.20: 2610–2617.

Qian, Jin, et al. 2017. "Perfluorooctane sulfonate adsorption on powder activated carbon: effect of phosphate (P) competition, pH, and temperature." *Chemosphere* 182: 215–222.

Radjenović, Jelena, et al. 2007. "Membrane bioreactor (MBR) as an advanced wastewater treatment technology." In D. Barcelo and M. Petrovic (Eds.) *Emerging contaminants from industrial and municipal waste*. Springer, Berlin, Heidelberg. 37–101.

Rayne, Sierra, Kaya Forest, and Ken J. Friesen 2008. "Congener-specific numbering systems for the environmentally relevant C4 through C8 perfluorinated homologue groups of alkyl sulfonates, carboxylates, telomer alcohols, olefins, and acids, and their derivatives." *Journal of Environmental Science and Health Part A* 43.12: 1391–1401.

Rodriguez-Freire, Lucia, et al. 2015. "Effect of sound frequency and initial concentration on the sonochemical degradation of perfluorooctane sulfonate (PFOS)." *Journal of hazardous materials* 300: 662–669.

Rodriguez-Freire, Lucia, et al. 2016. "Sonochemical degradation of perfluorinated chemicals in aqueous film-forming foams." *Journal of Hazardous Materials* 317: 275–283.

Ross, Ian, et al. 2018. "A review of emerging technologies for remediation of PFAS." *Remediation Journal* 28.2: 101–126.

Schaefer, Charles E., et al. 2017. "Electrochemical treatment of perfluorooctanoic acid and perfluorooctane sulfonate: Insights into mechanisms and application to groundwater treatment." *Chemical Engineering Journal* 317: 424–432.

Sekiguchi, Kazuhiko., et al. 2017. "Combined sonochemical and short-wavelength UV degradation of hydrophobic perfluorinated compounds." *Ultrasonics - Sonochemistry* 39: 87–92.

Singh, Raj Kamal. et al. 2019. "Rapid Removal of Poly- and Perfluorinated Compounds from Investigation-Derived Waste (IDW) in a Pilot-Scale Plasma Reactor." *Environmental Science and Technology* 53: 11375–11382.

Siriwardena, Dinusha P., et al. 2019. "Influence of groundwater conditions and co-contaminants on sorption of perfluoroalkyl compounds on granular activated carbon." *Remediation Journal* 29.3: 5–15.

Stasinakis, Athanasios S 2012. "Review on the fate of emerging contaminants during sludge anaerobic digestion." *Bioresource technology* 121: 432–440.

Stebel, Eva K., et al. (2019). "Absorption of short-chain to long-chain perfluoroalkyl substances using swellable organically modified silica." *Environmental Science: Water Research & Technology* 5.11: 1854–1866.

Stockholm Convention on Persistent Organic Pollutants 2007. *Guidelines on Best Available Techniques and Provisional Guidance on Best Environmental Practices relevant to Article 5 and Annex C of the Stockholm Convention on Persistent Organic Pollutants: Cement kilns firing hazardous waste* (Geneva, Switzerland: United Nations Environmental Programme). May. Available at: http://chm.pops.int/Portals/0/Repository/batbep_guideline08/UNEP-POPS-BATBEP-GUIDE-08-2.English.PDF (accessed November 15, 2020).

Storch, P.J. et al. 2018. "*Full-Scale Treatment of PFAs-Impacted Wastewater Using Ozofractionation*". *Presentation, 2018 Battelle Conference Proceedings*, Palm Springs, CA. Available at: https://www.battelle.org/docs/default-source/conference-proceedings/2019-sediments-conference-proceedings/a7.-characterization-and-remediation-of-pfas-contaminated-sediments-media/532.pdf?sfvrsn=75dc9522_2 (accessed November 15 2020).

Stratton, Gunnar R., et al. 2017. "Plasma-Based Water Treatment: Efficient Transformation of Perfluoroalkyl Substances in Prepared Solutions and Contaminated Groundwater" *Environmental Science and Technology* 51: 1643–1648.

Tang, Chuyang Y., et al. 2006. "Use of reverse osmosis membranes to remove perfluorooctane sulfonate (PFOS) from semiconductor wastewater." *Environmental Science & Technology* 40: 7343–7349.

Thompson, Jack, et al. 2011. "Removal of PFOS, PFOA and other perfluoroalkyl acids at water reclamation plants in South East Queensland Australia." *Chemosphere* 82: 9–17.

Trautmann, A. M., et al. 2015. "Electrochemical degradation of perfluoroalkyl and polyfluoroalkyl substances (PFAS) in groundwater." *Water Science and Technology* 71.10: 1569–1575.

Trojanowicz, Marek, et al. 2018. "Advanced oxidation/reduction processes treatment for aqueous perfluorooctanoate (PFOA) and perfluorooctanesulfonate (PFOS)–a review of recent advances." *Chemical Engineering Journal* 336: 170–199.

Ueno, Hiroyuki, et al. 1997. "Destruction of chlorofluorocarbons in a cement kiln." *Journal of the Air & Waste Management Association* 47: 1220–1223.

USEPA 2018 Method 537.1 Determination of Selected Per- and Polyflourinated Alkyl Substances in Drinking Water by Solid Phase Extraction and Liquid Chromatography/Tandem Mass Spectrometry (LC/MS/MS). EPA 600-R-18-352, Office of Research and Development, USEPA, Cincinnati, Ohio. Available at: https://cfpub.epa.gov/si/si_public_file_download.cfm?p_download_id=537290 (accessed November 15, 2020).

USEPA (2019). "*Per- and Polyfluoroalkyl Substances (PFAS): Incineration to Manage PFAS Waste Streams.*" Available at: https://www.epa.gov/sites/production/files/2019-09/documents/technical_brief_pfas_incineration_ioaa_approved_final_july_2019.pdf (accessed November 15, 2020).

USEPA 2019a. "Method 533: Determination of Per- and Polyfluoroalkyl Substances in Drinking Water by Isotope Dilution Anion Exchange Solid Phase Extraction and Liquid Chromatography/Tandem Mass Spectrometry". *EPA 815-B-19-020.* Washington, D.C. Available at: https://www.epa.gov/dwanalyticalmethods/method-533-determination-and-polyfluoroalkyl-substances-drinking-water-isotope (accessed November 15, 2020).

USEPA 2019b. "Validated Test Method 8327: Per-and Polyfluoroalkyl Substances (PFAS) Using External Standard Calibration and Multiple Reaction Monitoring (MRM) Liquid Chromatography/Tandem Mass Spectrometry (LC/MS/MS)". *Revision 0.* June 2019. Washington, D.C. Available at: https://www.epa.gov/hw-sw846/validated-test-method-8327-and-polyfluoroalkyl-substances-pfas-using-external-standard (accessed November 15, 2020).

USEPA 2020. "Interim Guidance on the Destruction and Disposal of Perfluoroalkyl and Polyfluoroalkyl Substances and Materials Containing Perfluoroalkyl and Polyfluoroalkyl Substances". *Interim Guidance for Public Comment.* December 2020. Washington, DC. Available at: https://www.regulations.gov/document/EPA-HQ-OLEM-2020-0527-0002 (accessed March 5, 2021).

Vecitis, Chad D., et al. 2008. "Kinetics and mechanism of the sonolytic conversion of the aqueous perfluorinated surfactants, perfluorooctanoate (PFOA), and perfluorooctane sulfonate (PFOS) into inorganic products." *The Journal of Physical Chemistry A* 112.18: 4261–4270.

Vecitis, Chad D., et al. 2010. "Sonochemical degradation of perfluorooctanesulfonate in aqueous film-forming foams." *Environmental science & technology* 44.1: 432–438.

Wang, Fei, et al. 2011. "Influence of calcium hydroxide on the fate of perfluorooctanesulfonate under thermal conditions." *Journal of hazardous materials* 192.3: 1067–1071.

Wang, Fei, et al. 2013. "Mineralization behavior of fluorine in perfluorooctanesulfonate (PFOS) during thermal treatment of lime-conditioned sludge." *Environmental science & technology* 47.6: 2621–2627.

Watanabe, Nobuhisa, et al. 2016. "Residual organic fluorinated compounds from thermal treatment of PFOA, PFHxA and PFOS adsorbed onto granular activated carbon (GAC)." *Journal of Material Cycles and Waste Management* 18.4: 625–630.

Westreich, Philippe, et al. 2018. "The removal of short-chain and long-chain perfluoroalkyl acids and sulfonates via granular activated carbons: A comparative column study." *Remediation Journal* 29.1: 19–26.

Wood, James, et al. 2020. "Ultrasonic degradation of perfluorooctane sulfonic acid (PFOS) correlated with sonochemical and sonoluminescence characterisation." *Ultrasonics - Sonochemistry* 68: 105–196.

Xiao, Feng, Matt F. Simcik, and John S. Gulliver (2013). "Mechanisms for removal of perfluorooctane sulfonate (PFOS) and perfluorooctanoate (PFOA) from drinking water by conventional and enhanced coagulation." *Water research* 47.1: 49–56.

Xiao, Leilei, et al. 2017a. "β-Cyclodextrin polymer network sequesters perfluorooctanoic acid at environmentally relevant concentrations." *Journal of the American Chemical Society* 139: 7689–7692.

Xiao, Xin, et al. 2017b. "Sorption of poly-and perfluoroalkyl substances (PFAS) relevant to aqueous film-forming foam (AFFF)-impacted groundwater by biochars and activated carbon." *Environmental science & technology* 51.11: 6342–6351.

Yamada, T., and P. H. Taylor 2003. "Laboratory Scale Thermal Degradation of Perfluoro-Octanyl Sulfonate and Related Precursors." Final Report, 3M Company. UDR-TR-03-00044, USEPA EDocket OPPT-2003-0012-0151.

Yang, Bo, et al. 2015. "Highly efficient electrochemical degradation of perfluorooctanoic acid (PFOA) by F-doped Ti/SnO2 electrode." *Journal of hazardous materials* 299: 417–424.

Yang, Bo, et al. 2016. "Efficient removal of perfluoroalkyl acids (PFAAs) from aqueous solution by electrocoagulation using iron electrode." *Chemical Engineering Journal* 303: 384–390.

Yang, Zi, et al. 2019. "A review on reverse osmosis and nanofiltration membranes for water purification." *Polymers* 11.8: 1252.

Yu, Qiang, et al. 2009. "Sorption of perfluorooctane sulfonate and perfluorooctanoate on activated carbons and resin: kinetic and isotherm study." *Water research* 43.4: 1150–1158.

Zaggia, Alessandro, et al. 2016. "Use of strong anion exchange resins for the removal of perfluoroalkylated substances from contaminated drinking water in batch and continuous pilot plants." *Water research* 91: 137–146.

Zeng, Chao, et al. 2020. "Removing per-and polyfluoroalkyl substances from groundwaters using activated carbon and ion exchange resin packed columns." *AWWA Water Science 2*. Available at: https://awwa.onlinelibrary.wiley.com/doi/10.1002/aws2.1172 (accessed November 15, 2020).

10 Effectiveness of Point-of-Use/Point-of-Entry Systems to Remove PFAS from Drinking Water

Craig Patterson, Jonathan B. Burkhardt
US Environmental Protection Agency, Cincinnati, Ohio

Donald Schupp, E. Radha Krishnan
Aptim Federal Services, Cincinnati, Ohio

Stephen Dyment
US Environmental Protection Agency, Office of Research and Development, Denver, Colorado

Steven Merritt
US Environmental Protection Agency Region 8, Denver, Colorado

Lawrence Zintek, and Danielle Kleinmaier
US Environmental Protection Agency Region 5, Chicago, Illinois

CONTENTS

10.1 Introduction....212
10.2 Methods and Materials....213
10.2.1 Test Water Preparation....213
10.2.2 Stability Study....215
10.2.3 Analytical Methods....215

10.2.4 Analytical Sampling Plan....215
10.2.5 RO System Design....216
10.2.5.1 iSpring RO System....216
10.2.5.2 HydroLogic RO System....216
10.2.5.3 Flexeon RO System....217
10.2.5.4 Booster Pumps....217
10.2.5.5 Modifications For Point-of-Entry Use....217
10.2.6 RO System Sampling Plan....218
10.2.7 GAC System Design....218
10.3 Results and Discussion....223
10.3.1 RO System Studies....223
10.3.1.1 RO #1 Test....223
10.3.1.2 RO #2 Test....223
10.3.1.3 RO #3 Test....225
10.3.1.4 RO Discussion....225
10.3.2 GAC Media Studies....226
10.3.2.1 GAC #1 Test....226
10.3.2.2 GAC #2 Test....227
10.3.2.3 GAC Media Discussion....228
10.3.3 RO and GAC Residuals Management....228
10.4 Modeling of GAC Results....229
10.4.1 Parameter Estimation Process....229
10.4.2 Pore Surface Diffusion Model....229
10.4.3 Application of the Model....230
10.5 Conclusions....231
Acknowledgments....233
References....233

10.1 INTRODUCTION

The widespread use of per- and poly-fluoroalkyl substances (PFAS) over the last half-century has resulted in these compounds being detected in environmental media. Although much is unknown, we do know that certain PFAS are toxic, resistant to degradation, and bioaccumulate in the food web and subsequently in animals and humans. At many sites, it is also likely that PFAS precursors and their degradation products are present; however, limited occurrence data are available due to the lack of comprehensive analytical method capabilities.

Recently, this has been a growing concern for EPA's Superfund program as these compounds have been identified in groundwater at many high-profile sites across the country, including the Widefield aquifer near Colorado Springs, Colorado. At this site, perfluorooctanoic acid (PFOA), perfluorooctane sulfonate (PFOS), perfluoroheptanoic acid (PFHpA), perfluorobutane sulfonate (PFBS), perfluorohexane sulfonic acid (PFHxS), and perfluorononanoic acid (PFNA) have been detected in the groundwater. Three large water utilities, seven smaller state regulated utilities, and hundreds of private wells are currently under investigation. At the three large water utilities, non-impacted surface water from the Pueblo Reservoir is being blended with impacted groundwater sources to meet EPA's health advisory level (HAL) for combined PFOA and PFOS concentrations. The seven smaller systems and numerous private well owners use the impacted aquifer as their sole source of drinking water. Based on concentrations found in the Widefield Aquifer and EPA's 2016 HAL of 70 nanograms (ng)/L of combined PFOA and PFOS concentrations (Federal Register 2016), these Widefield aquifer systems are at risk.

Private well owners and small municipalities, where the presence of PFAS are known, are subject to multi-year or even decade-long site-specific investigations, source determinations, risk assessments, and remedial designs and constructions. Additionally, many challenges exist at sites managing PFAS contaminants, including the lack of EPA analytical methods, for the detection of PFAS, their precursors and degradation products, and unknown toxicological impact of many PFAS, and remediation or treatment technologies.

Many private well owners are seeking information on commercially-available resources to treat PFAS-contaminated drinking water. Dissemination of the knowledge of the capabilities of commercially-available reverse osmosis (RO) and granular activated carbon (GAC) point-of-use (POU) and point-of-entry (POE) treatment systems on PFAS removal would be valuable to those well owners. To match their needs, the PFAS removal systems that were selected for this study were designated to be in accordance with the National Science Foundation (NSF) Standard P473 (NSF 2016).

10.2 METHODS AND MATERIALS

To assess the removal effectiveness of the target PFAS present in the Widefield Aquifer region, treatability studies were conducted using three commercially-available reverse osmosis (RO) treatment systems and two granulated activated carbon (GAC) media at the EPA Test and Evaluation (T&E) Facility in Cincinnati, Ohio. RO systems and GAC media were tested using test water simulating the PFAS composition and water quality in the Widefield Aquifer region. PFAS samples were analyzed by liquid chromatography/tandem mass spectrometry (LC/MS/MS) at the EPA Region 5 Chicago Regional Laboratory (CRL) in Chicago, Illinois.

Prior to RO system and GAC media tests, initial PFAS stability tests were conducted using the same test water to determine the stability of the PFAS compounds in the test water in the 5,000-gallon stainless steel RO feed tank and the 55-gallon stainless steel GAC media feed drum used in the study. Upon confirmation of stability, subsequent studies were completed on three RO POU/POE treatment systems were tested based on aspects of NSF Standard P473 (NSF 2016). Also, two GAC media were evaluated using Rapid Small-Scale Column Tests (RSSCTs) per ASTM D6586-03 (ASTM 2008).

10.2.1 Test Water Preparation

The feed water was prepared to simulate the PFAS contamination of the Widefield Aquifer by adding the PFAS concentrations shown in Table 10.1. These six PFAS target concentrations are worst case (maximum) concentrations compiled from

TABLE 10.1
Test water target PFAS composition.

CAS Number	PFAS Compounds	Carbon Chain Length	Target Concentration
375-95-1	Perfluorononanoic Acid (PFNA)	C9	200 ng/L
335-67-1	Perfluorooctanoic Acid (PFOA)	C8	800 ng/L
1763-23-1	Perfluorooctane Sulfonate (PFOS)	C8	1,600 ng/L
375-85-9	Perfluoroheptanoic Acid (PFHpA)	C7	200 ng/L
3871-99-6	Perfluorohexane Sulfonate (PFHxS)	C6	1,000 ng/L
375-73-5	Perfluorobutane Sulfonate (PFBS)	C4	300 ng/L

TABLE 10.2
Test water target water quality characteristics.

General Chemistry Water Parameters	
Temperature (°C)	RO: 25 ± 1°C, GAC: 20 ± 2.5°C
pH (pH Units)	8.2 ± 0.5
Turbidity (NTU)	<1 NTU
Free chlorine (mg/L)	<0.2 mg/L
TOC (mg/L)	RO: not specified (not adjusted) GAC: >1 mg/L (added as dehydrated NOM)
TDS (mg/L)	RO and GAC: 500 mg/L (added as NaCl)
Hardness (mg/L)	RO: 300 mg/L $CaCO_3$ (added as potassium chloride [KCl], magnesium sulfate [$MgSO_4$], sodium bicarbonate [$NaHCO_3$] and calcium sulfate [$CaSO_4 \cdot 2H_2O$]), GAC: not specified.

historic records (2013–2016) of Widefield Aquifer region water samples (CDPHE 2016). To align with the NSF P473 specified 2:1 PFOS:PFOA ratio, the PFOA feed concentration was increased from 200 ng/L to 800 ng/L.

Cincinnati tap water was added to a 5,000-gallon stainless steel tank equipped with a mixer, recirculation pump, and a chiller with a shell-and-tube heat exchanger to control feed water temperature. Sodium thiosulfate (1 mg/L) was added to dechlorinate the test water. The target water quality characteristics specified in Table 10.2 were maintained throughout the RO and GAC test runs to adhere to the requirements of NSF P473. The test water targets for pH (8.2), total dissolved solids (TDS) (500 mg/L) and hardness (300 mg/L) were maintained to align with historic maximum Widefield Aquifer well water sampling results (1992–2016) to determine their impact on the longevity and removal capabilities of RO membranes and GAC media.

After the feed water additives dissolved overnight, the target PFAS were measured, mixed, and dissolved in 1 L of deionized water. The PFAS were then added to the 5,000-gallon tank and allowed to mix for three hours for use for the RO treatability study. Before the start of the RO study, approximately 50 gallons of the feed water was drained into a 55-gallon stainless steel drum equipped with a mixer for use in the GAC RSSCT evaluations. Dehydrated Ohio River water natural organic matter (NOM) was rehydrated, filtered, and added to the 55-gallon drum to increase the total organic carbon (TOC) of the water to above 1 mg/L for the GAC media studies (Pressman et al., 2012).

10.2.2 Stability Study

A PFAS stability test was run without RO system and GAC adsorbents over an eight-day period to assess PFAS degradability and sorption in the test apparatus. The concentrations of the six target PFAS remained stable in both the 5,000-gallon tank and the 55-gallon drum over the eight-day test period. The PFOS concentration was significantly lower than the target concentration due to a lower spike amount inadvertently used when adding the quantity of PFOS to the test water. The PFOS spike amount was corrected in the following test runs. The PFBS concentration was below the detection limit possibly due to the stock solution used during the stability test. A different source of PFBS was secured for the following test runs. These adjustments resolved the concentration discrepancies with PFOS and PFBS during the actual test runs. Although no PFAS degradation or sorption was noted in the stability study, the initial target PFAS concentrations were increased by 25 percent for the treatability studies as a safety factor to ensure adequate PFAS concentrations for the tests.

10.2.3 Analytical Methods

PFAS analysis was performed following the EPA Region 5 Standard Operating Procedure for the Analysis of Polyfluorinated Compounds of Interest to Office of Superfund Remediation and Technology Innovation (OSRTI) in Water, Sludge, Influent, Effluent, and Wastewater by Multiple Reaction Monitoring LC/MS/MS (USEPA (U.S. Environmental Protection Agency), 2017). Per the EPA Region 5 PFAS Analytical SOP, 15-milliliter (mL) polypropylene sample containers were weighed before and after sample collection to allow the laboratory to determine the exact sample amount used during analysis. The target volume of sample per polypropylene container was 5 mL. Because the entire 5 mL sample volumes were analyzed, triplicate samples were collected from both influent and effluent for additional quality assurance/quality control (QA/QC) analyses. With a 28-day sample holding time, grab-water samples for PFAS analysis were shipped for overnight delivery to the EPA Region 5 CRL in Chicago, IL. All other water quality analyses were performed at the EPA T&E Facility in Cincinnati, OH.

10.2.4 Analytical Sampling Plan

Six target PFAS compounds were analyzed in both RO system and GAC media influent and effluent water samples to determine PFAS removal efficiency. The sample plan included analysis of 188 PFAS samples at the CRL in EPA Region 5 from one stability run and three test runs. Samples were collected from the 5,000-gallon stainless steel mix tank and the 55-gallon stainless steel drum and analyzed for PFAS, TOC, TDS, hardness, turbidity, pH, temperature, and free available chlorine. Only one free chlorine sample (influent and effluent) was collected at the beginning of each test to ensure that the test water was adequately dechlorinated.

TABLE 10.3
Summary of RO system specifications.

RO system	iSpring RCS5T	HydroLogic Evolution RO1000	Flexeon LP-700
Rated Capacity[a]	500 gpd (0.35 gpm)	1,000 gpd (0.7 gpm)	700 gpd (0.5 gpm)
Filters Included	Sediment filter Carbon pre-filter CTO filter RO membrane Carbon post-filter	Carbon pre-filter 2 RO membranes	Sediment filter Carbon pre-filter 2 RO membranes Carbon post-filter
System Recovery[a]	50%	50%, using 1:1 fitting	38%
Booster Pump	Yes	No	No
Connections	3/8″ Inlet ¼″ Outlet (tubing included)	½″ Inlet 3/8″ Outlet (tubing included)	3/8″ Inlet and Outlet (tubing not included)
Self-Supporting	Yes	Yes	No
Size (L x W x H)	8.5″ × 15″ × 18.5″	20.5″ × 11″ × 10″	18″ × 10.5″ × 32″
Weight	31 lbs	16 lbs	38 lbs

[a] The systems' rated pressure and efficiency are a function of the temperature and pressure of the feed water, gpd=gallons/day, gpm=gallons/minute.

10.2.5 RO System Design

The installation of each unit was very straightforward. Each of the RO systems was pre-assembled by the manufacturer prior to shipment and used quick connects to attach inlet and outlet tubing. The RO systems only required connections to feed water and discharge points for operation. Specifications for the three RO systems tested are summarized in Table 10.3. Table 10.4 shows the replacement part numbers and replacement frequency for the replacement filters and membranes in each of the three RO systems.

10.2.5.1 iSpring RO System

The five-stage iSpring RO System was free-standing and was pre-plumbed with a booster pump and color-coded feed and discharge tubing when it was removed from the box. Except for the RO membrane, the filters and cartridges described in Table 10.4 were pre-installed. Wrenches were included to install replacement membranes and filters.

10.2.5.2 HydroLogic RO System

The two-stage HydroLogic RO System was free-standing and included fittings for a 2:1 (green) or 1:1 (orange) waste to pure water ratio. The instructions stated that using the 1:1 fitting (chosen for this study) would provide a higher flow rate but could require more frequent replacement of the RO membranes. Three 90° elbow push fittings, the carbon pre-filter, and two RO membranes were installed with wrenches prior to use. Inlet and outlet tubing was color-coded and a ball valve at the end of the

TABLE 10.4
RO system replacement filters and membranes.

RO system	iSpring RCS5T	HydroLogic Evolution RO1000	Flexeon LP-700
Sediment filter	#FP15 (3–6 months)	Not Part of System	#200627 (12 months)
Carbon pre-filter	#FG15 (6 months)	#22043 (2,000 gallons of purified water)	#200658 (12 months)
Carbon block filter	#FC15 (6 months)	Not Part of System	Not Part of System
RO membranes	#MS5 (24 months)	#220445 (6 – 24 months) (requires 2)	#208802 (24 months) (requires 2)
Carbon post-filter	#FT15 (12 months)	Not Part of System	#200658 (12 months)

discharge line was provided for use when water was not needed. A booster pump had to be purchased separately and was installed in front of the HydroLogic RO system.

10.2.5.3 Flexeon RO System

The four-stage Flexeon RO System was pre-plumbed with filters when it was removed from the box. The Flexeon RO system was not self-supporting and did not come with a booster pump or inlet and outlet tubing. Except for the RO membrane, the filters and cartridges described in Table 10.4 were pre-installed. A booster pump was purchased separately and was installed in front of the Flexeon RO system.

10.2.5.4 Booster Pumps

The iSpring unit included an integrated booster pump to create a stable working environment for the RO membrane for situations when the water feed pressure was not sufficiently high (40–50 psi). A booster pump was purchased separately for the Hydrologic and Flexeon RO systems (HydroLogic Booster Pump Model No. HL29014). The booster pump was installed and included tubing, a pressure gauge, and switches that turned the pumps off when the water supply was stopped or when the discharge line was closed (by a faucet or valve). This would not be needed in a home setup if system pressure is available.

10.2.5.5 Modifications For Point-of-Entry Use

By themselves, the RO systems tested do not constitute a whole-house POE system; however, with some additional components, each of these RO systems could be a part of a POE system. The RO systems tested were designed to produce 500–1,000 gpd (0.35–0.70 gpm), a flow rate that is not suitable for many household activities (bath water, showers, washing machines, dishwashers). Also, the RO systems do not produce water at a significant pressure. However, it could serve as a POE system with the addition of a 200- to 500-gallon storage tank, a delivery pump, and a 20-gallon bladder tank. The storage tank would provide the necessary purified water to the home, and the delivery pump and booster tank would supply the necessary pressure. These RO systems would then be able to refill the storage tank to meet household water demands in real time. However, watering of lawns and

TABLE 10.5
RO system sampling plan.

Day #	Day of Week	Time of Day	Sample Hour	Time of Day	Sample Hour	Time of Day	Sample Hour
Day 1	Tues	AM	Startup*	Noon	4 hr	PM	8 hr
Day 2	Wed	AM	24 hr	Noon	30 hr	PM	36 hr
Day 3	Thurs	AM	48 hr	Noon	54 hr	PM	60 hr
Day 4	Fri	AM	72 hr	Noon	78 hr	PM	84 hr
Day 5	Sat	2 Day Stagnation Period*					
Day 6	Sun						
Day 7	Mon	AM	144 hr	PM	148 hr	PM	Shutdown*

* No samples collected

gardens and filling swimming pools with treated water would not be a viable option without a much larger GAC or RO system.

10.2.6 RO System Sampling Plan

The RO systems were originally planned to operate continuously for four days, shutdown for two days and then operate for another day as per NSF P473, as shown in Table 10.5 (NSF 2016). Due to the high use of water from the 5,000-gallon tank, the Flexeon and HydroLogic units were operated for two days, shutdown for four days and then operate for another day. Sampling occurred normally during the period when the units were operating.

The RO systems received test water from a 5,000-gallon stainless steel tank. The tank included a mixer and a recirculation pump that recycled the feed water from the bottom to the top of the tank. The feed to the RO system came off the recirculation line. A chiller with a shell and tube heat exchanger was located on the recirculation line to maintain the temperature of the tank at approximately 25°C. A schematic of the RO test system is shown in Figure 10.1. Figures 10.2, 10.3, and 10.4 show photographs of the iSpring RCS5T, HydroLogic Evolution RO1000, and the Flexeon LP-700 RO systems, respectively, at the EPA T&E Facility.

Grab samples were collected from each of the RO systems two or three times per day during the test period. Grab samples from the influent to the RO system were collected from the recirculation line on the 5,000-gallon tank. Grab samples from the effluent of the RO system were collected from the end of the permeate tubing. The flow rate of both the reject (concentrate) and the permeate water was measured throughout the test using a graduated cylinder. This allowed the calculation of the recovery, or efficiency, of the RO system.

10.2.7 GAC System Design

Two types of GAC media were evaluated using test water: (1) Calgon Filtrasorb 600 AR+, manufactured from select grades of bituminous coal; and (2) Evoqua 1230CX,

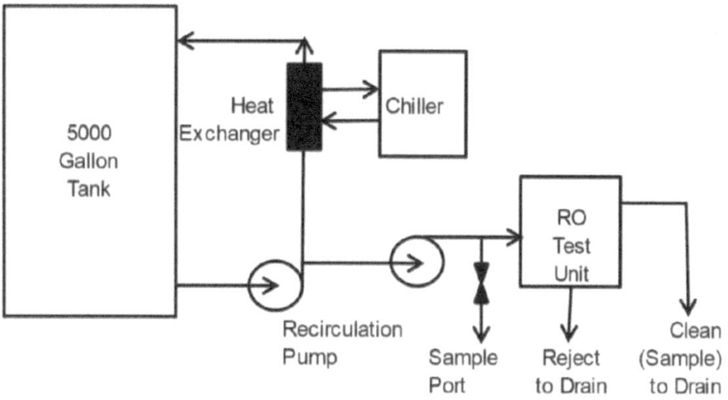

FIGURE 10.1 RO test schematic.

FIGURE 10.2 iSpring RCS5T RO system.

a coconut-based GAC as shown in Table 10.6. Two GAC media were tested separately using RSSCT methods described in ASTM Method D6586-03 (ASTM 2008). Samples were collected every 30 minutes over an eight-hour period.

The representative POE system for GAC filter was an H2O Filter Warehouse Model FT-1-GAC-20, 4–5 gpm Non-Backwashing Whole House Carbon Water Filter. This filter uses a 35″(H) × 9″(D) tank, 30 lbs (1 cubic foot) GAC, and is rated for 4 gpm with a treatment volume of 150,000 gallons. The RSSCT data were scaled up to approximate the use of GAC in this unit.

RSSCTs are based on fixed-bed mass transfer models. The mass transfer models are scaled down from the full-scale adsorber to a small column. Similarity of operation to that of large-scale adsorbers is assured by properly selecting the particle size,

FIGURE 10.3 HydroLogic Evolution RO1000 RO system (with optional booster pump).

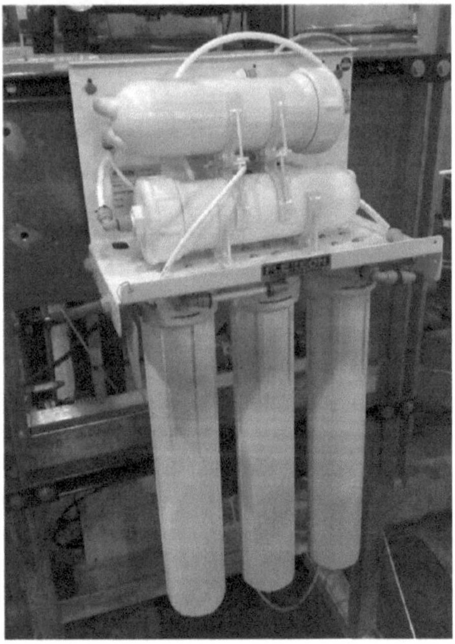

FIGURE 10.4 Flexeon LP-700 RO system.

TABLE 10.6
AC characteristics and RSSCT design parameters.

Parameter	Test 1	Test 2
GAC	Evoqua 1230CX	Calgon Filtrasorb 600 AR+
Source	Coconut	Bituminous Coal
Density	0.45 g/cm^3	0.62 g/cm^3
Porosity	0.47	0.39
Mesh Size	12 x 30	12 x 40
$EBCT_{LC}$	10 min	10 min
$d_{p,LC}$	1.150 mm	1.063 mm
$d_{p,SC}$	0.0825 mm	0.0825 mm
Scaling Factor	194.3	165.9
Q_{SC}	10 mL/min	10 mL/min
V_{SC}	0.515 mL	0.603 mL
M_{SC}	0.2294 g	0.3742 g

hydraulic loading, and empty bed contact time (EBCT) of the small adsorber. The RSSCT allows the tests to be conducted much more quickly and with less water than full-scale POE tests. Figure 10.5 shows a schematic of the RSSCT system. Figure 10.6 shows a photograph of the GAC RSSCT system at the EPA T&E Facility.

The EBCTs of the large-scale adsorber and the small-scale adsorber are related to the particle sizes of the GAC within the columns according to the following equation (Crittenden et al. 1991):

$$\frac{EBCT_{SC}}{EBCT_{LC}} = \left[\frac{d_{p,SC}}{d_{p,LC}}\right]^{2-x} = \frac{t_{SC}}{t_{LC}} \tag{10.1}$$

where *EBCT* is the empty bed contact time, d_p is the GAC particle size, and *t* is the elapsed time to conduct the tests for the RSSCT small columns (*SC*) and the POE

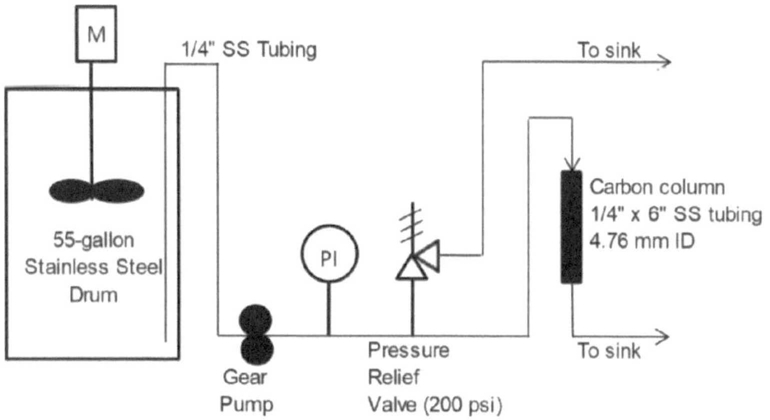

FIGURE 10.5 RSSCT schematic.

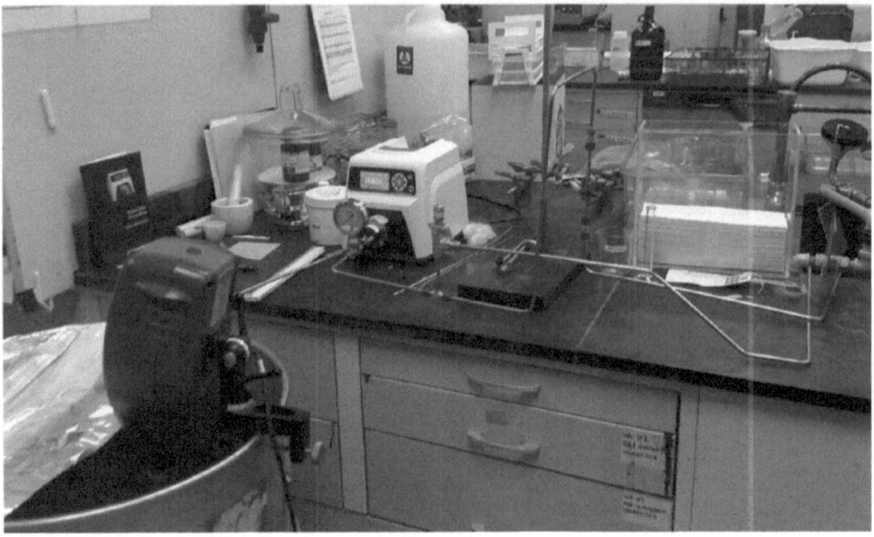

FIGURE 10.6 RSSCT system.

large columns (*LC*), respectively. *X* defines the dependence of the intraparticle diffusion coefficient on particle size. In this case, it is assumed that the intraparticle diffusivities do not change with particle size, so $X = 0$. Therefore, the time (minutes) of operation in the large column is proportional to the square of the ratio of the particle sizes in the large and small columns multiplied by the time of operation in the small column.

Other equations relating the large and small column operation include the following:

$$\frac{V_{SC}}{V_{LC}} = \frac{d_{p,LC}}{d_{p,\,SC}} \tag{10.2}$$

where *V* is the velocity in the *LC* and *SC* respectively, and

$$M_{SC} = EBCT_{LC}\left[\frac{d_{p.SC}}{d_{p,LC}}\right]^{2-x} Q_{SC} * \rho_{LC} \tag{10.3}$$

where *M* is the mass of GAC in the small column, *Q* is the flow rate in the small column, and ρ is the GAC density in the large column.

The RSSCT evaluation for this study used a 55-gallon stainless steel drum with a mixer as a holding tank for the feed water in a temperature-controlled laboratory. The feed water was pumped through the GAC column using a Masterflex® gear pump at approximately 10 mL/min. The flow rate was measured by collecting water in a graduated cylinder, and the flow rate was adjusted to maintain 10 mL/min throughout the study. One-quarter-inch stainless steel tubing was used throughout the system.

The inside diameter of the RSSCT column was 4.76 millimeter (mm). Each type of GAC was ground to a particle size between 170 and 200 mesh with a mortar and pestle. Fines were washed off the crushed GAC with deionized water, and the crushed GAC was dried in a 150°C oven for four hours. After weighing the crushed GAC for use in the RSSCT, the GAC was added to 40 mL of deionized water and boiled for 10 min to replace the air in the GAC's pores with water. Fiber glass wool was placed in the bottom of the column to keep the GAC inside the column.

Based on discussions with GAC vendors offering systems for PFAS removal, an EBCT of 10 min and a GAC RSSCT study duration of eight hours was selected with a 30-min sample collection frequency. An EBCT of 10 minutes was selected to provide adequate time for PFAS to absorb to GAC and to allow for reasonable GAC replacement frequencies for homeowners. Grab samples from the influent to the RSSCT were collected from the 55-gallon drum (Figure 10.5). Grab samples from the effluent of the RSSCT were collected from the end of the ¼-inch stainless steel tubing every 30 min over an eight-hour test period.

10.3 RESULTS AND DISCUSSION

10.3.1 RO System Studies

Three RO systems were tested: RO #1; iSpring RCS5T 500 gpd (0.35 gpm), RO #2; Hydrologic Evolution 1,000 gpd (0.7 gpm) and RO #3; Flexeon LP-700 gpd (0.5 gpm).

10.3.1.1 RO #1 Test

The first RO POU/POE system, the iSpring RCS5T (RO #1), was tested over a seven-day period at an average flow rate of 2285 mL/min. RO #1 was operated on Tuesday through Friday of the first week, shut down over the weekend, and restarted on Monday morning of the following week for the Day 7 samples.

All PFAS in the effluent were reduced below detection for the entire test period. RO #1 also reduced the turbidity, TOC, TDS, and hardness in the effluent. The average permeate flow rate from this unit was approximately 1020 mL/min, and the average concentrate (reject) flow rate from this unit was approximately 1265 mL/min. The recovery (flow rate of permeate/flow rate of influent water) of the RO #1 system was approximately 45.

10.3.1.2 RO #2 Test

The second RO POU/POE system, HydroLogic Evolution RO1000 (RO #2), was tested intermittently over a seven-day period. Because of the increased water volume necessary to test this unit at an average flow rate of 4895 mL/min, the unit was only operated on Tuesday and Wednesday of the first week before being shut down for the weekend. RO #2 was then restarted on Monday morning of the following week for the Day 7 samples.

Most of the PFAS samples were below the detection limit for the test. However, six of 42 RO #2 PFAS sample results were above the detection limit, as shown in Table 10.7 and Figure 10.7. Five of the PFAS were detected after startup on Day 7.

TABLE 10.7
Removal efficiency of RO #2 PFAS results greater than non-detect.

PFAS	Time (hr)	Influent Conc. (ng/L)	Effluent Conc. (ng/L)	Removal Efficiency (%)
PFOS	8	1100	22	98
PFOS*	144	1360	77	94.3
PFOA*	144	799	21	97.3
PFHxS	144	844	11	98.7
PFNA	144	210	49	76.7
PFOS	148	1330	20	98.5

* Exceeded the 70 ng/L PFOS+PFOA EPA Health Advisory Level

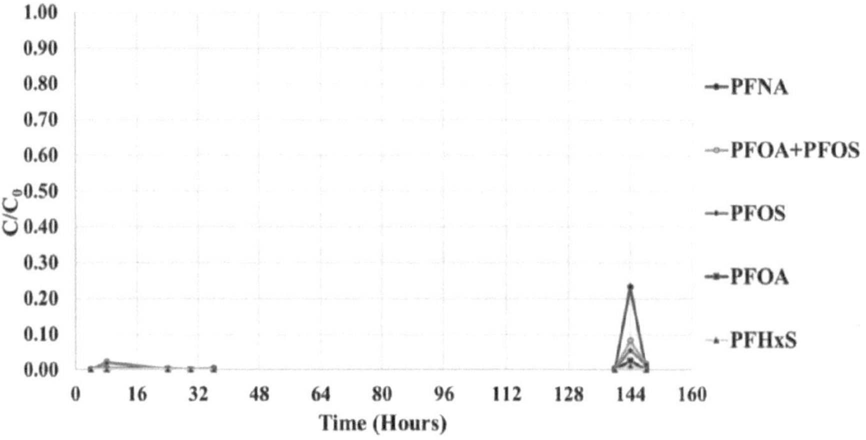

FIGURE 10.7 PFAS Removal vs. Time for RO #2 Test.

RO #2 reduced the turbidity, TOC, TDS, and hardness in the effluent, as shown in Table 10.8. Other than a slightly higher effluent turbidity after startup on Day 7, the RO membranes appear to be intact, as exhibited by the consistent removal of TOC, TDS and hardness. Due to the demineralization of the RO treated water shown in Table 10.8, a re-mineralization cartridge may be required to improve taste and to control corrosion in household premise plumbing.

The PFAS compounds found in the effluent may have resulted from the limited number of treatment barriers (activated carbon pre- and post- filters) in the two-stage Hydrologic RO system or a leaking membrane seal, because the breakthrough does not correlate to molecular size as might be expected with strictly a membrane system. However, even the breakthrough of the GAC from the activated carbon pre- and post-filters does not fully match the adsorption properties of the compounds. Also, it is unclear as to whether this was a one-time event, or whether it is to be expected for the days following Day 7.

TABLE 10.8
Water quality results from RO #2 test.

			Turbidity (NTU)		TOC (mg/L)		TDS (mg/L)		Hardness (mg/L)	
			Influent	Effluent	Influent	Effluent	Influent	Effluent	Influent	Effluent
Date	Day	Hr				Startup				
3/13/18	1	4	0.38	0.12	1.2	0.48	454	10	298	6
3/13/18	1	8	0.40	0.19	1.2	0.18	456	6	247	6
3/14/18	2	24	0.31	0.17	1.2	0.21	455	24	247	4
3/14/18	2	30	0.33	0.11	1.2	0.16	448	2	247	6
3/14/18	2	36	0.28	0.12	1.2	0.16	452	3	260	6
						Shutdown and Startup				
3/19/18	7	144	0.34	0.54	1.3	0.19	450	16	252	4
3/19/18	7	148	0.39	0.12	1.3	0.12	446	6	240	4

The average permeate flow rate from this unit was approximately 2165 mL/min, and the average concentrate (reject) flow rate from this unit was approximately 2730 mL/min. The recovery of the RO #2 system was approximately 44 percent.

10.3.1.3 RO #3 Test

The third RO POU/POE system, Flexeon LP-700 (RO #3), was tested over a seven-day period. Because of the increased water volume necessary to test this unit at an average flow rate of 4560 mL/min, the unit was only operated on Tuesday and Wednesday of the first week before being shut down for the weekend. RO #3 was then restarted on Monday morning of the following week for the Day 7 samples.

All PFAS in the effluent were reduced below the detection for the entire test. RO #3 also reduced the turbidity, TOC, TDS, and hardness in the effluent. The average permeate flow rate from this unit was approximately 1260 mL/min, and the average concentrate (reject) flow rate from this unit was approximately 3300 mL/min. The recovery of the RO #3 system was approximately 28%.

10.3.1.4 RO Discussion

The three RO systems were similar in many respects. They were each designed to be installed by a homeowner with basic "handyman" skills or a plumber, if necessary. The RO systems were all equipped with the required filters, membranes, and tools to install and replace these parts. However, the specifics of what was included varied (types of filters, tubing, booster pump, etc.). RO #1 and RO #2 (iSpring and HydroLogic) were self-supporting, while RO #3 (Flexeon) had to be mounted to a wall. Filters and membranes require replacement by the homeowner or plumber on a regular schedule and all three RO systems generated more reject water than purified water during this study.

Because RO systems tend to increase water corrosivity without additional processes, care should be made to evaluate potential lead and copper corrosion from the post-treatment premise plumbing for any POE systems. A re-mineralization cartridge may be required to improve the taste of RO-treated water and for corrosion control in household plumbing. Disposal of reject water via sewer or septic systems may be allowed under a homeowner exemption in many communities; however, special considerations should be made if there is a possibility of re-contaminating the aquifer.

10.3.2 GAC Media Studies

Two GAC media were tested using the test water: GAC #1; Evoqua 1230CX, a coconut-based GAC and GAC #2; Calgon Filtrasorb 600 AR+, manufactured from select grades of bituminous coal.

10.3.2.1 GAC #1 Test

The first RSSCT was conducted using Evoqua 1230CX (GAC #1), a coconut-based GAC. The target flow rate was 10 mL/min, and the flow rate was checked every 30 min before samples were collected. Each time the flow rate was adjusted to maintain 10 mL/min. The flow rate was maintained within 9.6–10.1 mL/min over the course of this test.

No PFAS were detected in the RSSCT effluent for the first 1.5 hours of testing. Individual PFAS began to break through the GAC at two hours and increased with time. The breakthrough time for each PFAS was dependent on the specific compound adsorption properties and its influent concentration. PFOA and PFOS were the first compounds detected in the RSSCT effluent, and PFNA was the last compound detected. All PFAS were detected within 5.5 hours of testing. By the end of the eight-hour testing period, the PFOS and PFBS had the highest concentrations in the effluent, and the PFNA had the lowest concentration in the effluent. Figure 10.8 shows the PFAS concentrations as a function of time for GAC #1. These results do not match the general results seen by others (Appleman et al. 2013; Dickenson and Higgins 2016; McCleaf et al. 2017) where the smaller PFAS (e.g., PFBS) generally break through before the longer chain PFAS (e.g., PFOA and PFOS). However, the PFBS concentration in the test water (260 ng/L) was much lower than the PFOS (1600 ng/L) and PFOA (800 ng/L) concentrations. The data suggest that in this case, the

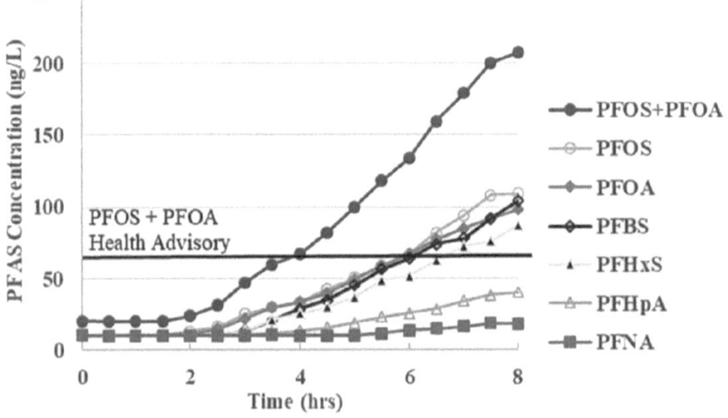

FIGURE 10.8 PFAS Concentrations vs. Time for GAC #1.

combined PFOS+PFOA results may be conservative for controlling the other PFAS. Care must be taken in translating these results to other waters and carbons.

Based on the EPA PFOS and PFOA HAL, GAC #1 lasted for approximately four hours before the total PFOS and PFOA concentration exceeded 70 ng/L. Using the RSSCT scaling factor for this test (194.3, as shown earlier in Table 10.6), the simulated life of a GAC filter using GAC #1 with 10 min EBCT would be approximately 800 hours (33 days). After this time, the total PFOS and PFOA concentration in the effluent would be expected to exceed the EPA HAL. GAC #1 treated approximately 4800 bed volumes of water prior to exceeding the EPA PFOS and PFOA HAL. These results suggest that the HAL level of 70 mg/L for combined PFOA and PFOS concentrations is protective when evaluating the breakthrough of the other PFAS studies; however, this is a function of the experimental design chosen in terms of water quality and PFAS studied. Others have shown different results, as mentioned above. Aside from a reduction in TOC, no other changes were evident in the water quality parameters for GAC #1.

10.3.2.2 GAC #2 Test

As a comparison to the coconut-based GAC from Evoqua, a second RSSCT was conducted using Calgon Filtrasorb 600 AR+ (GAC #2), a coal-based GAC. The target flow rate was 10 mL/min, and the flow rate was checked every 30 min before samples were collected. Each time the flow rate was adjusted to maintain 10 mL/min. The flow rate was maintained at 9.5–10.8 mL/min over the course of this test.

No PFAS were detected in the RSSCT effluent for the first hour of testing. Individual PFAS began to break through the GAC at 1.5 hours and increased with time. As with GAC #1, PFOA and PFOS were the first compounds detected in the GAC #2 effluent, and PFNA was the last compound detected. All PFAS were detected by 3.5 hours of testing. By the end of the eight-hour testing period, the PFOA and PFOS had the highest concentrations in the effluent, and the PFNA had the lowest concentration in the effluent. After eight hours of testing, compared to GAC #1, GAC #2 had higher concentrations of each of the PFAS except for PFBS. GAC #2 PFBS reached 70 ng/L 1.5 hours later than GAC #1 PFBS. Figure 10.9 shows the PFAS concentrations as a function of time for GAC #2. These results are similar to those seen for GAC #1 (Evoqua 1230CX). The caveat that the HAL level for PFOA and PFOS may not be protective for PFAS removal made in the GAC #1 discussion holds true for GAC #2 also.

Based on the EPA PFOS and PFOA HAL, the GAC #2 lasted for approximately three hours before the total PFOS and PFOA concentration exceeded 70 ng/L. Using the RSSCT scaling factor for this test (165.9, as shown earlier in Table 10.6), the simulated life of a GAC filter using GAC #2 with 10 min EBCT would be approximately 480 hours (20 days). After this time, the total PFOS and PFOA concentration in the effluent would be expected to exceed the EPA HAL. GAC #2 treated approximately 2900 bed volumes of water prior to exceeding the EPA PFOS and PFOA HAL. Aside from a reduction in TOC, no other changes were evident in the water quality parameters for GAC #2. The projected life of GAC #1 and GAC #2 is based on operation of the filters 24 hours/day and should be adjusted to account for actual household water usage patterns.

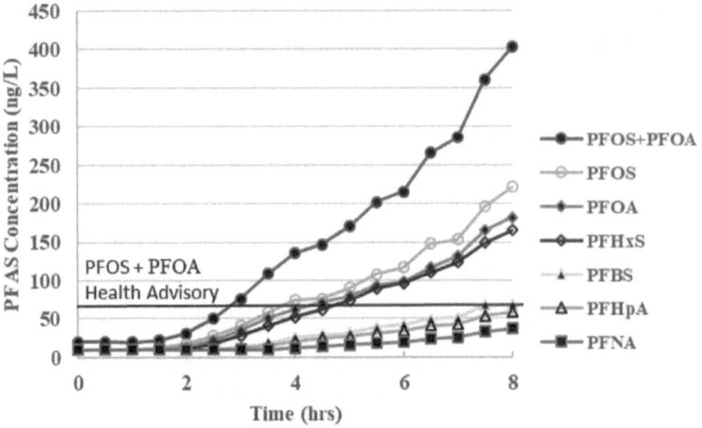

FIGURE 10.9 PFAS Concentrations vs. Time for GAC #2.

10.3.2.3 GAC Media Discussion

The representative POE system selected for the GAC filter used a 35"(H) x 9"(D) tank, 30 lbs (1 cubic foot) GAC, and was rated for four gpm with a treatment volume of 150,000 gallons. This size unit would only have an EBCT of two minutes, much smaller than a tank with an EBCT of 10 minutes (as recommended by GAC vendors for PFAS removal).

To meet the EBCT of 10 minutes, one alternative would be to use two tanks in series that are 65 inches high by 16 inches in diameter with approximately 240 lbs (8 cubic feet) GAC. However, scaling up the RSSCT data estimated that the coal-based Calgon F-600 GAC would have a lifetime of 20 days (2,900 bed volumes) compared to the coconut-based Evoqua GAC lifetime of 33 days (4,800 bed volumes) before requiring replacement based on maximum day PFAS concentrations tested before exceeding the EPA's HAL of 70 ng/L for PFOS and PFOA. Modeling the results for a lower concentration (average daily concentration) gave bed lives of 545 days (79,000 bed volumes) for the Calgon F-600 GAC and 791 days (115,000 bed volumes) for the Evoqua Coconut carbon before requiring replacement.

Increasing the size of the GAC filter tank or placing two filter tanks in series to lengthen the lifetime of the GAC media for PFAS removal would require additional plumbing and access to drainage lines for backwash wastewater. Replacing wet, contaminated GAC with dry GAC in the home is labor-intensive and requires backwashing to precondition and sort out GAC particle sizes and to remove air. The larger carbon filter tanks will also require additional space that may need to be heated depending on the local environment.

10.3.3 RO and GAC Residuals Management

Residual streams from home residences are exempt from Federal regulations as solid/hazardous wastes under the Resource Conservation and Recovery Act (RCRA). Therefore, homeowners can discharge RO reject water to the publicly owned treatment works (POTW). The RO reject wastewater concentrations from RO #1, RO #2 and RO #3 were calculated to range from 7 to 10 micrograms/liter. Therefore, care must be taken when discharging RO wastewaters to private septic or mound systems. Many GAC media manufacturers have carbon reactivation/disposal services. However, household RO and GAC cartridges containing PFAS are not considered hazardous waste and may also be disposed of in household trash.

10.4 MODELING OF GAC RESULTS

To investigate the impact of PFAS influent concentrations on GAC (bed volumes to breakthrough at 70 ng/L PFOS+PFOA), a model was utilized to predict GAC lifetime based on average PFOA (43 ng/L) and PFOS (137 ng/L) concentrations based on historic records (2013–2016) found in Widefield aquifer region water samples (CDPHE 2016). Adsorption Design Software (AdDesignS™) version 1.0, distributed as part of the ETDOT suite, developed by Michigan Technological University (Mertz et al. 1994; Hokanson et al. 1999), was used to model the adsorption of PFOA+PFOS compounds.

10.4.1 Parameter Estimation Process

Adsorptive capacity was estimated by calculating the area (integrating) between the influent and effluent concentrations and dividing the resulting value by the mass of carbon in the test apparatus. Any extrapolation of effluent concentration data needed to calculate the total breakthrough point was performed using a line fit for non-zero effluent concentrations and determining the intersection of that line with the average influent concentration. The capacity calculated in this step was used to estimate the Freundlich parameters. The Freundlich isotherm formula is:

$$q = K \cdot \bar{C}^{1/n} \tag{10.4}$$

where q is the adsorptive capacity (μg/g), $\bar{C}$ is the average influent concentration (μg/L) and K ((μg/g)(L/μg)^(1/n)) and n (unitless) are isotherm fitting parameters. The exponential term (1/n) was set to 0.45 for all models based on past experience with PFAS modeling. The K estimated in this step was used by the PSDM (Pore Surface Diffusion Model) code, as described below.

10.4.2 Pore Surface Diffusion Model

The PSDM module from AdDesignS™ was used to model the adsorption of PFAS onto each carbon (Crittenden et al. 1986; Mertz et al. 1994). AdDesignS™ uses input information about the adsorption column (diameter, length, flow rate, carbon mass), the adsorptive material/carbon (particle radius, apparent density, porosity) and the adsorbing species (molecular weight, molar volume, etc). The calculated Freundlich isotherm parameters as described above were used as inputs to AdDesignS™. Additional adsorption kinetics parameters were also required by the model. The film diffusion coefficient was calculated by the Gnielinski correlation (Mertz et al. 1994;

TABLE 10.9
PFAS AdDesignS™ Model Input Parameters.

GAC	PFAS	Kf	K	1/n	dp	ds
Evoqua 1230CX	PFBS	0.00613	19.354317	0.45	8.43E-06	1.31E-10
	PFHpA	0.00808	30.826032	0.45	7.94E-06	1.92E-12
	PFHxS	0.00797	105.445498	0.45	6.26E-06	1.16E-12
	PFNA	0.01077	36.955921	0.45	6.43E-06	1.07E-11
	PFOA	0.00634	91.528488	0.45	6.74E-06	2.34E-12
	PFOS	0.00726	207.547312	0.45	5.6E-06	1.41E-12
Calgon Filtrasorb 600 AR+	PFBS	0.00701	15.680191	0.45	9.25E-06	1.19E-11
	PFHpA	0.0052	12.703420	0.45	7.23E-06	8.68E-11
	PFHxS	0.00513	38.212900	0.45	7.51E-06	2.86E-11
	PFNA	0.00695	15.030018	0.45	6.93E-06	5.89E-11
	PFOA	0.00495	30.815503	0.45	5.28E-06	6.33E-11
	PFOS	0.00567	60.178352	0.45	3.65E-06	7.83E-11

Sontheimer et al. 1988). The pore diffusion coefficient was calculated from the Hayduk and Laudie correlation (Mertz et al. 1994; Sontheimer, et al. 1988). The surface diffusion coefficient was calculated by AdDesignS™ using the Sontheimer correlation (Mertz et al. 1994 Sontheimer et al. 1988). The surface and pore diffusion and film transfer coefficients were varied to produce a good fit to experimental data. The influent concentrations recorded during GAC RSSCTs 1 and 2 were input into the model. The surface and pore diffusion coefficients were varied to minimize the sum of squared (SSQ) difference between the modeled and experimental effluent concentrations. The minimum SSQ values were recorded with the associated pore and surface diffusion coefficients. PFAS AdDesignS™ Model input parameters are provided in Table 10.9.

10.4.3 Application of the Model

Adsorption kinetics parameters were recorded that resulted in the best match to experimental values. These were used to model configurations or operating schemes that were not directly tested in the GAC RSSCTs. Specifically, the influent concentrations were reduced from the worst-case (maximum) to average day concentrations seen in the Widefield aquifer (CDPHE 2016). Figures 10.10 and 10.11 show the model results of effluent concentrations for PFOS and PFOA for both RSSCT experiments using the worst-case (maximum, tested, shown in green) and average day concentrations (shown in blue) supplied to each carbon. The PFOS+PFOA concentration in the influent was approximately 3,000 ng/L for the worst-case scenario and 180 ng/L for the average day (a 16–17x reduction). For the maximum day, the model predicted an exceedance of the PFOS+PFOA HAL (70 ng/L) after approximately 3,400 bed volumes for Evoqua GAC #1 and approximately 2,700 bed volumes for Calgon GAC #2, which is consistent with the experimental values. Reducing the influent maximum day concentrations to the average day concentrations resulted in exceedances of the HAL after approximately 115,000 bed volumes (2.2 years of

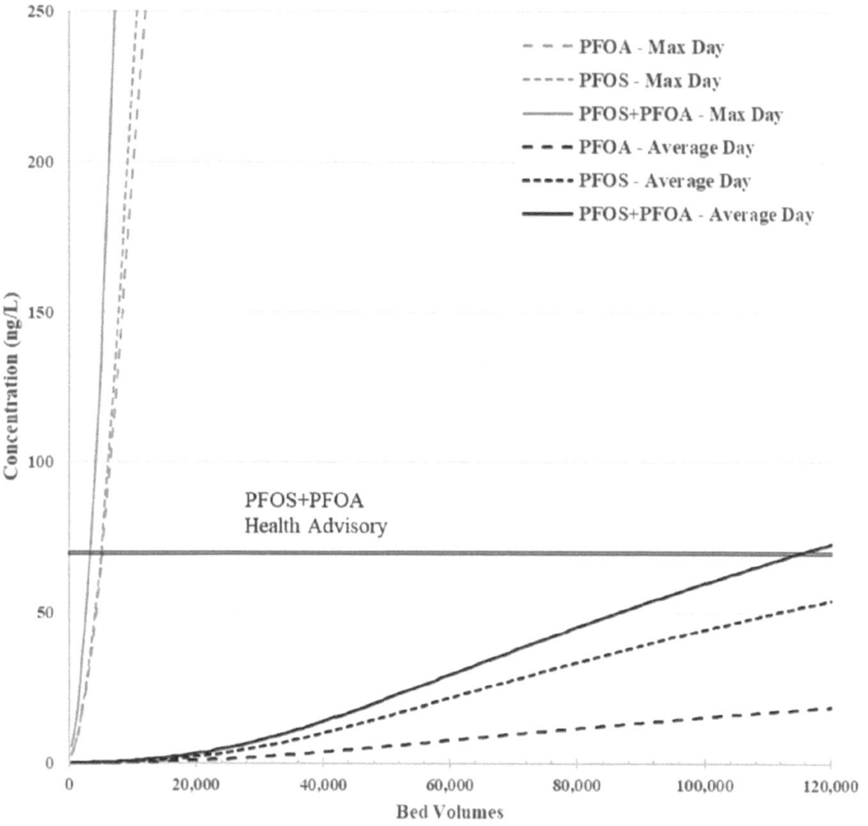

FIGURE 10.10 Model Results of PFOS+PFOA Drinking Water Concentrations vs. Bed Volumes for GAC #1.

operation) for GAC #1 and approximately 79,000 bed volume for GAC #2 (1.5 years of operation). For the 16–17x reduction in influent concentration, the model predicted about 34x (Evoqua) and about 29x (Calgon) improvements in number of treatable bed volumes before exceedance of the 70 ng/L HAL. It must be noted that this modeling has a number of assumptions (see Table 10.9) on the scale up approach to that made in the model fitting, and, therefore, it should not be relied upon by homeowners. GAC performance should be monitored to determine the bed volumes and time to PFOA+PFOS breakthrough based on the actual well water quality conditions that will vary based on distance from contamination source and source water hydrology.

10.5 CONCLUSIONS

In this study, RO and GAC technologies were evaluated to determine their ability as household POU/POE systems to remove PFAS from contaminated groundwater. The RO system and GAC media effluent results were compared for PFOA and PFOS at

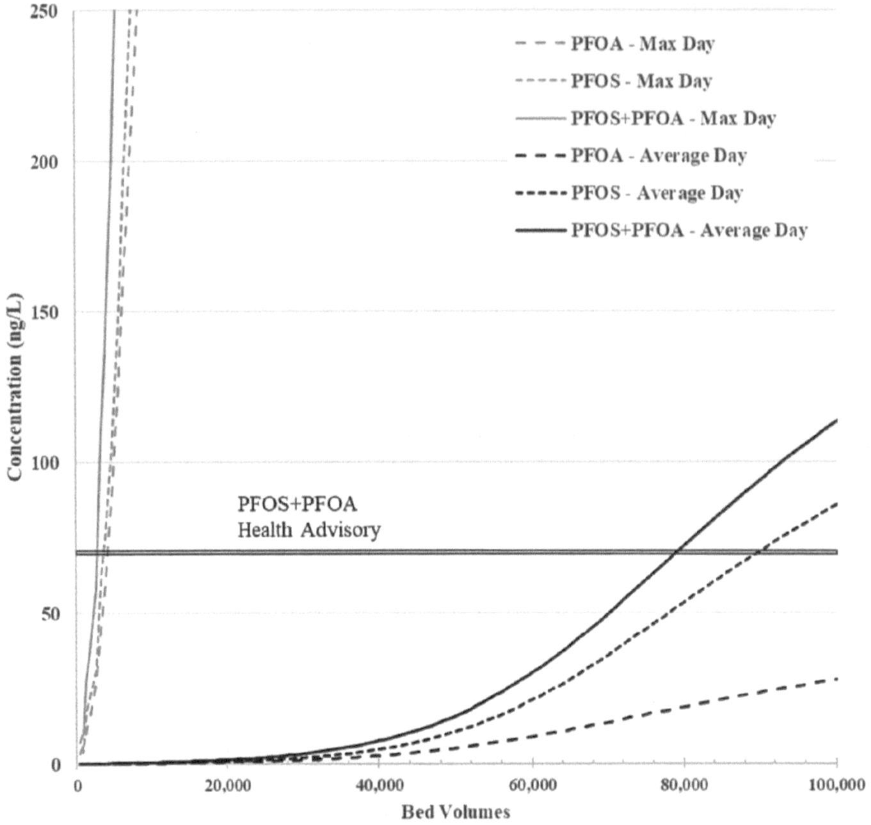

FIGURE 10.11 Model Results of PFOS+PFOA Drinking Water Concentrations vs. Bed Volumes for GAC #2.

individual and combined concentrations with the current 70 ng/L U.S. EPA HAL. Both GAC and RO POU/POE systems were shown to have potential to remove PFAS to below detections under specific water quality conditions, PFAS concentrations and operational conditions, studied herein. However, performance will vary as these conditions vary both spatially and temporally. In addition, long-term performance of these systems, particularly the membrane system, was not tested

Specifically, each of the three RO systems that was tested successfully removed PFAS from the influent water to below analytical detections for the majority of the sampling events. RO #1 and RO #3 (Flexeon and iSpring) did not show any PFAS in the treated water, while the RO #2 (HydroLogic) showed low levels of PFAS in the treated water after shutdown and startup.

The GAC POE systems were evaluated using RSSCT studies to save experimental time. Two different GAC materials were tested in this treatability study: one manufactured from coconut shells and the other manufactured from bituminous coal. While each of the GAC materials tested during this evaluation produced water with little or no PFAS in the effluent, it was found that the coconut-based carbon would treat more water than the coal-based carbon before requiring replacement. However, the GAC materials may react differently to variations in source water characteristics within the Widefield Aquifer region, and additional pilot-tests should be performed to ensure use of the best-performing GAC for each application. Also, more GAC materials are available that should be considered for testing.

Although this study suggests that utilizing the HAL of 70 ng/L for the combined concentrations of PFOA and PFOS was protective when compared to the breakthrough of the other PFAS, other water qualities may not yield the same results. Differences in water quality in terms of both PFAS concentrations and background water characteristics can change this result. Also, other PFAS are known to have weaker adsorption characteristics and would breakthrough well before PFOA and PFOS.

If a POU/POE system is correctly designed based on the source water's characteristics and the PFAS found therein, POU/POE water systems can provide relatively inexpensive treatment barriers for PFAS removal in the home. However, analysis of PFAS samples is costly for homeowners and can be a major hurdle in effective removal of PFAS from household water supplies. Proper operation and maintenance and conservative replacement of POU/POE components and media may be one way to circumvent the high cost of monitoring treated household drinking water.

ACKNOWLEDGMENTS

The U.S. Environmental Protection Agency, through its Office of Research and Development, funded and managed, or partially funded and collaborated in, the research described. It has been subjected to the Agency's administrative review and has been approved for external publication. Any opinions expressed in this chapter are those of the author(s) and do not necessarily reflect the views of the Agency, therefore, no official endorsement should be inferred. Any mention of trade names or commercial products does not constitute endorsement or recommendation for use.

REFERENCES

Appleman, T.D., Dickenson, E.R.V., Bellona, C., & Higgins, C.P., 2013. Nanofiltration and granular activated carbon treatment of perfluoroalkyl acids, *Journal of Hazardous Materials*, v. 260 (9) 740–746, September 15, 2013.

ASTM (American Society for Testing and Materials) D6586-03, 2008. *Standard Practice for the Prediction of Contaminant Adsorption on GAC in Aqueous Systems Using Rapid Small-Scale Column Tests*, ASTM International, Conshohocken, PA.

CDPHE (Colorado Department of Public Health and Environment), 2016. *PFC Levels in Drinking Water Distribution Samples Collected from the Widefield Aquifer Region*, https://environmentalrecords.colorado.gov/HPRMWebDrawer/RecordView/1208017 (Accessed December 21, 2016).

Crittenden, J.C., Reddy, P.S., Arora, H., Trynoski, J., Hand, D.W., Parram, D.L. & Summers, R.S., 1991. Predicting GAC performance with rapid small-scale column tests, *Journal AWWA*, v. 83 (1) 77–87, January 1991.

Crittenden, J.C., Hutzler, N.J., Geyer, D.G., Oravitz, J.L., & Friedman, G., 1986. Transport of organic compounds with saturated groundwater flow: model development and parameter sensitivity, *Water Resources Research*, v. 22 (3) 271–284, March 1986.

Dickenson, E.R. & Higgins, C., (2016). Treatment and mitigation strategies for poly- and perfluoroalkyl substances. WRF report 4322. Denver, CO: Water.

Federal Register, 2016. Lifetime health advisories and health effects support documents for perfluorooctanoic acid and perfluorooctane sulfonate, *Federal Register*, v. 81 (101), 33250, May 25, 2016. https://www.gpo.gov/fdsys/pkg/FR-2016-05-25/pdf/2016-12361.pdf

Hokanson, D.R., Hand, D.W., & Crittenden, J.W., 1999. "Adsorption Design Software™ (AdDesignS™)," *Copyright 1993–1999*. Michigan Technological University, Houghton, MI 49931. Commercial release: (1999).

McCleaf, P., Englund, S., Östlund, A., Lindegren, K., Wiberg, K., & Ahrens, L., 2017. Removal efficiency of multiple poly- and perfluoroalkyl substances (PFASs) in drinking water using granular activated carbon (GAC) and anion exchange (AE) column tests, *Water Research*, v. 120 (2017) 77–87.

Mertz, K.A., Gobin, F., Hand, D.W., Hokanson, D.R., & Crittenden, J.C., 1994. *Adsorption Design Software for Windows (AdDesignS)*, Michigan Technical University, Houghton, MI.

NSF (National Sanitation Foundation) International, 2016. *Draft NSF International Protocol, NSF* P473–2016, Drinking water treatment units – PFOA & PFOS, Ann Arbor, MI.

Pressman, J., McCurry, D., Parvez, S., Rice, G., Teuschler, L., Miltner, R., & Speth, T., 2012. Disinfection byproduct formation in reverse-osmosis concentrated and lyophilized natural organic matter from a drinking water source, *Water Research*, v. 46(16), 5343–5354.

Sontheimer, H., Crittenden, J.C., & Summers, R.S., 1988. Activated Carbon for Water Treatment, DVGW-Forschungsstelle, Engler-Bunte-Institut, Univ. of Karlsruhe, Fed. Rep. of Germany.

USEPA (U.S. Environmental Protection Agency), 2017. Standard Operating Procedure for the Analysis of Polyfluorinated Compounds of Interest to OSRTI in Water, Sludge, Influent, Effluent, and Wastewater by Multiple Reaction Monitoring Liquid Chromatography/Mass Spectrometry (LC/MS/MS), March 21, 2017, US EPA, Region 5 Chicago Regional Laboratory, 536 S. Clark Street (ML-10C), Chicago IL, 60605, CRL Document #OM021, Version #1, 68 pages.

11 Removing PFAS from Water

From Start to Finish

Caitlin Berretta, Thomas Mallmann, Kyle Trewitz and David M. Kempisty
Evoqua Water Technologies, USA

CONTENTS

11.1 Introduction....235
11.2 Technology Overview: Granular Activated Carbon and Ion Exchange Resin....236
11.2.1 Granular Activated Carbon (GAC)....237
11.2.2 Ion Exchange Resin (IX)....238
11.2.3 Granular Activated Carbon vs. Ion Exchange Resin....239
11.3 Predicting Performance....240
11.3.1 Piloting Initiatives....241
11.3.2 Bench-scale Testing....243
11.3.3 Modeling Capabilities....244
11.4 Permanent Treatment Solutions....245
11.4.1 Permanent GAC Solution - Kennebunk, ME....246
11.4.2 Permanent IX Solution - Stratmoor Hills, CO....249
11.5 Rental or Emergency Assets – Municipality in Vermont....250
11.6 The New and the Next: Emerging Technology and Applications....251
Acknowledgments....252
References....252

11.1 INTRODUCTION

Per- and polyfluoroalkyl substances (PFAS) are an expansive set of chemicals that the scientific community is largely still learning about every day. The removal of PFAS from source water of concern is affected in a dynamic fashion by the changing regulatory landscape, ever-improving analytical methods, new toxicological findings and evolving remediation technologies. All of these facets (e.g., regulation, analysis, toxicology, and remediation technology) are advancing along different paths and are influencing one another. In fact, Massachusetts recently finalized PFAS drinking

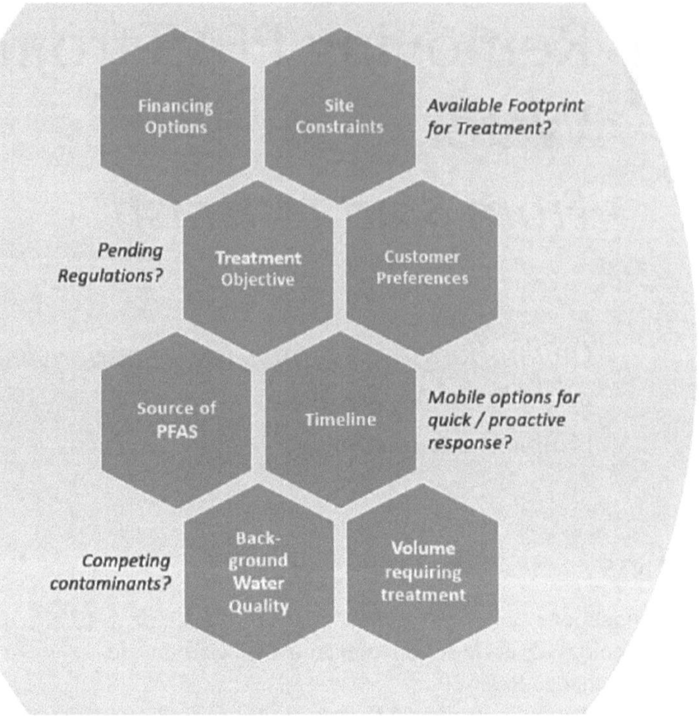

FIGURE 11.1 Important considerations when evaluating PFAS treatment options. Figure courtesy of Evoqua Water Technologies.

water standards which included a requirement for the Commonwealth to revisit the limits every three years due to the rapid growth in these areas (Commonwealth of Massachusetts, 2020). When corrective action is required, regardless of the application (e.g., municipal drinking water, industrial wastewater, military remediation, etc.), a whole host of questions are worth consideration. As an example, Figure 11.1 provides an abbreviated set of items to evaluate when considering PFAS treatment options. This graphic is not meant to be exhaustive but demonstrates the need for a multifaceted approach when selecting cost-effective PFAS solutions.

This chapter details the process applied by Evoqua Water Technologies LLC (Evoqua) to identify and provide optimal solutions for remediation of PFAS-impacted waters. Case studies are presented demonstrating a comprehensive approach employing the use of proven PFAS removal technologies, treatability testing, as well as both temporary and permanent installations.

11.2 TECHNOLOGY OVERVIEW: GRANULAR ACTIVATED CARBON AND ION EXCHANGE RESIN

Currently, the two most widely used and commercialized technologies for PFAS removal from water include granular activated carbon (GAC) and ion exchange (IX) resin. Membrane filtration, such as nanofiltration (NF) or reverse osmosis (RO), is another highly effective and commercialized technology for PFAS removal; however, the technology produces a PFAS-concentrated brine of significant volume (approx. 20% of treated water) that requires further treatment or management. Also, the capital cost of membranes is generally higher than GAC and IX. Other developing PFAS technologies, such as advanced reductive and oxidative processes, separation by foam fractionation, or the use of novel medias, have shown some promising results but often encounter challenges when employed at full scale due to cost, energy requirements, regulator certification, or meeting flow requirements (Ross et al., 2018). For these reasons, focus is given to the conventional and proven technologies of GAC and IX and the approach to ensure a comprehensive solution is identified and implemented.

11.2.1 Granular Activated Carbon (GAC)

GAC has a long history of use in the water treatment industry. Historical documents in Sanskrit text describe the use of GAC's precursor, charcoal, in the treatment of water (EPA, 2000). In the United States, activated carbons were used beginning in the 1930s for water treatment (Le Cloirec and Faur, 2006). Activated carbon is a highly porous, carbonaceous-based media with a high affinity for organic contaminants. It has an immense internal surface area; for example, a spoonful of activated carbon has an internal surface area equivalent to an entire football field. It can be manufactured from a wide variety of carbonaceous materials, but some common sources include various ranks of coal (e.g., bituminous, sub-bituminous, lignite, and anthracite), wood, and coconut shells. The EPA has named GAC the 'Best Available Technology' for organic contaminant removal (Pontius, 1995). As such, not only is GAC effective for PFAS removal, it is also effective at concurrently removing other regulated or soon-to-be regulated organic compounds (e.g., chlorinated solvents, pharmaceuticals, personal care products, etc.). To that end, the presence of other organics in the water can also impact performance as they compete with PFAS chemicals for adsorption sites. Therefore, it is important to understand the total organic carbon (TOC) concentration in the feed water when designing a GAC system.

While different GACs may look similar to the eye, performance varies between GAC types and can even be tailored for certain applications. Porosity is a significant parameter for carbon's performance for PFAS removal. While both pore and surface diffusion occur in the GAC particle, pore diffusion has been cited as the mass transfer controlling mechanism (Hand et al., 1989). Therefore, pore size distribution has been generally considered to affect the kinetics of adsorption, the capacity for the contaminants of interest, as well as the degree of carbon fouling by adsorption of other larger organics. Pore sizes are divided into three categories: macropores (> 500 Å), which are the largest pores; mesopores (20 Å – 500 Å), considered transport and adsorption pores; and micropores (< 20 Å), the smallest pores. The macropores function as feeder pores to the meso- and micropores while the mesopores also facilitate transport as well as serve as adsorption sites for larger adsorbates such as natural organic matter (NOM) (Rodríguez-Reinoso, 1997). Micropores make up a great portion of the pore volume of the GAC and are adsorption sites for smaller molecules (Snoeyink, 1990). The pore size distribution of mesopores to micropores needs to be

TABLE 11.1
Typical properties of common GAC feedstocks used for PFAS removal (approximate values). Data courtesy of Evoqua Water Technologies.

Carbon Feedstock	Iodine Number ($mg\ I_2/g$ GAC)	BET Surface Area (m^2/g GAC)	Apparent Density (g/cm^3 GAC)
Coconut-shell GAC	900–1100	950–1050	0.40–0.52
Bituminous GAC	900–1000	850–950	0.46–0.58
Low Density Sub-bituminous GAC	900–1000	850–950	0.34–0.46

considered to ensure the capacity can be properly utilized. For all applications (including PFAS applications), the background water quality should be understood when selecting a carbon's pore size distribution that provides adsorption capacity and reasonable kinetics while minimizing the adverse effects of competing organics (e.g., NOM). Treatability tests can be accomplished to empirically demonstrate efficient kinetics based on the ratio of mesopores to micropores and the background NOM characteristics and concentration (Kempisty and Summers, 2016).

Carbon properties such as iodine number and apparent density can provide an overall indication of the porosity and capacity of the carbon. Brunauer–Emmett–Teller (BET) surface area can also be used to provide a measure of capacity however is less reproducible due to a lack of an ASTM standard. Typical values for three common GAC properties based on carbon feedstock are included in Table 11.1. The iodine number is a measure of iodine adsorbed onto a 1-gram sample of GAC, reported as mg of iodine per gram of GAC. The results are an indicator of the microporosity of the GAC. The BET surface area measures the amount of a nitrogen gas being adsorbed onto the GAC surface. This is an indicator of the capacity of the carbon available for adsorption. Finally, the apparent density is measured as the GAC mass (g) divided by its total volume (including void spaces). Today's carbon market contains a range of different feedstocks and apparent densities. Apparent density is an important factor that is directly related to financial implications when purchasing media on a per pound basis and ensures the proper volumetric carbon fill within an adsorption vessel. Other properties, such as hardness and particle size, are also important to consider. A harder GAC will result in less attrition and breakdown of the product throughout its lifetime in use and during reactivation events. Particle size can be an important consideration to addressing head loss and the resulting effects on pumping costs and components of the treatment system (e.g., vessel, gaskets, and valves). When specified, these properties can help ensure the selection of a carbon with sufficient capacity and optimal fill. Ultimately, treatability testing is recommended prior to specification development.

11.2.2 Ion Exchange Resin (IX)

IX resins have also been proven as an effective media for PFAS removal and consist of beads that contain ion exchange sites for contaminant removal. The technology works by exchanging the targeted charged ion for a harmless ion within the bead.

Positively charged ion exchange sites attract negatively charged anions. For PFAS removal, the functional group is a quaternary amine with a positive charge to attract PFAS chemicals that form negatively charged anions when dissolved in water. IX resins are manufactured as durable, hydrated, spherical plastic beads. Bead size is typically between 16 and 50 mesh, which is smaller than GAC used for water treatment. There are two main physical structures for IX resin: macroporous beads have a physical pore structure similar to GAC and are opaque, while gel beads do not have physical pores and appear translucent. There are advantages and disadvantages to both structures. Macroporous resins have more surface area, which makes them more resistant to surface fouling. Gel resins have higher exchange capacity. The chemistry is quite similar, with both structures starting out as crosslinked polymers (e.g., polystyrene crosslinked with divinyl benzene). Functionalizing the polymer backbone gives the resin its ion exchange capacity and makes the resin more hydrophilic. Resins contain 20% to 80% water. This water is critical to facilitating transport of ions into and out of the resin beads. This is especially true for gel resins where little of the IX capacity is near the surface.

The media has a selectivity for certain ions more than others. Fundamentally, the resin will have a greater affinity for the contaminant of concern (e.g., PFAS) than for the exchangeable ion (e.g., chloride or sulfate), and an exchange for the preferred ion will occur thereby removing the contaminant from the water. IX resin manufacturers can alter the amine functional group as well as the polymer backbone to influence selectivity for PFAS. For example, most anion exchange resins used for demineralization are functionalized with trimethylamine (Type 1) or dimethylethanolamine (Type 2), while bulkier amines, such as tri-n-butylamine, are used in the manufacture of PFAS-selective resins (DuPont, 2020). However, it is important to understand the relative background anion concentrations that may interfere with the overall resin performance. By knowing the influent PFAS concentrations and typical background anion concentrations, estimated resin performance can be modeled.

Compared to GAC, the media requires a shorter residence time and can tolerate higher flow rates, which provides a smaller overall footprint (and lower upfront capital cost). However, the media is generally more expensive, which can result in greater service costs throughout the life of the system. From an operational perspective, backwashing is not recommended as it can result in premature leakage. As a result, often a particulate filter is employed in advance to protect the resin and avoid solids accumulation. If required, this additional step adds upfront costs, maintenance, and footprint to the overall system.

11.2.3 Granular Activated Carbon vs. Ion Exchange Resin

As discussed above, GAC has a larger system footprint and a higher upfront capital cost, but IX resin media is more expensive than GAC media. The lifecycle cost tradeoff ultimately comes down to the differences in total volume treated and the cost of service changeouts. Changeout frequency depends on several factors, including local site characteristics, water quality, and treatment goals. Each site is different. Even neighboring sites that draw water from the same aquifer may elect to install different technology based on slight differences in background water quality, treatment goals, available space, timing, upfront capital, comfort with the technology, and more. With the current patchwork of PFAS regulations, treatment goals can vary greatly from site to site. Future regulations are also expected to proceed in piecemeal, patchwork fashion. Often, despite the EPA's Health Advisory Levels (HALs) or promulgated regulations, the customer's desired treatment goals are "non-detect," or no detectable amount of any PFAS compound.

Generally, for both GAC and IX, the longer the carbon chain, the higher the affinity for removal. It is a common misconception that GAC does not remove short-chain PFAS (commonly defined as five or less fluorinated carbons for sulfonate-containing PFAS or six or less fluorinated carbons for carboxylate-containing PFAS). Short-chain PFAS are amenable to removal with GAC, but not as readily as the longer-chain PFAS. There are GAC systems that are designed to remove the short-chain PFBA (C_3F_7COOH) and maintain effluent concentrations below the detection limit. This affects operating costs as the GAC life is shorter than it would be if the treatment goal were to remove only PFOA ($C_7F_{15}COOH$). The same is true for IX. For example, breakthrough for PFBA (a 4-carbon chain) is typically observed before PFOA (an 8-carbon chain), but it is still possible to design an IX system to remove PFBA. Additionally, PFAS chains with a carboxylate functional group are generally more difficult to remove than those with a sulfonate group. For example, breakthrough for PFOA is seen before PFOS (both 8-carbon chain chemicals but with different functional groups, -COOH and SO_3, respectively). IX has a demonstrated higher affinity for the sulfonate functional group, often despite chain length. For example, IX typically has a higher affinity in removing PFOS (an 8-carbon chain) and PFBS (a 4-carbon chain) than PFOA.

As PFAS chemicals are typically detected in trace concentrations (parts per billion, μg/L, or parts per trillion, ng/L), the background water chemistry is an important design parameter in building an effective treatment system. Particularly in GAC treatment design, background TOC should be considered as it will compete with the PFAS chemicals for GAC adsorption sites. Large NOM compounds may also block access to carbon adsorption sites in the micropores. TOC can also affect IX resins, but often to a lesser extent than observed in GAC. Similarly, background anions present in the water such as chloride, nitrate, and sulfate will also compete with PFAS chemicals for the ion exchange sites (ITRC, 2020).

11.3 PREDICTING PERFORMANCE

To optimize treatment costs, upfront testing or modeling is recommended. The life-cycle costs of both GAC and IX technologies depends greatly on their changeout frequencies. As discussed, this will vary from site to site based on several factors. If the performance can be modeled or tested up front, the media that performs best on a particular water chemistry can be selected, thus optimizing operational costs over time. Depending on the available budget and timeline, there are several methods to predict the performance of GAC and IX medias, including bench-scale testing, on-site pilot testing, and modeling. An on-site pilot demonstration is often needed for regulatory approval of a new process or media. In certain cases, treatment studies must be conducted (e.g., piloting IX prior to implementation may be required in some states). Even if not required, the use of bench- and pilot-scale work can provide many advantageous insights. In the end, performance must be balanced against ease of implementation and operation, site constraints, and capital vs. operational costs, amongst other considerations.

11.3.1 Piloting Initiatives

Pilot-scale testing is conducted on site and consists of testing a fraction of the full-scale flow water through several proportionally-sized parallel media columns. These tests can be designed to test and compare several medias, including GAC versus IX resin. Due to the long throughputs typically observed with both medias in PFAS applications, it is not uncommon for pilot testing to take longer than six months in drinking water applications. The end point of the study can be defined as initial breakthrough or a certain effluent concentration. Although pilot testing takes more time than modeling or bench-scale testing, the results are more representative of full-scale performance as it considers the changes in water quality over time and uses full scale media (not ground).

As an example of piloting, Evoqua designed a large pilot system for a water utility on the US West Coast to compare different medias for PFAS removal from a variety of commercial producers. A portion of this pilot is shown in Figure 11.2. Fourteen different media were identified by the utility for testing, including eight GAC products, four IX products and two novel adsorbents. The test allowed for comparisons

FIGURE 11.2 One portion of a large pilot system designed to compare performance between eight different GAC products, four IX products, and two novel absorbents. Photo courtesy of Evoqua Water Technologies.

between the different media classes of GAC, IX resin, and novel adsorbent materials. Even in the same class, comparisons could be made between the many carbon or resin products on the market. Operating variables such as the empty bed contact time (EBCT) or pre-treatment could also be evaluated. Based on the results, the ideal media was selected. When extrapolated across the expected treatment duration of many years, even small improvements in media performance equate to significant savings in operational costs.

In another pilot project, Evoqua worked with a drinking water utility and engineering firm in the US Southeast to design a pilot comparing GAC and IX media performance in treating a groundwater well containing multiple PFAS compounds and having a background TOC concentration of 7.0 mg/L. The pilot results revealed that GAC was not an ideal media due to the high TOC concentration, while IX was able to meet water quality standards for an acceptable bed life. Following the pilot test, the utility and engineer purchased four full-scale vessels and selected four vendors to provide medias in a full-scale comparative study. To remove PFAS in the presence of elevated organics, each vendor selected media or media combinations they thought would perform best. Two vendors selected a resin configuration. Based on the initial pilot results, Evoqua selected to use a singular PFAS-selective IX media which had already demonstrated positive performance in the initial pilot test. The results, shown in Figure 11.3, demonstrate the value in piloting. The correct

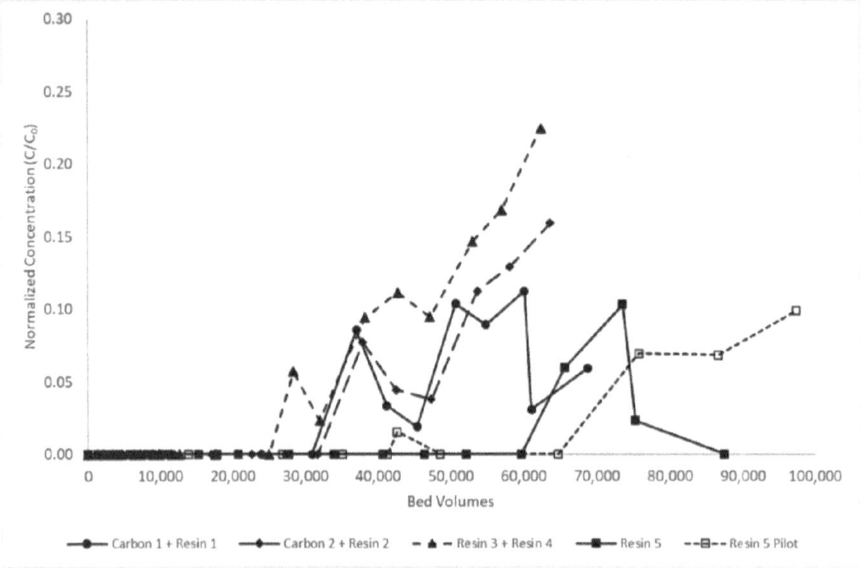

FIGURE 11.3 Comparing different configurations of media, full-scale piloting identified the ideal treatment configuration using ion exchange resin. Groundwater; PFOA: 12 ng/L, PFOS: 133 ng/L, TOC: 7 mg/L, EBCT: approx. 8 min. Prior to the full-scale pilot, the utility conducted a pilot comparing GAC and IX. The IX pilot data is included above and shows a similar performance for Resin 5 between pilot and full-scale performance. Pilot PFOA: 13 ng/L, PFOS: 72 ng/L, EBCT: 2 min. Data courtesy of Evoqua Water Technologies.

treatment can be selected, resulting in significant savings over the life of the treatment system. The single resin solution proved to provide the greatest volume of water meeting treatment objectives. Using these results and the price of the various treatment medias, the most cost-effective technology was selected.

11.3.2 Bench-scale Testing

While pilot-scale studies generally yield more representative results for full-scale systems, bench-scale tests can provide useful insights with lower costs and quicker turn-around times. Bench-scale work can be accomplished in a laboratory using a sample of site water. Rapid small-scale column tests (RSSCTs) can be used to evaluate treatment effectiveness of both GAC and IX medias. This testing method has historically been used to compare GAC media performance and is more recently being used with IX resins as well. It is important to note that the RSSCT methods for IX are under development and currently have limited data to correlate bench-scale to full-scale operation, as has been demonstrated for GAC. With the RSSCT, the principle of similitude is used to maintain the full-scale design hydraulic loading conditions while matching the controlling mass transfer process in a scaled-down bench-scale column. When conducting RSSCTs, it is important to ensure they are designed according to ASTM Standard D6586 for reproducibility and quality of results. Using ground media to achieve a smaller particle size, RSSCT results can be obtained in a fraction of the time (typically four to six weeks, mostly attributed to analytical turnaround) with a smaller volume of water as compared to full- or pilot-scale experimental setups that may take a year or longer to produce results. RSSCT breakthrough data provides direct comparisons between different media capacity and kinetics. Compared to pilot testing, however, the results are less representative of the full scale. This is largely based on its use of a single water sample which does not consider the changes in water quality over time.

To demonstrate, an RSSCT was performed to measure PFAS removal from a municipal well in the Midwestern US. This well had a total PFAS concentration of approximately 1000 ng/L. Using bituminous GAC, the water was filtered through an RSSCT column filled with the GAC ground and sieved to 170 x 200 US standard mesh. Samples were taken at regular intervals and sent to a third-party lab for PFAS analysis. The normalized concentration is plotted against the specific throughput measured in bed volumes in Figure 11.4.

Figure 11.4 illustrates each compound's affinity for adsorption onto the GAC. PFAS with greater tail lengths (e.g., 8-carbon PFOA vs. 6-carbon PFHxA) have greater hydrophobicity and exhibit greater affinity for the adsorbent. PFAS with sulfonate functional groups exhibit greater adsorption than PFAS containing carboxylate functional groups with the same number of carbon atoms. These observed trends can be further influenced by a variety of factors including other PFAS chemicals, PFAS pre-cursors, competing ions, natural organic matter, hydraulic loading conditions and contact times.

Another successful bench-top test involved investigation of a utility's concern about arsenic leaching from GAC at system startup. Depending upon the GAC media used, arsenic elution can occur and as a result, permit conditions for arsenic

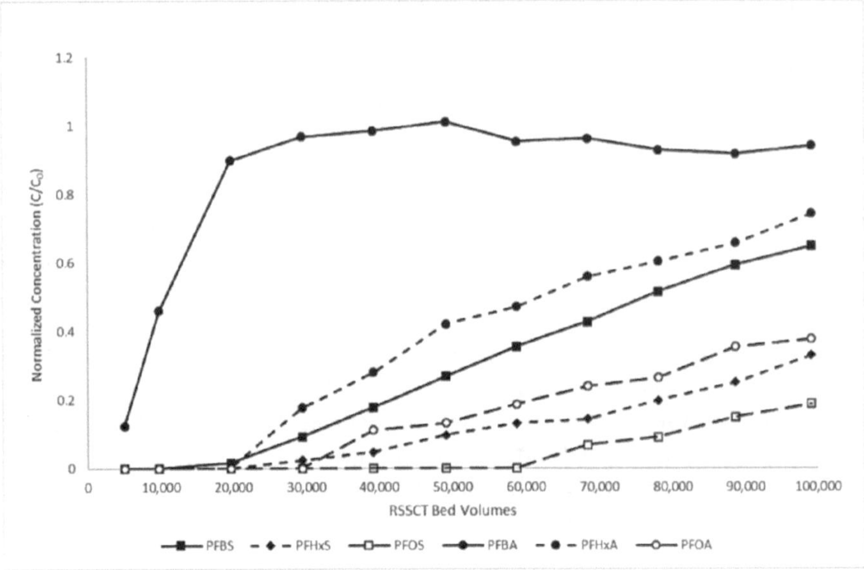

FIGURE 11.4 RSSCT (constant diffusivity) results demonstrating removal of six PFAS compounds from Midwest US municipality groundwater using bituminous GAC. PFAS concentration range: 1550 ng/L-1650 ng/L. EBCT: 8.3 min. Data courtesy of Evoqua Water Technologies.

discharges could be exceeded in the first few hundred bed volumes. Bench-scale testing was conducted on three GAC products. The results, included in Figure 11.5, demonstrated limited arsenic leaching from the coconut and sub-bituminous carbons with greater arsenic leaching observed in the bituminous product. Based on the results below, the utility opted for the sub-bituminous GAC as it could maintain discharges below the state drinking water maximum contaminant level for arsenic (5 ppb).

11.3.3 Modeling Capabilities

The quickest option for obtaining useful data to influence treatment decisions is from modeling simulations. This is especially true in the conceptual design phase when deciding to focus on one type of media over another or to estimate treatment costs. Models use isotherm data, media properties, water quality, and design operating conditions to predict full-scale operation. TOC interferences are difficult to model. Since GAC performance is more affected by background TOC, modeling estimates are more accurate when used to predict IX bed life or GAC performance in low TOC groundwater. Ideally, there would be a history of full-scale data supporting modeling output. Once demonstrated, the confidence in the modeling output is increased. Models allow rapid feedback of expected response due to process modifications. Using models, the technological solution can be optimized for ideal performance. As an example, two pilots could be run to determine differences observed when changing the EBCTs between 7.5 minutes and 15 minutes. If the model accurately

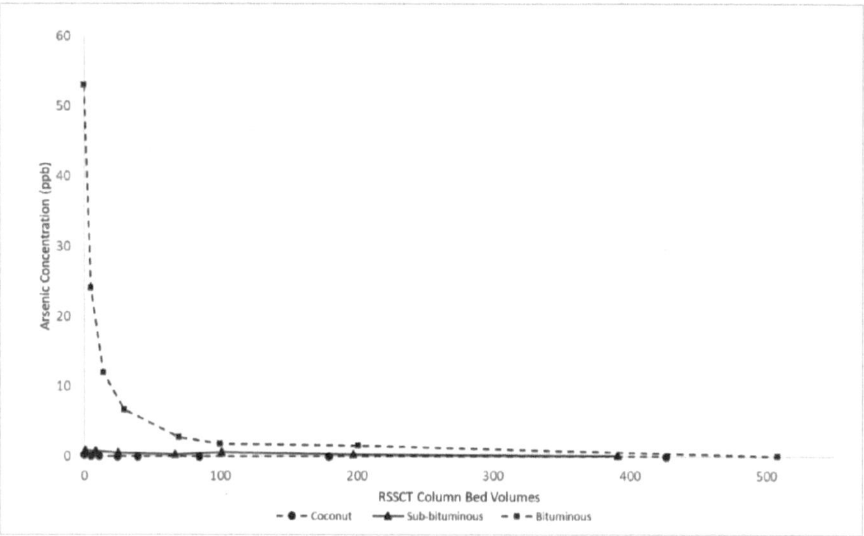

FIGURE 11.5 Bench-scale results investigating arsenic leaching from various carbon products upon treatment system start-up. Two of the three media tested had similar results, while the bituminous product had higher arsenic concentrations observed in the effluent. Data courtesy of Evoqua Water Technologies.

represents the two conditions, the ideal and specific EBCT could be predicted with the use of software. For example, the model might predict that 11.25 minutes is the ideal EBCT. Arriving at such a specific EBCT with piloting or bench-top experiments would be difficult and labor-intensive. And, as stated previously, when treating thousands or millions of gallons of water each day for years, even small improvements in efficiency result in significant financial savings. Certainly, there are large differences in performance observed from site to site and modeling estimates represent the best educated guess without conducting more expensive or labor-intensive options.

11.4 PERMANENT TREATMENT SOLUTIONS

Often the end goal for the site owner or responsible party is to receive an effective and efficient permanent treatment system to operate for a period of years. After considering the different treatment options, operating process control variables, and unique site characteristics (as well as the considerations in Table 11.1), selection of a final treatment solution occurs. Three full scale case studies highlighting this process for treatment implementation are included below.

When selecting a vendor for permanent treatment solutions, many factors should be considered. Selecting a vendor having experience implementing a variety of full-scale treatment solutions is important for several reasons. With experience and proper planning, complications with delivery and installation of the permanent system are less likely to occur. Experienced application engineers have a greater familiarity with approaches that yield successful results. In-house vessel fabrication is a valuable attribute and lends know-how to the overall solution. A capable and robust service arm of operators to install and perform media fills is important for a problem-free start-up and continuity of treatment. Finally, when unforeseen issues do arise, experience is an important attribute to provide quick resolution.

11.4.1 Permanent GAC Solution - Kennebunk, ME

As part of the EPA's third Unregulated Contaminant Monitoring Rule (UCMR3) testing, the Kennebunk, Kennebunkport, and Wells Water District (KKWWD) in Kennebunk, Maine, undertook PFAS sampling in their source wells. UCMR3 testing was required for all drinking water utilities serving over 10,000 people and included analysis for six PFAS compounds, including PFOA, PFOS, PFNA, PFHpA, PFHxS, and PFBS. Upon receiving the results, the utility was surprised to find levels of PFOA and PFOS in one of their groundwater wells. It is important to note that at that time (2016), there were not any PFAS regulations on a federal nor state-based level; there was only the EPA's non-mandatory HAL which is set at 70 ng/L for PFOA and PFOS combined. In addition, the analytical methods used for PFAS detection had a relatively higher detection limit than today's science, ranging from 10–90 ng/L (for comparison, today's detection limits are closer to 0.5–2 ng/L). KKWWD's results showed a combined level of PFOA and PFOS of 50 ng/L in one of KKWWD's wells. Although this result was lower than the EPA's HAL, the utility proactively decided to shut down the well and pursue options for remediation.

The utility wished to minimize any financial impact to ratepayers and sought to design the most cost-effective treatment solution. Their treatment goal was to remove PFOA and PFOS to non-detect levels (2 ng/L reporting limit). Only PFOA and PFOS were present; the other UCMR3 compounds (PFBS, PFHxS, PFHpA and PFNA) were all below reporting limits. KKWWD had a preference to use GAC since they had a high level of familiarity with the technology. The background water chemistry included low levels of TOC (1.3 mg/L), which would not be expected to significantly affect performance. Upon reaching out to Evoqua, an on-site pilot test was recommended to compare several types of GAC feedstocks to identify the product that provided the longest operational bed life to initial breakthrough.

To help the utility move quickly, an accelerated on-site pilot test was designed with halved beds (3 ft. beds in 3 in. diameter columns) maintaining the same hydraulic loading rate of a full-scale system. The EBCT was designed at 3.5 minutes in order to expedite breakthrough, and the test used operational-scale GAC mesh sizes (US Standard Sieve size 12x30 or 12x40). The commercial products tested included Evoqua and competitor material that represented four different carbon feedstock types: (1) standard coconut shell; (2) bituminous coal; (3) Evoqua's patented enhanced coconut shell; and (4) the reactivated version of Evoqua's enhanced coconut shell material. The results are shown in Figure 11.6.

KKWWD installed the enhanced coconut GAC product as it demonstrated the longest bed life, and therefore minimized the operational costs over time. The resulting full-scale design included the installation of two 12-foot diameter vessels, holding a total of 34,000 lbs. of media, in lead–lag configuration for built-in redundancy.

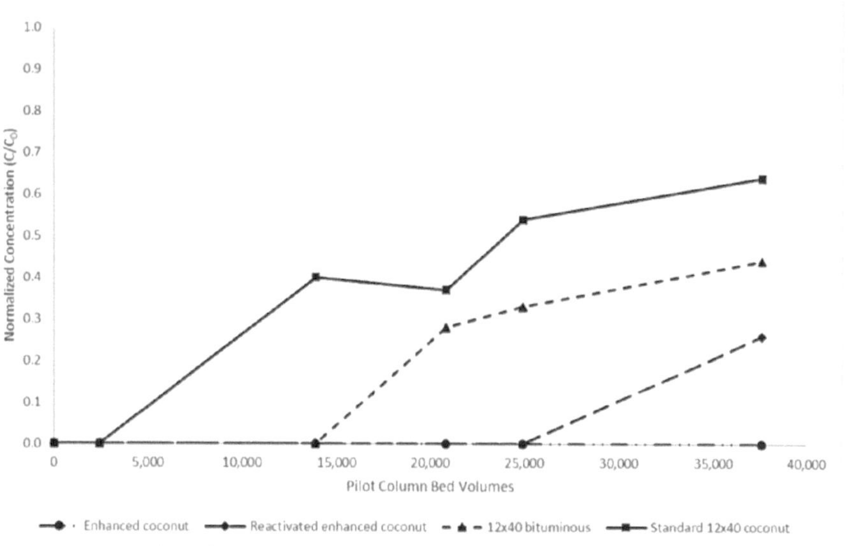

FIGURE 11.6 GAC Pilot on KKWWD water included four different carbons and identified the enhanced coconut product as having the lowest operational cost over time. Influent TOC: 1.3 mg/L. PFOA influent: 10 ng/L. EBCT: 3.5 min (to expedite breakthrough). Data courtesy of Evoqua Water Technologies.

Based on the presence of iron and manganese, a single 12 ft. diameter sand filter was added upfront to prevent accumulation of suspended solids in the GAC bed. The final installation, photo shown below as Figure 11.7, was located outdoors, although a permanent shelter was eventually built to prevent freezing and provide protection from the elements during the winter season.

The GAC vessel consisted of Evoqua's underdrain design, which was a slotted screen that protruded into the media bed (Figure 11.8). The Evoqua underdrain design provides the unique benefit of added built-in flexibility for future treatment modifications with the ability to hold multiple types of media sizes, including IX resin. PFAS chemistry, toxicology, and regulations are a dynamic and rapidly evolving science. At the time of installation, the treatment goals were specific to the EPA's HAL for PFOA and PFOS. Today, many states have drinking water or discharge limits in place for a growing list of PFAS chemicals which are subject to change as science progresses. As stated previously, neighboring Massachusetts' recently finalized PFAS drinking water standards included a requirement for the state to revisit the rule every three years. The regulatory environment will likely continue to be volatile for some time and this underdrain design allows the utility to limit investment in new capital improvements. An additional benefit to the underdrain design was reduced pressure drop compared to conventional vessel design. The reduced pressure drop resulted in lower pumping costs and less wear on the equipment from a maintenance perspective. As stated earlier, small improvements result in significant savings over the multi-year lifetime of the capital system.

FIGURE 11.7 The KKWWD full scale installation included one 12 ft. diameter sand filter followed by a 12 ft. diameter lead–lag GAC filtration system with enhanced coconut carbon. Photo courtesy of Evoqua Water Technologies.

FIGURE 11.8 The underdrain design was utilized in the GAC vessels at KKWWD. This design allowed for the flexibility to update the media type (GAC or IX resin). Photo courtesy of Evoqua Water Technologies.

Often, a probable source to PFAS contamination can be easily identified, such as a nearby military installation or chemical manufacturing site. In the KKWWD case, however, there were no obvious responsible parties. Following water sampling from a dozen monitoring wells surrounding the well site, the source was eventually identified as a neighboring dairy farm that had used land-applied biosolids containing PFAS. Biosolids historically used at the neighboring farm were used as soil enhancers and included waste from a nearby paper mill as well as solid waste from municipal wastewater (Buttarazzi, 2018). The responsible party being a farm was a concerning detail that speaks to the immense size of the PFAS contamination footprint. This is not just a water issue; PFAS chemicals are still used in a myriad of industrial processes today and land application of biosolids is still a common practice.

11.4.2 Permanent IX Solution - Stratmoor Hills, CO

Stratmoor Hills is another example of a municipality that installed a full-scale treatment unit for PFAS removal. Stratmoor Hills is a small Colorado township in El Paso county, Colorado with a population of 6900 people at the time of the last census. In 2016, PFAS was discovered in the township's Widefield aquifer and was widely attributed to the use of fluorinated fire-fighting foams at nearby Peterson Air Force base. Stratmoor Hills, along with several other communities, drew drinking water from this aquifer. Shortly after discovering PFAS contamination in the Widefield aquifer, Stratmoor Hills tested their drinking water wells and discovered one of their seasonal wells had elevated state-regulated PFAS chains of PFOA, PFOS, and PFHpA. Stratmoor Hills used this source during the summer months to assist in meeting the increased water demand.

Stratmoor Hills did not have the time to pilot technology as they needed a solution online in a matter of months before the seasonal well was expected to come online. However, a nearby utility pulling water from the same aquifer had recently piloted IX resin for PFAS removal, found it was a more economical solution than GAC, and successfully installed a full-scale system. This utility was willing to share their data with Stratmoor Hills included in Figure 11.9. The influent water quality for this

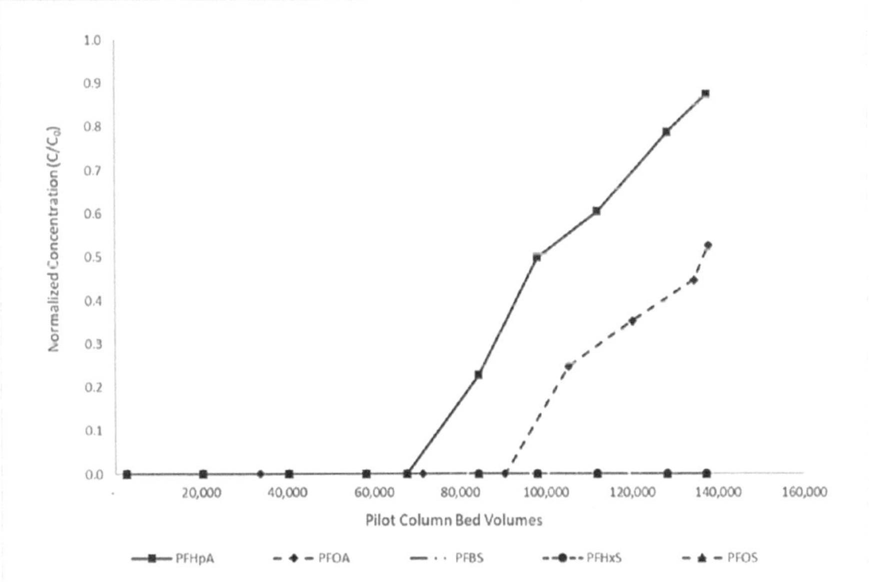

FIGURE 11.9 Pilot scale results shared with Stratmoor Hills from a nearby drinking water plant, pulling water from the same aquifer, using IX resin to remove PFAS. Total PFAS: approx. 270 ppt. TOC: 3.6 ng/L. EBCT: 2 min. Data courtesy of Evoqua Water Technologies.

FIGURE 11.10 The full-scale installation at Stratmoor Hills included two 4 ft. diameter lead–lag ion exchange filtration systems with a 5-micron pre-filter. Photo courtesy of Evoqua Water Technologies.

neighboring well was similar and contained the same PFAS chemicals; however in higher concentrations than Stratmoor Hills. As such, the shared pilot results were taken as a worst-case scenario to help Stratmoor Hills understand the estimated operational costs.

The results above included IX media performance exhibiting breakthrough for PFHpA first, followed by PFOA. The other present PFAS chains were sulfonates that demonstrated no breakthrough at 100,000 BV, which was the end point of the test. This supports the general trend that IX resin has a higher selectivity for the sulfonate PFAS chemicals, regardless of chain length. Based on these results, the need for a small footprint and a solution that could easily be moved to store for winter, the utility accepted the recommendation to use IX resin over GAC. A single-use IX system was implemented capable of treating 350 gallons per minute (Figure 11.10). The IX footprint for such a system was small, and occupies a space of 200 sq. ft. The system was versatile and able to be operated in both series and parallel treatment configurations. This size system allowed the use of municipally-owned forklifts to move the treatment units indoors when water demand was reduced during the winter off-season. Thus far, the system has operated during the high demand months for four summer seasons. Since the installation of the system, a second system has been procured by the municipality, doubling their treatment capacity. This second system was similarly designed to the first system. The redundancy in design maintained the capability to easily move the vessels while also providing operators product familiarity and ease of installation with similar equipment.

11.5 RENTAL OR EMERGENCY ASSETS – MUNICIPALITY IN VERMONT

There may be situations when time is not available to conduct performance-based testing. A rental solution may be preferred when a solution needs to be implemented quickly (e.g., a 'do not drink' order for a municipality), or on a temporary basis (e.g., construction de-watering or cleanup following firefighting efforts). A rental solution

FIGURE 11.11 Emergency rapid response installation of GAC system involved two of the stainless-steel tanks shown here in a lead–lag configuration with a coconut shell carbon. Photo courtesy of Evoqua Water Technologies.

can also be put in place as interim treatment while testing is underway to determine a more permanent design.

As an example, in early 2016, UCMR3 testing revealed elevated levels of PFOA in the drinking water of a small municipality in Vermont. The responsible party was identified as a former nearby industrial site that used PFAS chemicals as part of its processing. The State issued a 'do not drink' order for the drinking water supplying the 450 residents of the small community. Incorporating a solution as quickly as possible was a top priority for the town.

The town hired an engineer to work with a manufacturer that had emergency rental vessels on hand. Within two weeks of the initial order, a mobile treatment system consisting of two high-pressure lead-lag 2000 lb. GAC systems (four vessels total) was deployed quickly (Figure 11.11). Due to the low TOC content in the groundwater, a coconut shell-based GAC was used. Although it was a temporary system, the carbon installed at start-up lasted 2.5 years prior to changeout. A permanent treatment system was designed and is in operation today without any interruption to water quality or quantity.

11.6 THE NEW AND THE NEXT: EMERGING TECHNOLOGY AND APPLICATIONS

Moving forward, research and development efforts will continue testing GAC and IX media to understand how these products can be used in a changing regulatory environment. Exploration of pre-treatment steps in front of adsorbent media (both GAC and IX) to minimize the adverse effects of competing organics is ongoing. While IX performance is proven for PFAS treatment, the technology is far from a mature product. New suppliers are creating PFAS-selective resins, while those already in the market are improving their existing products. In addition to GAC and IX, novel adsorbents are being developed from cyclodextrins and bentonite-based clays. These media are promising, particularly for applications with elevated TOC and TSS, where carbon and resin usage rates can be costly. However, these new media have the added burden of building case studies and obtaining the regulatory approvals required for use in some applications. Treatability testing such as bench-scale RSSCTs and pilot demonstrations will continue to be the methods used for selecting the best technology and media for each application.

As PFAS testing increases across many industries, new applications for PFAS treatment continue to develop. These developing applications include, among others: military and commercial airport remediation; construction dewatering; landfill leachate treatment; treatment of process water from metal plating; food and beverage; and the dairy industries. Control and understanding of air emissions of PFAS is another area receiving increased regulatory and public attention. Each application brings its own set of needs that continue to shape new PFAS treatment solutions. For example, food and beverage applications need to consider water chemistry consistency across a product line and tight restrictions for bio-growth. Metal platers are typically low-flow applications and often run into an added disposal issue of chrome loading onto the media, creating a listed hazardous waste.

As we opened this chapter, PFAS science continues to rapidly evolve across many fronts, including analytical capabilities, newly identified industrial sources, toxicology, regulations, and the treatment technologies to remove it from the environment. Seemingly, the only constant with PFAS is simply that it is constantly evolving. For all these reasons, the most important note of this chapter is to ensure the reader has a strong, technology-agnostic partner with diversified experience to help navigate this dynamic landscape.

ACKNOWLEDGMENTS

The authors would like to thank John Lombardo and Mike Stock for compiling information and photos for this chapter.

REFERENCES

Buttarazzi, D. 2018. *Dairy Farm Contaminated KKWWD's Kimball Lane Well.* https://www.seacoastonline.com/news/20180201/dairy-farm-contaminated-kkwwds-kimball-lane-well (accessed October 5, 2020).

Commonwealth of Massachusetts, 2020. *D310 CMR 22.07F: Per- and Polyfluoroalkyl Substances (PFAS) Monitoring and Analytical Requirement.* https://www.mass.gov/doc/pfas-mcl-revisions-to-310-cmr-2200-clean-version-9-16-2020/download (accessed October 24, 2020).

Dupont, 2020. *Technical guidance for removal of PFAS using ion exchange resins.* https://www.dupont.com/content/dam/dupont/amer/us/en/corporate/PFAS/Guide%20for%20PFAS%20Removal%20Using%20Ion%20Exchange%20ResinsFIN.pdf (accessed November 17, 2020).

Environmental Protection Agency, 2000. *The History of Drinking Water*. https://nepis.epa.gov/Exe/ZyPDF.cgi/200024H9.PDF?Dockey=200024H9.PDF (accessed October 5, 2020).

Hand, D.W., Crittenden, J.C., Arora, H., Miller, J.M., Lykins, B.W., Jr., 1989. Designing fixed-bed adsorbers to remove mixtures of organics. *Journal - American Water Works Association*, 66–77; doi: 10.1002/j.1551.8833.1989.tb03324.x accessed (November 18, 2020).

ITRC (Interstate Technology Regulatory Council), 2020. *PFAS - per- and polyfluoroalkyl substances: treatment technologies*. https://pfas-1.itrcweb.org/12-treatment-technologies/ (accessed November 17, 2020).

Kempisty, D.M. and Summers, R.S.; 2016. Effect of influent groundwater quality on adsorption of low concentrations of 1,2 dichloroethane by granular activated carbon. *Journal of Environmental Engineering*. doi: 10.1061.(ASCE)EE.1943-7870.0001142 (accessed October 5, 2020)

Le Cloirec, P., Faur, C.. 2006. Adsorption of organic compounds onto activated carbon - applications in water and air treatments. *Interface Science and Technology*. 7: 376–419. doi: 10.1016/S1573-4285(06)80017-7 (accessed October 5, 2020).

Pontius, F.W., 1995. An update of the federal drinking water regs. *Journal AWWA* 87:2, 48–58. doi: 10.1002/j.1551-8833.1995.tb06316.x (accessed November 17, 2020).

Rodríguez-Reinoso, F., 1997. Activated Carbon: Structure, Characterization, Preparation, and Applications. *Introduction to Carbon Technologies*. Printer of the Univ. of Alicante, Alicante, Spain.

Ross, I., McDonough, J., Miles, J., Storch, P., Kochunarayanan, P. T., Kalve, E., Hurst, J., Dasgupta, S., Burdick, J., 2018. A review of emerging technologies for remediation of PFASs. *Remediation*. 28: 101–126 doi: 10.1002/rem21553 (accessed November 13, 2020)

Snoeyink, V.L., 1990. *Adsorption of Organic Compounds. Water Quality and Treatment*. 4th ed. McGraw-Hill Inc., New York.

12 Rapid Removal of PFAS from Investigation-derived Waste in a Pilot-scale Plasma Reactor

Raj Kamal Singh, Nicholas Multari, and Chase Nau-Hix
Clarkson University, USA

Richard H. Anderson
Air Force Civil Engineer Center, USA

Stephen D. Richardson
GSI Environmental Inc., USA

Thomas M. Holsen, and Selma Mededovic Thagard
Clarkson University, USA

CONTENTS

12.1 Introduction....256
12.2 Materials and Methods....257
 12.2.1 Chemicals....257
 12.2.2 Site Selection and Groundwater Collection....258
 12.2.3 Plasma Reactor....258
 12.2.4 Experimental Procedures....258
 12.2.5 Analytical Procedures....259
12.3 Results and Discussion....260
 12.3.1 Initial PFAA Concentrations....260
 12.3.2 Initial PFAA Precursor Concentrations....261
 12.3.3 Plasma Treatment of IDW Samples....262
 12.3.4 Effect of Co-contaminants on PFAS Removal Efficiency....266

12.3.5 Energy Requirements of Plasma Treatment for PFOA and PFOS Removal....267
Acknowledgments....268
References....268

12.1 INTRODUCTION

Poly- and perfluorinated alkyl substances (PFAS) have been widely used in numerous industries and commercial applications and were present in aqueous film-forming foam (AFFF) formulations used for firefighting and training activities. PFAS are highly persistent, can undergo long-range transport in groundwater (Hu et al. 2016), and are often detected in groundwater (Schultz et al. 2004; Backe et al. 2013), surface water and biota (Oakes et al. 2010; Awad et al. 2011; Lam et al. 2017; Ghisi et al. 2019). Thousands of polyfluorinated precursors along with perfluorinated alkyl acids (PFAAs) have been commercially synthesized and used throughout commerce. PFAAs only contain C–F bonds in their hydrophobic chain, unlike their precursors which contain both a fluorinated and non-fluorinated portion (Conder et al. 2008). Although the fluorinated portion of PFAAs molecules is highly resistant to microbial and chemical transformation, the non-fluorinated portion in precursors can be degraded and reduced, leaving the perfluorinated portion to result in the formation of PFAAs (Liu and Avendano 2013).

The U.S. Defense Department (DoD), along with numerous other agencies and entities around the world, are currently undertaking expansive portfolio-wide investigations of the extent of PFAS contamination, in particular at AFFF-impacted sites (Anderson et al. 2016; Anderson and Hunter, 2019). These investigations will continue for decades and result in copious amounts of liquid investigation-derived waste (IDW) primarily from development and purge water from the installation and sampling of monitoring wells. Current disposal options for liquid IDW at sites where existing treatment systems are not available are limited and costly due to off-site transportation costs. Mobile treatment options are desperately needed to optimize IDW management on-site during large-scale PFAS characterization efforts.

To date, liquid IDW is either disposed of at permitted facilities or opportunistically treated on-site by pump-and-treat (P&T) using granulated activated carbon (GAC) or ion exchange (IX) resin, which are the most common approaches to remove PFAAs in groundwater (Woodard and Berry, 2017). While GAC and IX successfully remove some PFAAs, the processes are limited by groundwater chemistry and adsorption capacity, particularly for short-chained PFAAs. In addition, both processes produce a residual (spent GAC and IX brine) that requires disposal or further treatment. Nanofiltration and reverse osmosis have been shown to remove PFAAs from water; however, complete destruction of the resulting concentrate requires incineration which is costly (Appleman et al. 2013). Alternative PFAAs destructive treatment technologies such as advanced oxidation processes (AOPs) do not rely on sorption and disposal and include photochemical oxidation (Giri et al. 2012), thermally-induced persulfate oxidation (Liu et al. 2012), electrochemical oxidation (Zhuo et al., 2011), and sonochemical degradation (Campbell et al. 2009). Each of these, while still in development, is currently ineffective and/or prohibitively expensive.

In plasma-based water treatment, highly reactive oxidative (e.g., $^{\bullet}OH$, O, $HO_2^{\bullet}$, $O_2^{\bullet-}$, O_2, H_2O_2) and reductive species (e.g., free and aqueous electrons e_{aq}^{-}, $H^{\bullet}$, etc.) are generated by the formation of an electrical discharge between a high voltage and a grounded electrode in the vicinity of liquid water (Sunka et al. 1999; Mededovic Thagard et al., 2009). The reductive species such as solvated and free electrons have been shown to play an important role in PFAAs degradation in plasma-based water treatment process (Mededovic Thagard et al., 2016a).

The best-performing plasma reactor for treating surfactant-like compounds such as PFAAs is the gas discharge reactor with argon bubbling (Mededovic Thagard et al., 2016a; Stratton et al. 2017). In this reactor argon is pumped through a submerged gas diffuser to form a layer of foam containing surfactant-like contaminants (e.g., PFAS) at the plasma–liquid interface where chemical reactions occur. Our previous study compared several performance indicators of PFAS degradation among plasma treatment (our bench-scale gas discharge reactor), sonolysis, UV-activated persulfate, electrochemical treatment, and DC plasma in O_2 bubble processes (Stratton et al. 2017). Our plasma-based process was four, eight and 57 times more efficient than the electrochemical, persulfate and sonolytic processes, respectively, and two orders of magnitude more efficient than the DC plasma system. The same study also showed that the effectiveness of the plasma process in treating PFAS is not diminished by the presence of co-contaminants. This work demonstrated that increasing the plasma reactor treatment volume from 1.4 L (bench-scale system) (Stratton et al. 2017) to 4 L (this work) and redesigning the plasma reactor elements while maintaining the same features of the bench-scale reactor (i.e., plasma is formed in argon gas and is contacting liquid surface) increased the PFAS treatment time by an order of magnitude.

In this study, a 4 L pilot-scale plasma reactor was used to treat IDW in order to reduce PFAAs concentrations below United States Environmental Protection Agency's (USEPA's) health advisory concentration level (HAL). Before the treatment, PFAAs and their precursors in all IDW samples were characterized. The effect of co-contaminants (volatile organic compounds (VOCs) and heavy metals) and precursors on PFAAs degradation was studied, and the most significant operational parameters and IDW's physicochemical characteristics influencing the treatment determined. The operational energy cost was also determined.

12.2 MATERIALS AND METHODS

12.2.1 Chemicals

Linear perfluoroalkyl carboxylic acids, sulfonates, and labeled internal standards consisting of perfluorobutanoic acid (PFBA), perfluoropentanoic acid (PFPeA), perfluorohexanoic acid (PFHxA), perfluorohexane sulfonate (PFHxS), perfluoroheptane sulfonate (PFHpS), perfluorooctanoic acid (PFOA), perfluorooctane sulfonate (PFOS), perfluorononanoic acid (PFNA), perfluorodecanoic acid (PFDA), perfluoroundecanoic acid (PFUnA), and perfluorododecanoic acid (PFDoA) were purchased from Wellington Laboratories (Guelph, ON). Individual calibration and labeled internal standard for precursor fluorotelomer sulfonates (10:2 FTS, 8:2 FTS, 6:2 FTS, and 4:2 FTS), perfluorooctane sulfonamide (FOSA-1), perfluorooctane sulfonamide acetate

(FOSAA), N-ethylperfluorooctane sulfonamide (EtFOSA), N-ethyl perfluorooctane sulfonamido acetic acid (EtFOSAA), N-methylperfluorooctane sulfonamide (MeFOSA), and N-methyl perfluorooctane sulfonamido acetic acid (MeFOSAA) were also purchased from Wellington Laboratories (Guelph, ON). Methanol and acetonitrile (LC–MS grade) were purchased from Thermo Fisher Scientific (Waltham, MA).

12.2.2 Site Selection and Groundwater Collection

PFAS-contaminated groundwater was collected from monitoring wells in fire training areas, burn pits, and various other water sources as part of ongoing field investigations at U.S. Air Force installations. In total, 13 samples were collected in 18.9 L high-density polyethylene (HDPE) containers, secured in coolers, and shipped to Clarkson University packed on ice. Samples were stored at 4°C upon receipt. To maintain confidentiality of sample locations, in this study, samples were labelled as IDW 1 through IDW 13.

12.2.3 Plasma Reactor

Two plasma reactors were installed into an 8 × 20 ft. mobile trailer. Each plasma reactor was a rectangular Delrin tray (66.0 cm × 48.3 cm × 1.27 cm) with an acrylic lid that contained six tungsten saw tooth high-voltage (HV) electrodes for the generation of plasma and seven stainless steel strips (62.2 cm × 2.54 cm) as grounded electrodes (Mededovic et al. 2016b). Six gas diffusers (61.0 cm × 6.35 cm × 2.00 cm, 0.4 µm pore size and 50 psi maximum pressure; Pentair, USA) were positioned on the bottom of the reactor and arranged alternately to the grounded electrodes. The HV electrodes were suspended in the headspace region, a distance of 1.2 cm from the surface of the water (the thickness of the treated water layer was 1.27 cm and the total reaction volume ~4 L). The schematic of the overall experimental setup and the actual plasma reactor in operation are shown in Figure 12.1. Argon gas was continuously bubbled through the diffusers during the experiments at a flow rate of 16–18 L/min and recirculated. The recirculation did not compromise treatment efficiency (Singh et al. 2019) and allowed any contaminants present in the gas phase to be retreated by the plasma. A high-voltage power supply (TDK-Lambda, 102A-40 kV-NEG) was used to charge a set of 32 nF capacitors which were discharged via a rotating spark gap into the plasma reactor at the discharge frequency of 60 Hz. A voltage of (−)35 kV was used for all the experiments. Stratton et al. (2015) provides the schematic for the electrical circuit.

12.2.4 Experimental Procedures

The plasma reactor was designed to be operated in a continuous mode (up to 2 gpm); however, in this study, the limited quantities of available IDW (~12 L per IDW sample) required that the reactor be operated in a batch mode. The IDW solution electrical conductivities ranged between 22 and 26,300 µS/cm, and the pH between 5.3 and 8.0.

Before treatment, sediments in the IDW samples were allowed to settle in their original containers; no other pre-treatment was performed. Clear IDW samples (supernatant) were treated between 5 and 60 minutes, depending on the initial

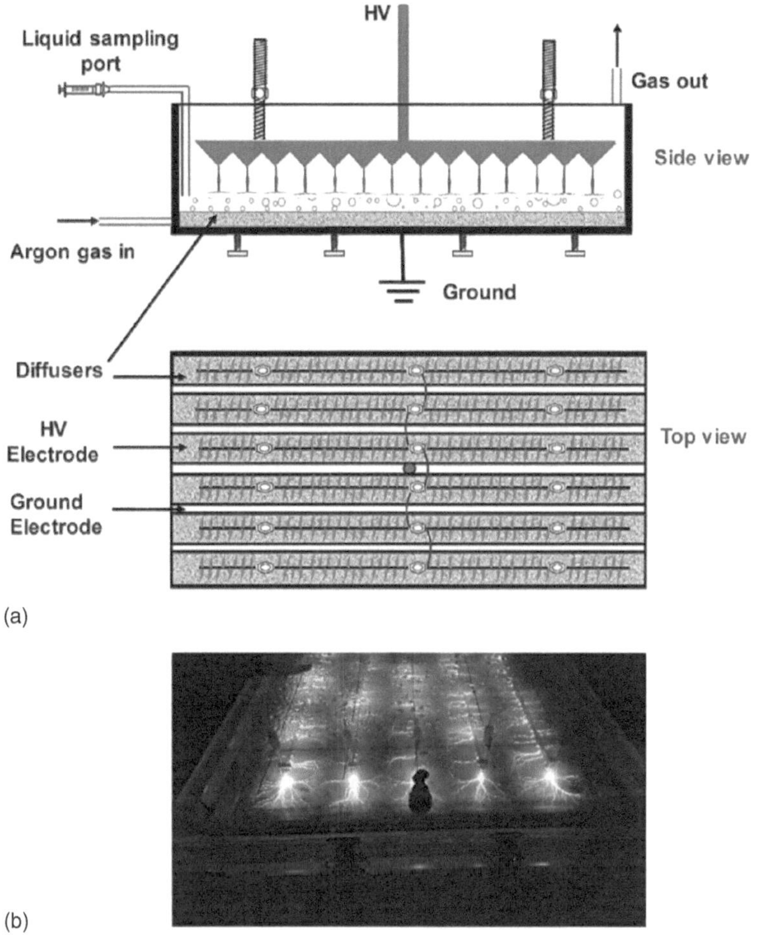

FIGURE 12.1 The pilot-scale plasma reactor: (a) schematic and (b) reactor in operation. No erosion of the electrodes was observed during and after the experiments.

sample characteristics (initial PFAS concentration and solution conductivity). Total oxidizable precursors were measured using a modified persulfate oxidation method developed by Houtz and Sedlak (2012). A control experiment was performed to validate TOP analysis at our precursor concentration range using synthetic lab water that contained a mixture of total identified PFAS precursors (TIP) (10:2 FTS, 4:2 FTS, 6:2 FTS, 8:2 FTS, EtFOSA, EtFOSAA, FOSA-1, FOSAA, MeFOSA, and MeFOSAA) each at a concentration of 0.5 μg/L. Good recoveries were found (96.7 ± 4.3%).

12.2.5 Analytical Procedures

Twelve PFAAs and their 10 precursor concentrations were measured in negative ionization mode using either a UPLC-QToF-HRMS (Xevo G2-XS, Waters Corp.) or UPLC-MS–MS (Thermo Scientific, Vanquish-TSQ ALTIS) equipped with Waters

Acquity HSS T3 column (2.1 mm x 100 mm, 1.8 µm) (samples analyzed on both instruments were not statistically different). Samples were diluted with methanol (3:1 ratio), then sonicated and centrifuged prior to injection (20 µL). For QA/QC, all samples were spiked with 2 ng of labeled internal standards. Six-point calibration in the range of 10 and 5000 ng/L was used for quantification using C-13 isotopic dilution or internal standard methods. Quantification was performed only when the standards yielded a regression fit of $r^2 > 0.98$ and deviation < 30% and surrogate recovery was between 70 and 120%. Detection limits were approximately 9 ng/L. Singh et al. (2019) discussed a detailed description of the analytical methods, QA procedures, and individual detection limits. Fluoride ion concentration was measured using an Accumet Excel XL60 meter kit (Fisher Scientific) as per EPA Method 9214.

The total fluorine concentration was measured using CIC (Metrohm, Switzerland). This instrument combined ion chromatography (930 Compact IC Flex and 920 Absorber Module with conductivity detector) and an automatic combustion module (Analytik Jena) which enabled trace-level determination of fluoride at sub ppb (µg F/L) levels. Heating of the 1 mL sample that was placed into the boat converted organofluorines into HF which, upon transferring into an absorption unit, dissociated into H^+ and F^-. The concentration of total F^- (from organofluorine and F^- from plasma treatment of PFAS) was measured using IC. A Metrosep A Supp 5 150/4.0 column with a Metrosep A Supp 4/5 guard column was used for the separation of anions. Operational conditions were: combustion temperature 1050 °C; argon as carrier gas (100 mL/min); oxygen as combustion gas (300 mL/min); Milli-Q water absorber liquid; 0.32 M/0.10 M sodium carbonate/bicarbonate solution mobile phase (1.0 mL/min); sample volume 1000 µL; sample loop 1000 µL; column temperature 30 °C. A five-point calibration curve was prepared at 0 to 500 µg/L using PFOA (Sigma-Aldrich, USA). All solutions were prepared in Milli-Q water (0.056 µS/cm). The minimum detection limit (MDL) was 5 ppb (µg F/L).

Samples were analyzed with methanol in a ratio of 3:1 water:methanol solution. The sample boat, elution lines, and combustion module were frequently cleaned by injecting 2 or 3 Milli-Q water samples after every three samples. The samples were analyzed after assuring the blank concentration was below the MDL.

VOCs, iron (Fe), and manganese (Mn) were analyzed by a commercial analytical lab (TestAmerica Laboratories Inc., Denver, CO, USA). TOC concentration was measured using a TOC analyzer (Shimadzu, TOC-VCPH). Solution electrical conductivity and pH were measured using probes (YSI, MultiLab IDS 4010–2). Turbidity was measured using a turbidity meter (Hach, 2100 N). Total alkalinity and hardness were measured following standard methods (APHA, 2017).

12.3 RESULTS AND DISCUSSION

12.3.1 Initial PFAA Concentrations

Of the 12 PFAAs analyzed for, PFOS, PFHpS, PFOA, PFHxS, PFHpA, PFHxS, PFBS, PFPeA, and PFBA were found in all of the samples. PFDS, PFDA, and PFNA were below detection limits (approximately 9 ng/L) in IDW samples 11, 4 and 3, respectively (Figure 12.2). Total PFAAs concentrations ranged from 2.7 to 1440 µg/L

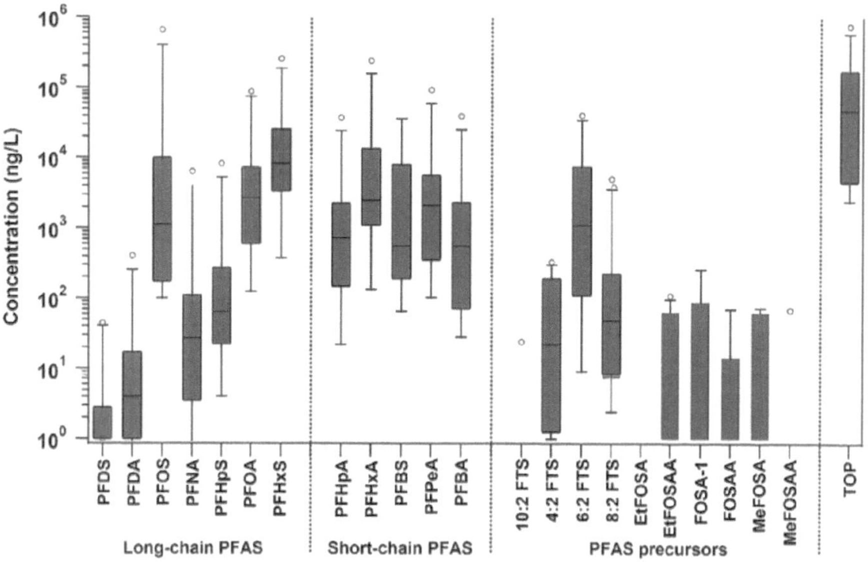

FIGURE 12.2 Concentrations of PFCAs, PFSAs, precursors and TOP in 13 IDW samples. Ends of the boxes represent the first and third quartiles, horizontal lines marked inside the box represent median, whiskers represent minimum and maximum values, and small hollow circles represent the outliers (determined with F test using the interquartile range). TOP concentration = Total PFAAs concentration after oxidation – Total initial PFAAs concentration.

(mean 151 ± 392 μg/L); ΣPFCAs ranged from 0.3 to 500 μg/L (mean 55 ± 136 μg/L); and ΣPFSAs ranged from 0.3 to 950 μg/L (mean 96 ± 256 μg/L). Individual PFAA concentrations ranged from 3 to 650,000 ng/L (Figure 12.2). The mean concentration was the highest for PFOS (54.3 ± 179 μg/L) and the lowest for PFDS (5.3 ± 13.9 μg/L). For the other PFAAs, concentrations decreased in the order of PFHxS (33 ± 69 μg/L), PFHxA (25 ± 65 μg/L), PFOA (13 ± 27 μg/L), PFPeA (9 ± 25 μg/L), PFBS (7 ± 13 μg/L), PFBA (4 ± 10 μg/L), PFHpA (4 ± 10 μg/L), PFHpS (0.8 ± 2 μg/L), PFNA (0.5 ± 1.7 μg/L), and PFDA (0.04 ± 0.11 μg/L). The sum of PFOA (13.2 ± 26.6 μg/L) and PFOS (54.3 ± 179 μg/L) concentrations was 5 to 10,500 times above the USEPA lifetime HAL in all the samples. One IDW sample (IDW 13) had extremely high concentrations of PFOA (87 μg/L) and PFOS (650 μg/L). PFNA, PFHxS, PFHpA, and PFBS, which fall under the USEPA's third Unregulated Contaminant Monitoring Rule (UCMR3), were also detected at mean ratios of 25, 1100, 400, and 78 times higher than their minimum reporting levels of 0.02, 0.03, 0.01, and 0.09 μg/L, respectively.

12.3.2 Initial PFAA Precursor Concentrations

Average TOP concentrations were high in all of the samples (74 ± 105 μg/L with a range of 20 to 750 μg/L). The ratio of average TOP concentrations to average total PFAA concentrations varied from 0.11 to 5.75. Total fluorine concentration (organic fluorine + fluoride) measured by CIC ranged from 98 to 4900 μg F/L, which on average was 90 times higher than organic fluorine measured in the TOP assay. Of the 10 individual precursors investigated, fluorotelomer sulfonates (6:2 FTS, 8:2 FTS and 4:2 FTS) were detected in 11 samples with mean concentrations of 7.2 ± 12.7, 0.5 ± 1.5, and 0.1 ± 0.1 μg/L, respectively. Other precursors such as FOSA-1, EtFOSAA, EtFOSA, MeFOSA, and FOSAA were detected in a few samples (2 to 6) at very low concentrations (120 ± 110, 60 ± 30, 120 ± 10, 60 ± 20 and 50 ± 30 ng/L, respectively). MeFOSAA was detected at a concentration of 68 ng/L in only one sample (Figure 12.2). TOP concentrations were on average ten times higher than the total directly measured precursor concentrations indicating the presence of numerous undefined precursors (the ratio of TOP/ΣTIP concentration varied from 2.5 to 513).

12.3.3 Plasma Treatment of IDW Samples

During treatment of the 4 L IDW samples, PFOA and PFOS concentrations decreased rapidly and were below HAL concentrations within one minute for IDW 1– 9, 10 minutes for IDW 10 and 11, 15 minutes for IDW 12, and 50 minutes for IDW 13 (Figure 12.3 (a) and (b)).

For long-chain PFAAs (PFOA and PFOS), plasma treatment achieved a removal efficiency > 93% within five minutes for IDW 1– 9, 30 minutes for IDW 10– 12, and 60 minutes for IDW 13 (Figure 12.4). Removal efficiencies were also higher for long-chain PFAAs than for short-chain PFAAs. For PFDS, PFDA, PFOS, PFNA, PFHpS, PFOA, and PFHxS, removal efficiencies were 93 ± 11, 94 ± 12, 94 ± 13, 95 ± 8, 100 ± 0.1, 98 ± 4, and 96 ± 11%, respectively. For the short-chain PFAAs (PFHpA, PFHxA and PFBS), removal efficiencies were 98 ± 2, 85 ± 18, and 56 ± 38%, respectively at the end of the treatment; however their concentrations were continuing to decrease (see Figure 12.5). In general, PFSAs were more rapidly removed than PFCAs of similar chain-length and similar initial concentrations, a result that agrees with Stratton et al. (2017). This is likely due to PFCAs' higher surface activity due to the presence of an extra CF_2 group in the molecule.

For the ten identified precursors (10:2 FTS, 4:2 FTS, 6:2 FTS, 8:2 FTS, EtFOSA, FOSA-1, FOSAA, MeFOSA, MeFOSAA, and EtFOSAA) removal efficiencies were 40 ± 73, 92 ± 17, 94 ± 20, 90 ± 32, 91 ± 12, 92 ± 16, 95 ± 9, 6 ± 117, 79 ± 5, and 65 ± 64 %, respectively. These removal efficiencies were higher than those observed for TOP (65 ± 20%, Figure 12.4) which should be considered for all PFAS destruction technologies since any TOP not removed can be broken down to PFAAs in the environment (Houtz and Sedlak 2012). Negative removal (increasing concentrations) were observed for PFPeA, PFBA and MeFOSA (−14, −562 and −159%, respectively) which is likely due to the formation of short-chain PFCAs (PFPeA and PFBA) from the degradation of long-chain PFAAs and the oxidation of their precursors, as discussed in the subsequent section. MeFOSA could be a degradation product of other precursors such as MeFOSAA, EtFOSA, EtFOSAA, and N-methyl perfluorooctane sulfonamidoethanol (MeFOSE) and formed by the terminal cleavage of $-CH_3COO^-$, $-CH_3$, and/or $-CH_3OH$ group from the parent molecules or from the degradation of unknown precursors (Liu and Avendano 2013). The lower

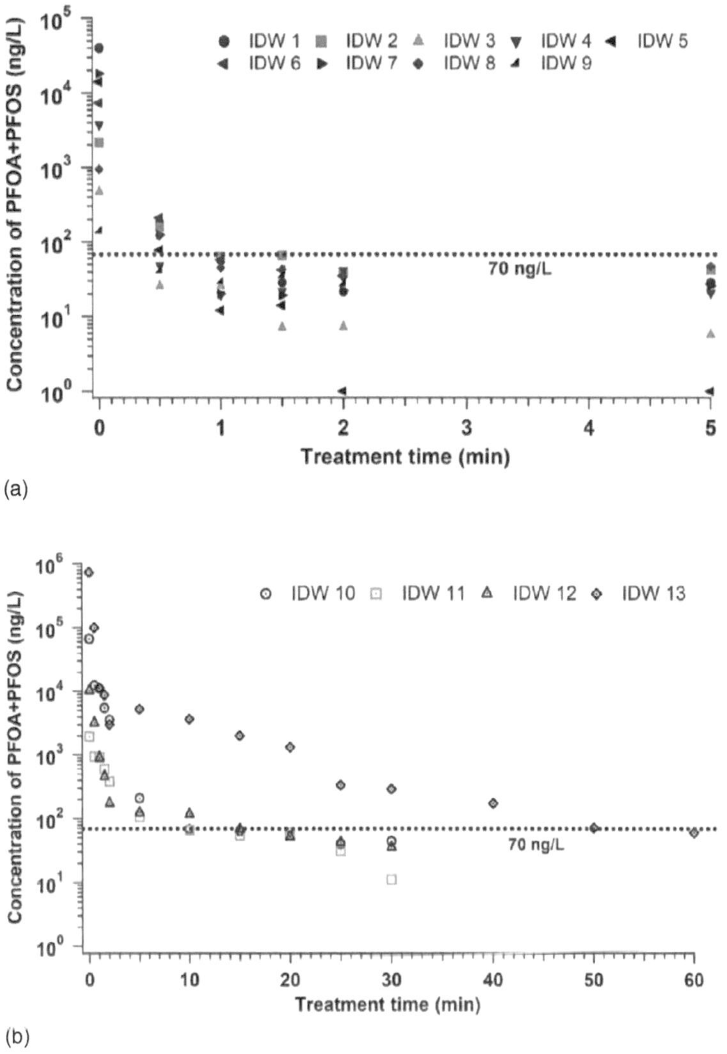

FIGURE 12.3 Degradation profiles of combined PFOA and PFOS concentrations in: (a) IDW samples 1 to 9 plotted for treatment time of five minutes and (b) IDW samples 10 to 13 plotted for treatment time of 60 minutes. The IDW treatment volume is 4 L. Blue horizontal dotted lines represent USEPA's HAL. Detection limits were approximately 9 ng/L.

degradation efficiency for some precursors could be attributed to their lower initial concentration relative to most other PFAAs as the degradation followed pseudo first order kinetics. Also, the high TOP concentrations measured indicate the presence of many unknown precursors which could also be the potential source of the identified precursors.

The time-dependent concentration profiles of PFAAs in IDW 12 (Figure 12.5) are typical of what was observed for all the samples. In general, the concentrations of the

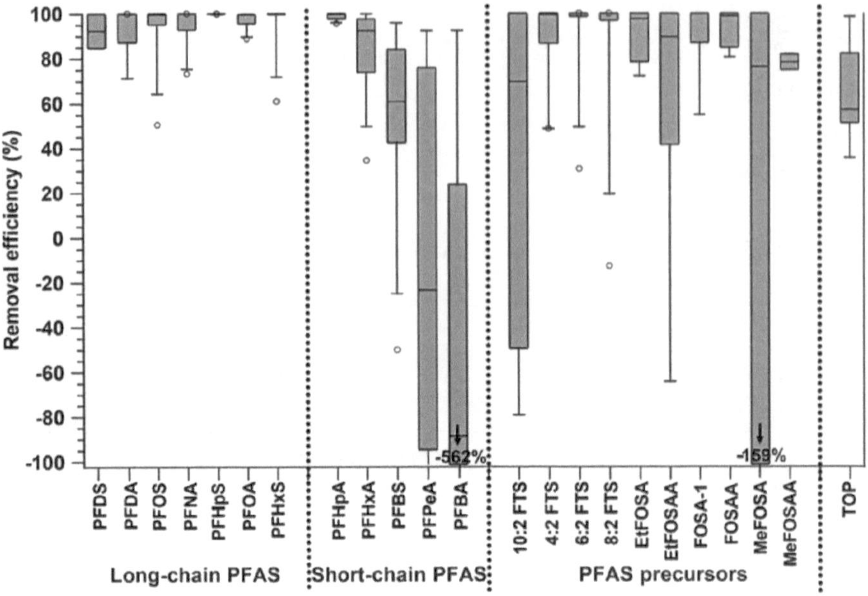

FIGURE 12.4 Box and Whisker plot showing the plasma removal efficiency (%) of long-chain PFAAs, short-chain PFAAs, PFAAs precursors and TOP in 13 different IDW samples. Removal efficiency is between +100 and −100%, where negative removal efficiency is due to the formation of short-chain PFAAs from degradation products of long-chain PFAAs. PFBA and MeFOSA had negative removal efficiencies of −562 % and −159 %, respectively, as indicated by the downward-pointing arrows. Treatment time used were 5 minutes for IDWs 1– 9, 30 minutes for IDW 10– 12, and 60 minutes for IDW 13.

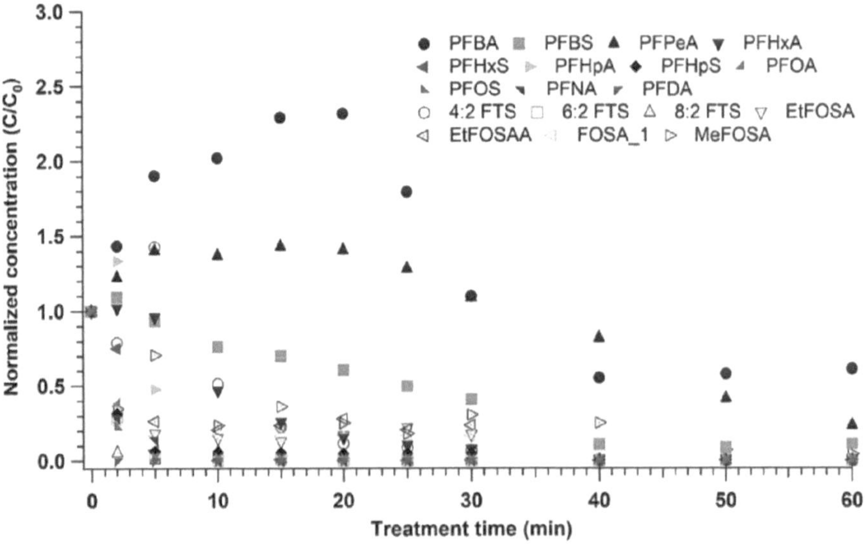

FIGURE 12.5 Degradation profiles of different PFAAs and precursors with treatment time in IDW 12 (representative) sample.

PFAAs decreased with treatment time. The exceptions were short-chain PFAAs (PFHpA, PFHxA, PFBS, PFPeA and PFBA) whose concentrations increased early in the treatment process. A decrease in the removal rate with a decrease in the PFAAs chain length (Figure 12.5) was likely due to several factors. First, short-chain PFAAs are formed from the degradation of long-chain PFAAs (Singh et al. 2019) and also from the oxidation of PFAAs precursors (Liu and Avendano 2013). Second, short-chain PFAAs exhibit decreased surface activity relative to longer-chain PFAAs which decreases their concentration (i.e., surface excess) at the plasma–liquid interface. In our previous study, we showed that surfactant-like molecules (e.g. PFAAs) accumulate at the gas–liquid interface where they are rapidly degraded by plasma-generated species (Mededovic Thagard et al., 2016a).

A Pearson's correlation analysis was performed to investigate the impact of IDW characteristics on pseudo first order removal rate constants for PFOA and PFOS ($k_{PFOA + PFOS}$) and explain the results shown in Figure 12.3. The integration of the kinetic rate expression was carried out between time zero and the time required for the concentration of PFOA+PFOS to drop below 70 ng/L. It should be noted that in some experiments the removal rates were so high that only three data points were used in the regression analysis. In those cases, the rate constants may have been underestimated since actual treatment times were shorter that those used in the calculation.

The impact of total PFAAs concentration (ng/L), combined initial PFOA and PFOS concentration (ng/L), TIP concentration (ng/L), TOP concentration (ng/L), PFAS+TOP concentration (ng/L), PFAAs+TIP (ng/L), solution conductivity (μS/cm), TOC (mg/L) and total fluorine (μg/L) on combined PFOA and PFOS (PFOA+PFOS) removal were investigated. Of these variables, the removal rate of PFOA+PFOS was significantly correlated only with TIP ($r = -0.55$, $p = 0.049$) and marginally correlated with total fluorine ($r = -0.49$, $p = 0.087$). Other parameters such as total PFAAs+TIP ($r = -0.45$, $p = 0.127$), total PFAAs ($r = -0.44$, $p = 0.131$), conductivity ($r = -0.42$, $p = 0.152$), PFAAs+TOP ($r = -0.39$, $p = 0.186$) and total PFOA+PFOS ($r = -0.39$, $p = 0.188$), were also negatively correlated with PFOA+PFOS removal rates; however, they were not statistically significant. TIP was the most influential parameter suggesting that the concentrations of precursors can decrease the PFAAs degradation rate by scavenging reactive species generated by the plasma and also by generating PFOA and PFOS. In addition, no correlation with TOC was observed ($r = -0.18$, $p = 0.55$), indicating that the presence of natural organic matter such as VOCs, humic acid, and fulvic acid does not affect the removal rates of PFAS. This is because the removal of PFOA and PFOS is taking place at the gas–liquid interface where (non-surfactant) co-contaminants are least likely to interfere, as Stratton et al. (2017) explained in detail. Some studies also reported the co-occurrence of non-PFAS surfactants such as hydrocarbons with PFAS in groundwaters (Richardson and Ternes 2018, Du et al. 2017). Although these hydrocarbons might compete with PFAS for the sites and reactive species at the gas–liquid interface, their presence is not expected to lower the PFAS treatment effectiveness because surfactant compounds (PFAS) are degraded directly in the plasma via "above-surface" reactions which proceed at extremely high rates; rates of reactions that occur beneath or at the plasma–liquid interface (e.g., hydrocarbons) are significantly lower (Mededovic Thagard et al., 2016a).

Longer times required for treating IDW 12 and 13 were the result of high TIP concentration (in addition to extremely high initial PFAA concentrations). IDW 11 required longer treatment times not because of TIP concentrations, but because the high concentration of ions reduced the contact area of the plasma and the treated liquid (Mededovic et al. 2016a; Wang et al. 2019). Due to a complex relation among total fluorine, TIP, and initial PFOA and PFOS concentrations, IDW 10 also required longer treatment times.

Singh et al. (2019) identified and quantified gaseous and liquid byproducts of PFAS degradation following a plasma treatment. Herein, the concentration of only inorganic fluoride ions before and after the treatment were measured. Of all the samples, only IDW samples 1, 3, 10, 12, and 13 were above the fluoride detection limit of 0.1 mg/L following the treatment. For those samples there was a significant increase in fluoride concentrations post-treatment (from 260 to 2490 μg/L) which accounted for 33 ± 14% of the organically bound fluorine in the measured PFAAs and their precursors. Also, unlike in Singh et al. (2019), the fluoride mass balance in this study was difficult to perform because the measured total PFAS concentration was low (ranged between tens and few hundreds of μg/L) compared to minimum quantification concentration of fluoride measurement required by both F electrode and IC (low ppm level). However, we measured total fluorine (TF) (organic and inorganic) before and after plasma treatment using combustion ion chromatography. A slight decrease in TF concentration (0 to 32%) was observed after the plasma treatment which could be attributed to sorption of some PFAS fraction onto the diffusers or the reactor walls.

12.3.4 Effect of Co-contaminants on PFAS Removal Efficiency

The fate of different VOCs and heavy metals comprising the solutions was investigated by measuring their concentrations before and after the treatment. The IDW samples contained measurable concentrations (0.26 to 28 μg/L) of 1,1-dichloroethane, 1,2-dichloroethane, bromodichloromethane, trichloromethane, chloroform, benzene, ethyl benzene, methyl tert-butyl ether, acetone, and xylene along with iron (0 to 2600 μg/L) and manganese (9 to 5000 μg/L), and had TOC concentrations in the range of 0.1 to 13 mg/L. The Pearson's correlation analysis did not reveal any significant correlations between initial VOCs ($r = -0.29$), iron ($r = -0.12$), and manganese ($r = -0.09$) concentrations and PFAA removal rates. Following plasma treatment, VOC concentrations in all samples were below detection limits likely due to stripping by the argon bubbling and simultaneous plasma treatment. Since argon was continuously purged and recycled, any stripped VOCs were continuously exposed to the plasma where they were most likely degraded, although their ultimate fate was not evaluated in this study. Following treatment, iron and manganese concentrations either increased or decreased compared to their initial concentrations. The mechanism(s) for this result is unknown but may involve bulk liquid redox reactions of metal ions formed by the electrolysis of stainless-steel ground electrodes.

12.3.5 Energy Requirements of Plasma Treatment for PFOA and PFOS Removal

The energy requirements for the treatment were calculated using the Electric Energy per Order (EE/O, kWh/m^3) parameter which represents the energy in kWh needed to degrade a contaminant by an order of magnitude in a unit volume (m^3) of contaminated water (Nzeribe et al. 2019):

$$EE/O = \frac{P \times t \times 1000}{V \times 60 \times \log\left(C_i / C_f\right)} \tag{12.1}$$

In Equation (12.1), P (kW) is total power consumed by the high voltage supply and the rotating spark gap motor (1.4 kW for all IDW samples), t (min) is the time required to achieve 90% degradation of contaminant (= 2.3035851/k where k is the pseudo-first order reaction rate constant (min^{-1})), V is the treatment volume in the reactor (4 L), and C_i and C_f are the initial and final concentrations of PFOA+PFOS (in µg/L).

The EE/O for the treatment of IDW 1–9 ranged from ~1.7–6.3 kWh/m^3 (IDW 1–9). Extremely high energy requirements were observed for IDW 10–13 (14.3 to 56.0 kWh/m^3) due to the longer treatment times (> 10 mins) required to reduce the initial concentrations of PFOA+PFOS.

Our recent review paper evaluated and compared the energy consumption (EE/O) and treatment costs of prepared PFOA-contaminated water and PFOS-contaminated water among plasma, advanced reduction processes (ARPs), electrochemical oxidation, sonolysis, heat-activated persulfate, and photocatalytic processes (Nzeribe et al. 2019). The reader is reminded that in this study, EE/O was calculated for the treatment of both PFOA and PFOS in actual groundwater. As reported by Nzeribe et al. (2019), for the similar initial concentrations of PFOA and PFOS, the energy demands for the treatment of PFOA are ~5000 kWh/m^3 for persulfate (Lee et al. 2010), photochemical oxidation (Chen et al. 2015), and sonolytic processes (Moriwaki et al. 2005). For the treatment of PFOS, the sonochemical treatment becomes prohibitively expensive (>20,000 kWh/m^3) (Moriwaki et al. 2005), whereas energy demands for photochemical oxidation decrease to non-competitive 377 kWh/m^3 (Jin et al. 2014). Electrochemical oxidation for PFOA treatment is significantly less energy intensive (EE/O = 132 kWh/m^3) than the other three technologies, but that cost is significantly increased for PFOS treatment (500 kWh/m^3) (Schaefer et al. 2017). With the EE/O value of 24 kWh/m^3 for the removal of PFOA in a prepared solution, ARPs fall within the same energy requirement range as the plasma process when treating extremely high concentrations of PFOA and PFOS and their precursors in groundwater (Qu et al. 2010). Therefore, electrical discharge plasma appears to be one of the most efficient technologies for the removal of PFOA and PFOS.

To the best of our knowledge this is the first study reporting large volume (4 L) treatment of PFAS-containing liquid IDW. The high concentrations of PFAAs and precursors were effectively and efficiently reduced by one to three orders of magnitude in the plasma reactor to below USEPA's HAL. The efficient removal of high levels of legacy PFAS and precursors makes the plasma treatment a promising solution for remediation of PFAS-contaminated IDW.

ACKNOWLEDGMENTS

This project was funded by the U.S. Air Force (Contract FA8903-17-C-0015). The authors would like to thank Monique Nixon, program officer of AFCEC, for her support. We sincerely acknowledge the help of Mohammad Rahman and Junda Ren for analysis of total fluorine using CIC.

REFERENCES

Anderson, R. H.; Long, G. C.; Porter, R.; Anderson, J. K. Occurrence of select perfluoroalkyl substances at US Air Force aqueous film-forming foam release sites other than fire-training areas: Field-validation of critical fate and transport properties *Chemosphere* 2016, 150, 678–685.

Anderson, R.; Hunter, D. T.; Adamson, Hans F. S. Partitioning of poly-and perfluoroalkyl substances from soil to groundwater within aqueous film-forming foam source zones *J. Contam. Hydrol.* 2019, 220, 59–65.

Appleman, T. D.; Dickenson, E. R. V.; Bellona, C.; Higgins, C. P. Nanofiltration and granular activated carbon treatment of perfluoroalkyl acids. *J. Hazard. Mat.* 2013, 260, 740–746.

Awad, E.; Zhang, X.; Bhavsar, S. P.; Petro, S.; Crozier, P. W.; Reiner, E. J.; Fletcher, R.; Tittlemier, S. A.; Braekevelt, E. Long-term environmental fate of perfluorinated compounds after accidental release at Toronto airport. *Environ. Sci. Technol.* 2011, 45, 8081–8089.

Backe, W. J.; Christensen, K. E.; Field, J. A. Zwitterionic cationic, and anionic fluorinated chemicals in aqueous film forming foam formulations and groundwater from U.S. Military bases by nonaqueous large-volume injection HPLC-MS/MS. *Environ. Sci. Technol.* 2013, 47, 5226–5234.

Campbell, T. Y.; Vecitis, C. D.; Mader, B. T.; Hoffmann, M. R. Perfluorinated surfactant chain-length effects on sonochemical kinetics *J. Phys. Chem. A* 2009, 113, 9834–9842.

Chen, M. J.; Lo, S. L.; Lee, Y. C.; Huang, C. C. Photocatalytic decomposition of perfluorooctanoic acid by transition-metal modified titanium dioxide. *J. Hazard. Mat.* 2015, 288, 168–175.

Conder, J. M.; Hoke, R. A.; DeWolf, W.; Russell, M. H.; Buck, R. C. Are PFCAs bioaccumulative? A critical review and comparison with regulatory criteria and persistent lipophilic compounds. *Environ. Sci. Technol.* 2008, 42, 995–1003.

Du, Z.; Deng, S.; Zhang, S.; Wang, W.; Wang, B.; Huang, J.; Wang, Y.; Yu, G.; Xing, B. Selective and fast adsorption of perfluorooctanesulfonate from wastewater by magnetic fluorinated vermiculite. *Environ. Sci. Technol.* 2017, 51, 8027–8035.

Ghisi, R.; Vamerali, T.; Manzetti, S. Accumulation of perfluorinated alkyl substances (PFAS) in agricultural plants: A review. *Env. Res.* 2019, 169, 326–341.

Giri, R.R.; Ozaki, H.; Okada, T.; Taniguchi, S.; Takanami, T. Factors influencing UV photodecomposition of perfluorooctanoic acid in water *Chem. Eng. J.* 2012, 180, 197–203.

Houtz, E. F.; Sedlak, D. L. Oxidative conversion as a means of detecting precursors to perfluoroalkyl acids in urban runoff. *Environ. Sci. Technol.* 2012, 46, 9342–9349.

Hu, X. C.; Andrews, D. Q.; Lindstrom, A. B.; Bruton, T. A.; Schaider, L. A.; Grandjean, P.; Lohmann, R.; Carignan, C. C.; Blum, A.; Balan, S. A.; Higgins, C. P.; Sunderland, E. M. Detection of poly- and perfluoroalkyl substances (PFASs) in U.S. drinking water linked to industrial sites, military fire training areas, and wastewater treatment plants *Environ. Sci. Technol. Lett.* 2016, 3, 344350.

Jin, L.; Zhang, P.; Shao, T.; Zhao, S. Ferric ion mediated photodecomposition of aqueous perfluorooctane sulfonate (PFOS) under UV irradiation and its mechanism. *J. Hazard. Mat.* 2014, 271, 9–15.

Lam, N. H.; Cho, C.-R.; Kannan, K.; Cho, H.-S. A nationwide survey of perfluorinated alkyl substances in waters, sediment and biota collected from aquatic environment in Vietnam: Distributions and bioconcentration profiles. *J. Hazard. Mat.* 2017, 323, 116–127.

Lee, Y.; Lo, S.; Chiueh, P.; Liou, Y.; Chen, M. Microwave-hydrothermal decomposition of perfluorooctanoic acid in water by iron-activated persulfate oxidation. *Wat. Res.* 2010, 44, 886–892.

Liu, C. S.; Higgins, C. P.; Wang, F.; Shih, K. Effect of temperature on oxidative transformation of perfluorooctanoic acid (PFOA) by persulfate activation in water *Sep. Purif. Technol.* 2012, 91, 46–51.

Liu, J.; Avendano, S. M. Microbial degradation of polyfluoroalkyl chemicals in the environment: A review. *Environ. Int.* 2013, 61, 98–114.

Mededovic Thagard, S.; Takashima, K.; Mizuno, A., Chemistry of the positive and negative electrical discharges formed in liquid water and above a gas–liquid surface Plasm. *Chem. Plasm. Process.* 2009, 29, 455–473.

Mededovic Thagard, S.; Stratton, G. R.; Dai, F.; Bellona, C. L.; Holsen, T. M.; Bohl, D. G.; Paek, E.; Dickenson, E. R. Plasma-based water treatment: development of a general mechanistic model to estimate the treatability of different types of contaminants. *J. Phys. D: Appl. Phys.* 2016a, 50, 014003.

Mededovic Thagard, S.; Stratton, G.; Bellona, C.; Holsen, T. M. Enhanced contact electrical discharge plasma reactor for liquid and gas processing, 2016b, application number 15/018780.

Moriwaki, H.; Takagi, Y.; Tanaka, M.; Tsuruho, K.; Okitsu, K.; Maeda, Y. Sonochemical decomposition of perfluorooctane sulfonate and perfluorooctanoic acid. *Environ. Sci. Technol.* 2005, 39, 3388–3392.

Nzeribe Blossom, M.; Crimi, M.; Mededovic Thagard, S.; Holsen, T. M. Physico-chemical processes for the treatment of per- and polyfluorinated alkyl substances (PFAS): A review. *Critic. Rev. Env. Sci. Tech.* 2019, 49, 866–915.

Oakes, K. D.; Benskin, J. P.; Martin, J. W.; Ings, J. S.; Heinrichs, J. Y.; Dixon, D. G.; Servos, M. R. Biomonitoring of perfluorochemicals and toxicity to the downstream fish community of Etobicoke Creek following deployment of aqueous film-forming foam. *Aquatic Toxicol.* 2010, 98, 120–129.

Qu, Y.; Zhang, C.; Li, F.; Chen, J.; Zhou, Q. Photo-reductive defluorination of perfluorooctanoic acid in water. *Wat. Res.* 2010, 44, 2939–2947.

Richardson, S. D.; Ternes, T. A. Water analysis: emerging contaminants and current Issues, *Anal. Chem.* 2018, 90, 398–428.

Schaefer, C. E.; Andaya, C.; Burant, A.; Condee, C. W.; Urtiaga, A.; Strathmann, T. J.; Higgins, C. P. Electrochemical treatment of perfluorooctanoic acid and perfluorooctane sulfonate: Insights into mechanisms and application to groundwater treatment. *Chem. Eng. J.* 2017, 317, 424–432.

Schultz, M. M.; Day, T. C.; Field, J. A. Quantitative determination of fluorotelomer sulfonates in groundwater by LC MS/MS. *Environ. Sci. Technol.* 2004, 38, 1828–1835.

Singh, R. K.; Fernando, S.; Baygi, S. F.; Multari, N.; Mededovic Thagard, S.; Holsen, T. M. Breakdown products from perfluorinated alkyl substances (PFAS) degradation in a plasma-based water treatment process. *Environ. Sci. Technol.* 2019, 53, 2731–2738.

Stratton, G. R.; Bellona, C. L.; Dai, F.; Holsen, T. M.; Mededovic Thagard, S. Plasma-based water treatment: Conception and application of a new general principle for reactor design. *Chem. Eng. J.* 2015, 273, 543–550.

Stratton, G. R.; Dai, F.; Bellona, C. L.; Holsen, T. M.; Dickenson, E. R. V.; Mededovic Thagard, S. Plasma-based water treatment: Efficient transformation of perfluoroalkyl substances in prepared solutions and contaminated groundwater. *Environ. Sci. Technol.* 2017, 51, 1643–1648.

Sunka, P.; Babicky, V.; Clupek, M.; Lukes, P.; Simek, M.; Schmidt, J.; Cernak, M., Generation of chemically active species by electrical discharges in water Plasm. *Sourc. Sci. Tech.* 1999, 8, 258–265.

Wang, H.; Wandell, R. J.; Tachibana, K.; Voráč, J.; Locke, B. R. The influence of liquid conductivity on electrical breakdown and hydrogen peroxide production in a nanosecond pulsed plasma discharge generated in a water-film plasma reactor. *J. Phys. D: Appl. Phys.* 2019, 52, 075201.

Woodard, S.; Berry, J.; Newman, B. Ion exchange resins for PFAS removal and pilot test comparison to GAC. *Remed.* 2017, 27, 19–27.

Zhuo, Q. F.; Deng, S. B.; Yang, B.; Huang, J.; Yu, G. Efficient electrochemical oxidation of perfluorooctanoate using a Ti/SnO2-Sb-Bi anode *Environ. Sci. Technol.* 2011, 45, 2973–2979.

13 Recent Advances in Oxidation and Reduction Processes for Treatment of PFAS in Water

Yaal Lester
Azrieli College of Engineering, Israel

CONTENTS

13.1 Introduction 272
13.2 Advanced Oxidations 272
 13.2.1 Activated Persulfate 273
 13.2.2 Electrochemical Oxidation 274
 13.2.3 Ultrasonication 275
13.3 Chemical Reduction 276
13.4 Advanced Reductions 277
 13.4.1 Examined ARPs 278
 13.4.2 Reaction Mechanism 279
 13.4.3 Structure–Reactivity Relationship 279
13.5 Impact of Environmental and Operational Parameters on ARPs 280
 13.5.1 Dissolved Oxygen 280
 13.5.2 Dissolved Organic Matter and NO_3^- 281
 13.5.3 Water pH 281
 13.5.4 Operational Parameters: Sensitizer Concentration 282
13.6 Plasma Technologies 282
13.7 Full-Scale Considerations 283
 13.7.1 Removal of PFAS During Conventional Full-Scale Treatments 283
 13.7.2 Specific Applications: Treating Concentrated Waste Streams 283
13.8 Summary and Conclusions 284
 13.8.1 Oxidation Processes 285
 13.8.2 Reduction Processes 285
References 286

13.1 INTRODUCTION

Contaminated drinking water has been recognized by regulators and the scientific community as an important human exposure path to perfluoroalkyl substances (PFAS). Situations where locally contaminated drinking water resources were linked with increased PFAS blood levels have been widely documented in Germany (Wilhelm et al. 2009), Japan (Saito et al. 2003), and the U.S. (Emmett et al. 2006; Hoff 2009; Mohr 2010; NY Department of Health 2019; U.S. Department of Health and Human Services 2019).

A key barrier in the reduction of human exposure to drinking water PFAS is the development of effective water treatment technologies. PFAS are extremely stable, hence resist most conventional water treatment technologies such as coagulation/flocculation/sedimentation, filtration, biofiltration, and membrane filtration (Appleman et al. 2014; Merino et al. 2016; Rahman et al. 2014). A large number of studies showed that concentration of PFAS remain constant (and even increase) during full-scale water and wastewater treatments, even when relatively advanced tertiary treatments are applied (Blackbeard et al. 2016). Other advanced treatments, such as activated carbon, high-pressure membranes and ion exchange, showed greater success (Rahman et al. 2014); however, these technologies only concentrate the contaminants and transfer them from aqueous to the solid phase (which needs further treatment). More importantly, these techniques are much less effective for the removal of short-chain compounds, which are gaining increased attention in the regulatory community (Mueller and Yingling 2018).

Oxidation and reduction (redox) processes are considered highly attractive alternatives for PFAS mitigation, principally due to their destructive nature. In other words, these treatments do not produce liquid waste streams requiring additional treatment, nor spent adsorbent material for disposal. Notably, advanced oxidation and reduction processes (AOPs and ARPs respectively) employ powerful radicals such as •OH and hydrated electron (e_{aq}^-) to transform organic contaminants into harmless products. In the previous book of this series, Kempisty et al. (2018) elaborated on the mechanisms of different oxidation and reduction approaches and their application for PFAS degradation (Chapter 12). In this chapter we present recent studies and developments (published during 2018–2020) and underline processes which show higher potential than others.

13.2 ADVANCED OXIDATIONS

Due to their high chemical stability, PFAS are generally not degraded by conventional oxidants such as chlorine species, ozone, potassium permanganate, and hydrogen peroxide. In these cases, advanced oxidation processes (AOPs) are often employed. During AOPs, oxidation of contaminants takes place through their reaction with reactive free radicals, mostly hydroxyl •OH. Radicals are typically generated *on-site* by activating chemical oxidants such as ozone (O_3) and hydrogen peroxide (H_2O_2) with different activation techniques (e.g., UV, ultrasound, microwave). Popular AOPs include the combination of O_3 and H_2O_2 (O_3/H_2O_2), Fenton reactions, UV/titanium dioxide (UV/TiO_2), UV/H_2O_2, and activated persulfate.

Others, less conventional AOPs, are ultrasonication, electro-oxidation and plasma technologies.

Although AOPs are considered highly effective in degrading recalcitrant organic contaminants, their successful application for treating PFAS in water is questionable. Applying AOPs for PFAS mitigation aims to destroy the C–F bonds to form F^- ions, which can combine (for example) with Ca^{2+} to form non-toxic, insoluble CaF_2 (Hori et al. 2004). This reaction, however, is unfavorable due to the chemical inertness of the C–F bond and the high electronegativity of fluorine. The •OH attack of organic compounds usually takes place through H-abstraction or electron transfer; the lack of available hydrogen on PFAS disables H-abstraction, while fluorine electronegativity hampers electron transfer. The upper limit for the second-order rate constant of •OH reaction with PFAS is in the range of 10^6 1/Ms (for $C \geq 2$), which is three orders of magnitude lower than common •OH reaction rates with more conventional water contaminants (Lutze et al. 2018). These relatively low rate constants often lead to poor defluorination under typical AOP conditions (Glover et al. 2018; Nzeribe et al. 2019). Several AOPs were tested over the last years for treating PFAS in water and wastewater, as described below.

13.2.1 Activated Persulfate

Activated persulfate is commonly applied for environmental remediation and the degradation of organic contaminants (Bruton and Sedlak, 2017). The process involves the production of sulfate ($SO_4^{\bullet -}$) and •OH radicals by activating sodium persulfate ($Na_2S_2O_8$) with thermal, transition metal, or UV activation methods. In their previous book, Kempisty et al. (2018) showed that activated persulfate can decompose perfluorinated carboxylic acids (PFCA) (and to a lesser extent perfluorinated sulfonic acids - PFOS); however, this would only be under the conditions of high persulfate doses, high temperature or extreme pH.

Different hypothetical reaction mechanisms were presented for PFCA decomposition, such as the initial attack of the compound by sulfate radicals and the cleavage of the carbon–carbon (C–C) bond between the perfluorinated backbone and the carboxyl functional group (COOH). This is followed by reaction of the unstable perfluoroalkyl radicals with water to form •OH and thermally unstable alcohol ($C_nF_{2n+1}OH$). The alcohol undergoes HF elimination and hydrolysis, resulting in a CF_2 unit shortened PFCA (Yin et al. 2016). This series of reactions continues until all PFCA are mineralized to fluoride and carbon dioxide (Figure 13.1) (Hori et al. 2010, Lee et al. 2012, Yin et al. 2016).

In a more recent study, Lutze et al. (2018) confirmed the basic ability of sulfate radicals to degrade and mineralize PFCAs, and provided new insights into the mechanism, kinetics, and cost-effectiveness of the activated sulfate process. The researchers determined that PFCA (C4-C8) reacts with sulfate radicals at a kinetics range of 10^4 1/Ms, independent of chain length. Using the calculated reaction rates, Lutze and his coworkers concluded that treating PFCA with sulfate radicals, generated by UV/$S_2O_8^{2-}$, is, however, costly and impractical, with energy demand as high as 55 kW h m^{-3} for 90% degradation in pure water.

A different approach was taken by Wang et al. (2019), who developed a mechanochemical method to treat PFAS. The method used a ball milling reactor, with alumina

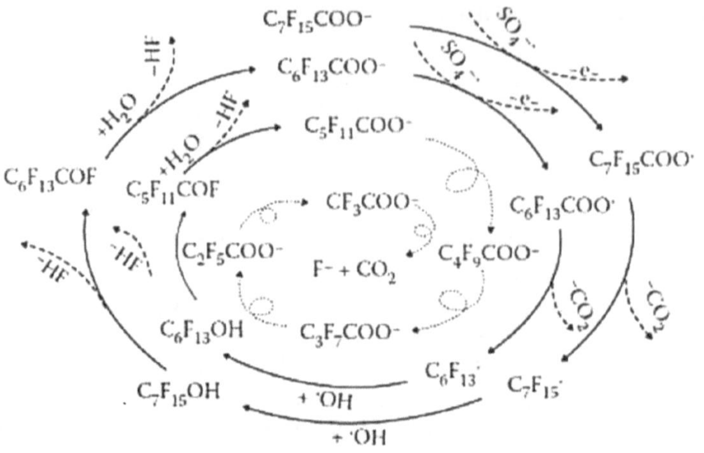

FIGURE 13.1 Proposed degradation pathway of PFOA by heat-activated persulfate. Reprinted with permission from Yin et al. 2016.

(Al_2O_3) and persulfate as co-milling agents. The mechanical effects activated the persulfate and produced $SO_4^{\bullet -}$ radicals. Complete (100%) defluorination was achieved after 2 h of treatment for different chain-length PFCA, whereas PFOS was also reduced but to lesser extent (56% degradation and 9.7% defluorination).

13.2.2 Electrochemical Oxidation

Electrochemical oxidation is considered a relatively effective AOP, as it combines direct oxidation of (adsorbed) contaminants by electron transfer at the anode and indirect degradation of contaminants in the bulk liquid, through reaction with •OH produced by anodic oxidation of water (Radjenovic & Sedlak, 2015). Kempisty et al. (2018) summarized some of the principal proposed mechanisms for PFAS degradation by electrochemical oxidation. These mechanisms are typically initiated by direct electron transfer from the contaminant to the anode, followed by the formation of PFAS radical which is subsequently defluorinated in a series of reactions.

While electrochemical oxidation was successful in degrading PFAS in large numbers of lab-scale studies (Barisci and Suri 2020; Kempisty et al. 2018; Le et al. 2019; Wang et al. 2020b), its large-scale application is still limited. The reasons are mainly associated with the cost and efficacy of full-scale electrodes and the high energy consumption (Schaefer et al. 2015). Previous and recent studies all focused on lab-scale systems, evaluating different electrodes such as titanium suboxide (Y. Wang et al. 2020b) and boron-doped diamond (Barisci and Suri 2020), analyzing the effect of background water constituents (Gomez-Ruiz et al. 2017) and developing potential applications or combining electrochemical oxidation with other technologies.

For example, Soriano et al. (2017) presented a strategy for removing and degrading perfluorohexanoic acid (PFHxA) from industrial wastewater, comprised of nanofiltration (NF) separation followed by electrochemical degradation of the NF concentrate. Destruction of PFHxA, which is a short-chain PFAS (C-6), is extremely interesting, since this compound appears to be one of the popular alternatives to long-chain PFAS. The examined NF membrane (NF270, Dow/Filmtech) rejected up to 99.4% of PFHxA without any noticeable membrane fouling, while electro-oxidation (with boron-doped diamond electrodes) degraded as high as 98% of PFHxA retained in the NF concentrate, with more than 95% reduction of total organic carbon. Gomez-Ruiz (2017) used electro-oxidation with a boron-doped diamond (BDD) anode to degrade PFAS in industrial wastewater. Current density of 50 mA/cm^2 and 10 h of treatment resulted in 99.7% PFASs removal, high TOC removal of > 90%, and extended defluorination.

13.2.3 Ultrasonication

Ultrasonication was also presented in the previous book of this series as a potential mean for PFAS treatment. Ultrasonication relies on the generation of cavitation bubbles in water by acoustic waves, at frequencies ranging from 20 to 1000 kHz (Furuta et al. 2004). The collapse of these micro-bubbles releases transient high temperature and pressure, and generates reactive oxygen species (ROS) which further decompose organic contaminants through oxidation or pyrolysis (Rayaroth et al. 2016). Kempisty et al. (2018) summarized some of the plausible PFAS degradation mechanisms during ultrasonication, which are mostly initiated by the adsorption of PFAS onto the bubble–water interface, followed by the bubble collapse and thermal cleavage of a C–C or the C bond connecting the chain with the functional head-group. Other aspects presented earlier relate to the impact of operational (frequency) and environmental parameters (pH, temperature), and the combination of the process with other oxidants. The general conclusion was that ultrasonication may successfully degrade PFAS at bench-scale systems; however, its severe scaling up problems presently prevent its wider implementation.

Combining ultrasonic with other methods/chemicals may nevertheless offer a path for cost-effective applications, as was recently demonstrated by Sekiguchi et al. (2017) and Hu et al. (2018). Sekiguchi and his coworkers (2017) examined the decomposition of hydrophobic PFAS by ultrasonic irradiation combined with short-wavelength UV light (low pressure UV lamp emitting at 185 and 254 nm). This treatment sequence attempted to solve the fact that ultrasonication is mostly limited to hydrophobic PFAS with low efficacy in degrading hydrophilic compounds. Hydrophobic PFAS are sorbed to the cavity and thermally decomposed in the cavities itself or at the cavity–water interface. Degradation of hydrophobic PFAS often lead to the formation of shorter, hydrophilic PFAS, which remain intact in solution. Sekiguchi et al. (2017) showed that, indeed, ultrasonication can decompose hydrophobic PFAS at a relatively high rate, which generates hydrophilic PFAS intermediates. Some of these intermediates, mostly ones with light-absorbing carboxyl head-groups, can be subsequently decomposed by UV irradiation. Hu et al. (2018) investigated the combination of ultrasonication with permanganate to degrade perfluorooctanoic acid (PFOA) in aqueous solutions. They showed that the addition of permanganate increases PFOA defluorination by more than 5-fold (compared to ultrasonication alone). They concluded that MnO_2 nanoparticles were generated during treatment and catalyzed PFOA degradation by enhancing pyrolytic decomposition and oxidation by permanganate in the vicinity of the particle's surface.

As mentioned above, scalability is a key barrier for the wider application of ultrasonication in the treatment of PFAS, and studies employing large-scale systems are extremely scarce. Constructing a cost-effective large-scale sonochemical reactor requires the design and optimization of large numbers of parameters, including the number and locations of transducers, frequency of operation, geometries of reactors, power dissipation per unit volume, and cavitational yield. Gole et al. (2018) used a 91-L multi-transducer sonochemical reactor to treat two commercially available aqueous film-forming foams (AFFFs), and conducted an extended parametric study. They showed that defluorination was better for dual frequency (1 MHz and 500 kHz) than a single one (1 MHz), due to the synergic effects of dual frequencies on the size and life span of the cavity. This leads to improved contact of PFAS and ions with the bubble cavity, which ultimately increase the rate of defluorination (Prabhu et al. 2004). Other operational parameters included pH and dilution factor, which both showed inverse correlation to the rate of defluorination (Gole et al. 2018).

13.3 CHEMICAL REDUCTION

Reductive degradation of PFAS typically involves direct electron transfer from the reducing agent to the contaminant, and the release of fluoride ions. Although this mechanism is highly desirable (results in non-fluorinated product), the kinetics are often too slow or adversely affected by dissolved oxygen and other background water constituents. The chemical reductants frequently investigated for PFAS mitigation are zero valent metals, specifically, zero valent iron (ZVI).

Zero valent iron (ZVI) is mostly used for in-situ groundwater remediation. It is relatively inexpensive and abundant, and was successful in degrading a wide range of environmental contaminants, including halogenated compounds such as trichloroethylene - TCE (Phillips et al. 2010; Truex et al. 2011). Hori et al. (2006) showed that ZVI could degrade and defluorinate PFOS and other shorter-chain (C2–C6) perfluoroalkyl sulfonates in water, albeit under extreme temperature conditions of 350°C. The specific method was successfully applied to decompose PFOS in an antireflective coating agent used in semiconductor manufacturing.

To improve ZVI and enable its use for PFAS mitigation under environmental conditions, two approaches were recently examined: (i) developing nano-scale ZVI (nZVI); and (ii) combining ZVI with other materials. Nanomaterials in general and nZVI in particular are increasingly employed for water decontamination. The nano size of these materials results in high specific surface area, leading to increased reactivity (as heterogenic reactions take place on the surface). Arvaniti et al. (2015) evaluated the use of Mg-aminoclay coated nZVI for PFAS mitigation in water under environmental conditions. Coating of nZVI is required to prevent their aggregation and loss of effectiveness, which often take place due to the small size of the particles. The tested nZVI was much more efficient in degrading large numbers of PFAS (up to 96% removal) than commercial nZVI powder and uncoated nZVI. Degradation mechanisms included adsorption as a first step followed by (partial) reduction and defluorination.

Liu et al. (2020) demonstrated the use of ZVI coupled with biochar for treating seven PFAS of different headgroups and chain-lengths in water under environmental conditions. They showed that the combination ZVI and biochar was more effective than ZVI alone, and the removal rate decreased with a decrease of chain length. Removal mechanism was a combination of sorption and reductive defluorination, as was implied by the increase in aqueous fluoride ions during treatment. The treatment is considered cost-effective for on-site remediation of PFAS-contaminated water due to the low cost of the material (ZVI and biochar).

13.4 ADVANCED REDUCTIONS

Advanced reduction processes (ARPs) are a relatively new, underexplored class of water treatment processes, with high PFASs degradation potential. This technology is analogous to the above-mentioned AOPs, which typically combine an activation method and a chemical agent to generate highly reactive radicals. However, while AOPs produce highly reactive oxidants, ARPs generate reducing radicals such as hydrated electrons (e_{aq}^-) and sulfite radical anions ($SO_3^{\bullet-}$). This is commonly done by applying a source of energy (e.g., UV light, gamma-rays) to activate conventional reductants (e.g., sulfite, iodide) (Vellanki et al. 2013):

$$SO_3^{2-} + h\nu \rightarrow SO_3^{\bullet-} + e_{aq}- \tag{13.1}$$

$$I^- + H_2O + h\nu \rightarrow I\bullet + e_{aq}- \tag{13.2}$$

$$SO_3^{\bullet-} / e_{aq} - + PFAS \rightarrow Product \tag{13.3}$$

It was generally accepted that e_{aq}^- is the dominant radical responsible for PFAS degradation during ARPs. This radical is a strong reducing agent which may lead to the cleavage of the C–F bonds, the elimination of fluoride, and the eventual generation of non-fluorinated compounds (schematically presented in Figure 13.2).

Relatively limited data is available on the application of ARPs for PFAS mitigation, with fewer than ten scientific publications mentioned in Kempisty et al. (2018). These early studies demonstrated the degradation of PFAS by UV-KI (e.g., Qu et al. 2010), UV-sulfite (e.g. Vellanki et al. 2013) and UV-dithionite, focusing on

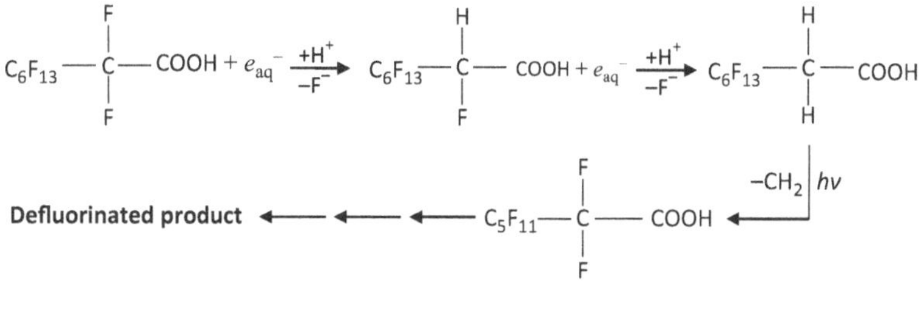

FIGURE 13.2 Schematic defluorination mechanism of PFOA by e_{aq}^- adapted from Song et al. 2013.

degradation mechanisms and the impact of different parameters such as the type of PFAS (ionic headgroup and chain length), concentration of chemical reductant, and the presence of oxygen (a well-known reducing radicals' scavenger).

Recent studies published during 2018–2020 evaluate alternative ARPs (other than conventional UV-KI and UV-sulfite), further elucidate the mechanism of PFAS degradation by e_{aq}^-, and establish a structure–reactivity relationship to enable prediction of degradation of a wide spectrum of PFAS (Bentel et al. 2019, Cui et al., 2020).

13.4.1 Examined ARPs

As mentioned above, most of the early work on ARPs focused on conventional processes, principally UV-KI and UV-sulfite. Attempts to improve these processes have been carried out in recent years. Yu et al. (2018) proposed the combination of sulfite/iodide/UV as an efficient method for producing e_{aq}^-. While sulfite and iodide had a positive synergetic effect on e_{aq}^- production, iodide contribution dominated due to its much higher absorptivity and the higher quantum yield of e_{aq}^-. The treatment's efficiency increases with pH, but the process was still effective under weakly acidic conditions and in field surface water. Bao et al. (2019) used UV/sulfite to treat commercial chlorinated polyfluorinated ether sulfonate (trade name F-53B), an alternative chemical to PFOS, frequently used in the manufacturing of chrome mist suppressants in China. Their study showed that chlorinated polyfluorinated ether sulfonate is much more degradable than PFOS, with complete degradation (below detection limit) after 1 min in ultrapure water and 60 min in chrome plating industrial wastewater. Stoichiometric fluoride recovery was also achieved, confirming significant defluorination.

Recently, Trojanowicz and his coworkers (2019, 2020) evaluated the decomposition of PFOA and PFOS by gamma-rays and electron beam (EB) processes. The basic mechanism of these processes involves irradiation of the water matrix with an electron beam from electron accelerators or gamma-rays from different isotope sources (usually ^{60}Co or ^{137}Cs). This leads to high yields of free radicals (principally •OH, •H and e_{aq}^-), which subsequently attack and degrade water contaminants (Equation 13.4).

$$H_2O + \gamma / EB \rightarrow \bullet OH, e_{aq}^-, \bullet H \quad (13.4)$$

Trojanowicz et al. (2019) found that degradation of PFOA under environmental conditions was ineffective for both gamma-rays and EB processes. To improve degradation, conditions had to be modified to create a reductive environment: deaeration and highly alkaline pH (where e_{aq}^- is predominant). Subsequently, 90% of PFOA was decomposed during gamma-irradiation at a 10 kGy radiation dose (below detection limit at 25 kGy) in Ar-saturated solution and $pH \geq 12.5$. A lower removal rate of 50% at 80 kGy was obtained for EB under similar conditions (Trojanowicz et al. 2019). Degradation of PFOS by gamma-rays and EB was generally less effective than PFOA with removal rates of < 65% for gamma-rays and < 40% for EB in most cases (Trojanowicz et al. 2020). Cleavage of C–C bonds by e_{aq}^- was identified as the dominant mechanism of PFOA radiolysis with only minor contribution from direct attack on the strong C–F bond.

In an attempt to improve PFAS degradation under ambient conditions, formate was added to the aqueous solution (Trojanowicz et al. 2019, 2020). Formate may react with free radicals to generate the highly reactive carbonate radicals, known to degrade organic contaminants:

$$HCOO^- + \bullet OH \rightarrow CO_2^{\bullet -} + H_2O \; k = 3.2 \times 10^9 \, 1/Ms \quad (13.5)$$

$$HCOO^- + \bullet H \rightarrow CO_2^{\bullet -} + H_2 \; k = 2.1 \times 10^8 \, 1/Ms \quad (13.6)$$

Indeed, the addition of formate increased the degradation rate of PFAS for both gamma-radiation and EB in aerated solution (albeit at pH 12). This emphasizes the potential of this process as a means of avoiding an expensive deaeration phase.

13.4.2 Reaction Mechanism

Recently, Bentel et al. (2019) have confirmed that two parallel reaction pathways take place during degradation of PFOA by e_{aq}^-: (i) exchange of fluorine by hydrogens (H/F); and (ii) chain shortening. In the first path, two H/F exchange takes place on the α-position adjacent to the carboxyl group ($C_{n-1}F_{2n-1}$–CF_2–COO^-), yielding the less-fluorinated compound $C_{n-1}F_{2n-1}$–CH_2–COO^-. Additional H/F exchange is possible for long-chain PFOA. The second path involves the generation of unstable perfluorinated alcohols (C_nF_{2n+1}–OH) through decarboxylation and hydroxylation, and the subsequent elimination of fluoride ion. The formed acyl fluoride undergoes hydrolysis, release a second fluoride and transforms into a shorter-chain PFOA ($C_{n-1}F_{2n-1}$–COO^-). This decarboxylation–hydroxylation–elimination–hydrolysis cycle (DHEH) continues until the compound is fully degraded (Bentel et al. 2019, Cui et al. 2020).

PFOS degradation takes place via three putative mechanisms, slightly different than PFOA (owing to their different head-groups): (i) Desulfonation; (ii) H/F exchange; and (iii) chain shortening through C–C cleavage. During desulfonation, the C–S bond between the headgroup and the perfluoroalkyl chain is attacked and broken by e_{aq}^-, likely producing PFOA as an intermediate. The C–S bond has relatively low energy (272 kJ/mol) compared to the C–C bond (346 kJ/mol) and longer length: 4.463Å compared to 1.529 Å –1.627Å for C–C and 1.651Å for S–O. H/F exchange may take place at the PFOA by-product or at one of the central C–F bonds of the parent PFOS, which have a relatively low dissociation energy (446.31 kJ/mol) compared to other C–F bonds in the chain (Cui et al., 2020). Cleavage of C–C bonds on PFOS may take place following the attachment of e_{aq}^- on the middle –CF_2–, which facilitates the breakdown of the adjacent C–C bond.

13.4.3 Structure–Reactivity Relationship

Early and recent studies have shown that PFAS degradation mechanism and kinetics depend, in most cases, on the headgroup and the fluoroalkyl chain length. Bentel et al. (2019) concluded that the two headgroup types impact the rate of PFAS defluorination differently. For PFCA (n ≥ 2), the carboxylic headgroup accelerates defluorination, as implied by the fact that defluorination is slower than the disappearance of the parent compound during ARP (hence, removal of the headgroup slowed down defluorination). An opposite phenomenon was observed for PFSA, where the sulfonic acid headgroup seemed to limit defluorination rate.

Chain length impacts mostly PFSA degradation, with little effect on PFCA. This observation is valid for the degradation of the parent compound as well as for defluorination (Bentel et al. 2019; Park et al. 2009). For PFAS, increase in chain length significantly increases degradation and defluorination rates. For example, Bentel et al. (2019) measured marginal defluorination of CF_3SO_3 by e_{aq}^-, compared to PFAS with longer alkyl chain. The decrease in defluorination rate at decreasing chain lengths is ascribed to the high strength of primary and secondary C–F bonds in a short chain and the more "available" middle $-CF_2-$ bonds in long chains.

13.5 IMPACT OF ENVIRONMENTAL AND OPERATIONAL PARAMETERS ON ARPs

One of the principal drawbacks of ARPs, which limit their wider use at full-scale, is the impact of environmental parameters, more specifically: dissolved oxygen (DO) and other water constituents, and pH.

13.5.1 Dissolved Oxygen

Dissolved oxygen (DO) is a well-known scavenger of reducing radicals, the most important one in natural waters. Oxygen readily reacts with e_{aq}^- and other reducing radicals at a near diffusion rate (Buxton et al. 1988; Guo et al. 2013) which highly reduces available reductants and PFAS degradation rate. In fact, due to the adverse effect of DO, ARPs are rarely applied to treat aerated water (open to the atmosphere) with no nitrogen purge (Cui et al. 2020). For example, Song et al. (2013) observed a more than 10-fold decrease in defluorination of PFAS, when changing conditions of treated water from nitrogen-purged to air–purged (Equations 13.7–13.9).

$$e^- + O_2 \rightarrow O_2^{\bullet-} \quad k = 1\text{x}10^{10}\ 1/\text{Ms} \tag{13.7}$$

$$e^- + O_2^{\bullet-} \rightarrow O_2^- \quad k = 1.3\text{x}10^{10}\ 1/\text{Ms} \tag{13.8}$$

$$SO_3^{\bullet-} + O_2 \rightarrow SO_5^{\bullet-} \quad k = 1\text{x}10^{9}\ 1/\text{Ms} \tag{13.9}$$

Methods to mitigate the impact of DO are rare. Gu et al. (2016) employed high photon flux UV-sulfite to promote e_{aq}^- production and overcome the scavenging of DO (and other scavengers). A high-pressure mercury UV lamp which emits a much higher flux and broader spectrum than the more conventional low- and medium-pressure UV lamps was used as a light source. Experiments showed efficient degradation of PFOS, even in the presence of 5 mg/L DO and 7 mg/L NO_3^-- N (another e_{aq}^- scavenger). Sun et al. (2018b) used a different approach, applying nitrilotriacetic acid (NTA) as e_{aq}^- sensitizer in combination with low-pressure mercury UV lamp. The sensitivity of this system to DO was relatively low: PFOS degradation and defluorination were decreased by merely 12% and 23%, respectively, when shifting from N_2- to air-saturated water. The low DO sensitivity was attributed to NTA and its UV-excited species (NTA*), which react with oxygen and other e_{aq}^- scavengers, decreasing their concentration and, subsequently, enhancing the degradation rate of PFOS.

13.5.2 Dissolved Organic Matter and NO_3^-

Other water constituents known for their potential impact on ARPs are dissolved organic matter (DOM) and NO_3^-, both ubiquitous in natural waters. These solutes may act as light absorbers (screening e_{aq}^- sensitizer) or radical sensitizers and scavengers. This multi-mechanism effect often leads to different, sometimes even opposing, kinetics of PFAS degradation in the presence of DOM and NO_3^-. For example, numerous studies showed that DOM (specifically humic acid – HA) has an accelerating effect on ARP degradation of PFAS at low HA concentrations, whereas high HA levels decreased PFAS degradation rate (Cui et al. 2020, Guo et al. 2019). HA is believed to enhance the degradation of PFAS directly through photogeneration of e_{aq}^-, or indirectly through enhancing e_{aq}^- generation by other sensitizers (Cui et al. 2020). An example of an important indirect path is the HA reaction with I-containing intermediates in the UV-iodide process (e.g., HOI, IO_3) and the recycling of iodide into the treatment (Sun et al. 2017). The inhibiting mechanisms of DOM are principally light absorption and blocking from other sensitizers, and scavenging of e_{aq}^- either directly or indirectly by photo-generated reactive species (•OH, $^1O_2$, H_2O_2, and excited triplet state DOM) (Cui et al. 2020).

Nitrate (NO_3^-) is an effective e_{aq}^- scavenger with reaction rate constant in the 10^{10} 1/Ms range (Cui et al. 2020). Hence, its presence typically decreases ARP degradation rate of PFAS, especially at high NO_3^- concentrations of 30–60 mg/L (Equations 10–12) (Gu et al. 2017).

$$NO_3^- + e_{aq}^- \rightarrow (NO_3)^{\bullet 2-} \quad k = 1\text{x}10^{10}\ 1/\text{Ms} \tag{13.10}$$

$$(NO_3)^{\bullet 2-} + H_2O \leftrightarrow (NO_3H)^{\bullet -} + OH^- \tag{13.11}$$

$$(NO_3H)^{\bullet -} \rightarrow NO^{2\bullet} + OH^- \tag{13.12}$$

13.5.3 Water pH

Water pH has a significant influence on PFAS degradation and defluorination by e_{aq}^-. It is generally accepted that PFAS degradation is slower at acidic pH, mainly due to e_{aq}^- scavenging by H^+. Therefore, successful degradation of PFAS by ARPs is mostly achieved at alkaline pH (Equation 13) (Cui et al. 2020).

$$e_{aq}^- + H^+ \rightarrow H \bullet k = 2.3\text{x}10^{10}\ 1/\text{Ms} \tag{13.13}$$

Another issue related to pH is its variation during treatment, which may occur in an ARP system with limited buffering capacity. Multiple factors may contribute to pH variability: H^+ consumption by e_{aq}^- and other side reactions may lead to increase in pH, whereas the release of F^- and the formation of hydrofluoric acid (HF, pKa = 3.45) may lower the pH of the treated water.

13.5.4 Operational Parameters: Sensitizer Concentration

Concentration of the chemical sensitizer (e.g., SO_3^{2-} or I^-) is essential for the degradation of PFAS, since it directly impacts the production of e_{aq}^-. Typically, increasing sensitizer concentration will enhance the degradation of PFAS until an optimal concentration. Beyond this point, additional sensitizer will either be ineffective or even decrease the PFAS degradation rate. This phenomenon was previously demonstrated both for UV/SO_3^{2-} (Sun et al. 2018a) and UV/I^- (Qu et al. 2010). The accepted explanation for this ball-shape kinetics is that at low concentrations any additional sensitizer will lead to enhanced photoproduction of e_{aq}^-, whereas at higher sensitizer concentrations there is a growing risk for scavenging by other photo-produced intermediates (e.g., $S_2O_6^{2-}$, different I species).

13.6 PLASMA TECHNOLOGIES

Plasma-based systems are not considered classical oxidation or reduction processes, since they simultaneously produce oxidants (e.g., •OH and H_2O_2) and reductive species (e.g., e_{aq}^-), as well as UV light and heat, all of which may contribute to PFAS degradation (Kempisty et al. 2018). This brief section describes plasma technology and its place in both the oxidative and reductive categories of PFAS destruction. The reader is referred to Chapter 12 of this book (by Singh et al.) for a more comprehensive discussion on its use for PFAS water treatment.

Depending on the energy applied and the temperature of the plasma, the process can be classified as thermal or nonthermal. Nonthermal plasma, carried out at gas temperatures of up to 1000°K, is more popular for water treatment. In their earlier book, Kempisty et al. (2018) showed that PFAS degradation and defluorination by plasma take place by both direct thermal decomposition and reaction with generated ions and reactive species (identified and unidentified). Additionally, plasma was less sensitive than other oxidation/reduction processes to the presence of background water constituents, since reaction primarily takes place at the plasma–liquid interface.

Recently, Singh et al. (2019a) used a low-intensity plasma system to detect and quantify all transformation products of PFOA and PFOS degradation and determine the degradation mechanism. They concluded that PFAS degradation by plasma occurs through a stepwise reduction of the parent compound to shorter-chain PFAS and organic acids, followed by defluorination and the release of F^-. Other studies focused on the design and optimization of lab- and pilot-scale plasma reactors (Chen et al. 2019; Mahyar et al. 2019; Saleem et al. 2020). The various lab reactors tested differ in their plasma type (thermal and nonthermal) and generation method, electrodes material and configuration, volume of water treated, power supply, and more. The pilot was designed as a 4 L Aragon gas discharge reactor and was successfully applied to treat PFAS in liquid investigation-derived waste, obtained from 13 groundwater monitoring wells at U.S. Air Force installations (Singh et al. 2019b). This study is one of the rare cases demonstrating plasma degradation of PFAS at field scale.

13.7 FULL-SCALE CONSIDERATIONS

In this section we will describe efforts to apply oxidation and reduction processes to treat PFAS at full-scale. The section is divided into: (i) PFAS removal in conventional water treatment facilities; and (ii) potential full-scale applications specifically designed for PFAS mitigation.

13.7.1 Removal of PFAS During Conventional Full-Scale Treatments

Over the last decade, a large number of studies examined the removal of PFAS in full-scale water and wastewater treatment facilities. Some extended surveys include the ones conducted by Appleman et al. (2014) and Rahman et al. (2014), and more recent ones by Pan et al. (2016), Chen et al. (2018), Coggan et al. (2019), and others (Crone et al. 2019; Glover et al. 2018; Sun et al. 2018a). The wide majority showed that, when employed, redox processes did not contribute to PFAS removal. On the other hand, anion exchange, granular activated carbon, and RO/NF membranes could remove PFAS compounds, with greater capacity for longer-chain PFAS than for shorter ones. For example, Glover et al. (2018) analyzed four full- and pilot-scale potable reuse treatment facilities from different areas of the U.S. They showed that neither ozone nor AOP (UV/H_2O_2) could reduce PFAS concentration when applied under typical full-scale conditions: ozone dose of 1–2 mg-O_3:mg-TOC and UV/H_2O_2 at 900 mJ/cm^2; and 10 mg/L of hydrogen peroxide. Moreover, numerous studies showed that redox processes may even increase the concentrations of PFAS during full-scale wastewater (Blackbeard et al. 2016; Coggan et al. 2019) and water (Sun et al. 2018a) treatments, as a result of their reaction with PFAS precursors.

13.7.2 Specific Applications: Treating Concentrated Waste Streams

The large volume of research described in this chapter regarding PFAS mitigation by redox processes may lead to two key conclusions. The first is that many redox processes may effectively degrade and defluorinated PFAS. In fact, these destructive processes are the only alternative for complete defluorination and detoxification of PFAS. Second, redox processes are currently not competitive at full-scale water- and wastewater treatment facilities (compared to GAC and ion exchange), mainly due to their high cost and technical upscaling problems. As a result, recent work has focused on the adaption of redox processes for specific applications, where they are more advantageous. In this context, treating PFAS concentrated waste streams, specifically waste from sorbent regeneration and membrane brine applications, has emerged as an important, high-potential application. In this application, a relatively small volume of highly concentrated water is treated; therefore, the redox reactors used can be relatively small and economical. In addition, the common handling alternative for these waste streams is relatively expensive and usually consists of disposal into a permitted landfill or incineration. Both of these options have their own concerns and drawbacks.

Regeneration of ion exchange sorbents and RO desalination produces liquid waste that contains high concentrations of PFAS and must be treated, especially if the water is reused or discharged into the environment. For this purpose, the use of electro-oxidation showed high treatment potential. Liang et al. (2018) applied electro-oxidation with titanium electrodes to degrade a mixture of PFOA and PFOS in ion exchange resin regenerant with high organic carbon content (9,880–15,800 mg/L). They showed degradation rate between 77% to > 99% after 16–17 h of treatment, with three orders of magnitude increase in fluoride concentration in the treated water. This study was later translated to a patent (US 2019 / 0185352 A1), proposing electro-oxidation as a valid treatment for a variety of concentrated effluents as long, as the concentration of salt is at least 10 mM. This concentration makes it suitable, for example, for desalination brine and other concentrated streams. Soriano et al. (2017) used electrochemical oxidation to degrade perfluorohexanoic acid in NF concentrate. Degradation reached 98% with more than 95% mineralization (measured through reduction of total organic carbon). It is important to note, however, that electrochemical oxidation of chloride-containing waters can generate chlorate and perchlorate as toxic by-products, especially if boron-doped diamond (BDD) electrodes are used (Wang et al. 2020).

Wu et al. (2019) proposed the use of hydrothermal alkaline conditions (condensed water, 200°C–350°C, 2–16.5 MPa, pH $\geq$ 9) for destroying PFOS in concentrated liquid waste. They found complete conversion of C–F bonds to F^- within 40 min of reaction in a lab-scale reactor; however, upscaling tests are still required. One of the rare field-scale studies was published by Gole et al. (2018), who used a sonochemical 91 L reactor to treat highly concentrated commercial aqueous film-forming foams (ANSUL and 3M) with PFAS concentration of 1%–5%, w/w. Approximately 90.5% and 26.6% reduction of PFSA and PFCA was achieved, respectively, in 13 h. The estimated energy cost of the treatment was \$0.015–\$0.019 per liter.

13.8 SUMMARY AND CONCLUSIONS

Presently, effective removal of PFAS from water and wastewater at full-scale is accomplished with separation techniques, such as anion exchange, granular activated carbon and RO/NF membranes. These processes, however, do not eliminate the compounds and generate a concentrated waste stream that requires further treatment. On the other hand, oxidation/reduction (redox) processes may provide an attractive treatment alternative due to their destructive nature. This chapter summarizes recent knowledge and advancement in the use of redox processes for the treatment of PFAS, and provide insights into drawbacks and future study.

The principal challenge in degrading PFAS by redox processes is the high stability of the carbon–fluorine bond which resists attacks from most oxidants and reductants. Degrading PFAS simply through chain shortening (i.e., cleavage of C–C bonds) will not necessarily eliminate the compound's toxicity; hence C–F cleavage and defluorination is required for an efficient redox treatment.

13.8.1 Oxidation Processes

Conventional oxidants such as chlorine species, ozone, potassium permanganate, and hydrogen peroxide are typically ineffective in degrading and defluorinating PFAS, due to their relatively low redox potential. Some oxidizing radicals produced during AOPs are more successful, albeit depending on the target PFAS and the treatment conditions. The hydroxyl radical (•OH), the principal oxidant in most AOPs, reacts relatively slowly with the PFAS C–F bonds—three orders of magnitude slower than its reaction with common water contaminants. This means that the vast majority of AOPs are not cost-effective for PFAS mitigation. Alternatively, some AOPs employ other oxidation mechanisms, such as sulfate radicals ($SO^{\bullet -}_4$) or direct electron transfer, which shows higher potential for PFAS degradation. For example, activated persulfate, which involves the production of $SO^{\bullet -}_4$, decomposes PFOA with some degree of successes; however, this process was much less effective in degrading PFOS. Electrochemical oxidation and ultrasonication also degrade PFAS, mostly by reactions which initiated with direct oxidation by electron transfer (electro-oxidation) or thermal bond cleavage (ultrasonication). Application of these two technologies at full-scale is, however, limited, due to technical scalability problems which reduces the treatment cost-effectiveness at large-scale.

13.8.2 Reduction Processes

Reductive degradation of PFAS is typically more favorable than oxidation; it involves electron transfer from the reducing agent to the contaminant and the release of a fluoride ion (i.e., defluorination). The kinetics of this mechanism is, however, often too slow or adversely affected by dissolved oxygen and other background water constituents.

Reducing agents are either conventional chemicals, mostly zero valent iron (ZVI), or reducing radicals such as hydrated electrons (e_{aq}^-). Simple application of ZVI is typically ineffective in degrading PFAS. However, improved results were obtained when ZVI was applied under extreme operational conditions, combined with other materials (such as activated carbon) or when using nanoparticle ZVI. Reducing radicals, principally e_{aq}^- generated during ARPs, are more effective in degrading PFAS (than ZVI). Degradation (and defluorination) takes place via two parallel reaction pathways for PFCA: exchange of fluorine by hydrogens (H/F) and chain shortening. For PFOS, the above reactions are preceded by desulfonation and the cleavage of the C–S bond between the headgroup and the perfluoroalkyl chain.

A large number of ARPs were tested for degrading PFAS; effective degradation could be achieved principally under anaerobic lab conditions. The main limiting factor is dissolved oxygen (DO) which readily reacts with e_{aq}^- at a near diffusion rate, decreasing its availability in aerated water and reducing the rate of PFAS degradation. Only limited attempts were carried out to mitigate the impact of DO on ARPs' effectiveness. They include the use of high photon flux UV/sulfite to promote e_{aq}^- production and the application of an alternative sensitizer (i.e., nitrilotriacetic acid) with improved e_{aq}^- observed yield. Adapting ARPs to environmental conditions should be in the heart of future research, aiming to develop full-scale reducing processes for treating PFAS in water.

REFERENCES

Appleman, T. D., Higgins, C. P., Quiñones, O., Vanderford, B. J., Kolstad, C., Zeigler-Holady, J. C., & Dickenson, E. R. V. (2014). Treatment of poly- and perfluoroalkyl substances in U.S. full-scale water treatment systems. *Water Research*, 51, 246–255. doi: 10.1016/j.watres.2013.10.067.

Arvaniti, O. S., Hwang, Y., Andersen, H. R., Stasinakis, A. S., Thomaidis, N. S., & Aloupi, M. (2015). Reductive degradation of perfluorinated compounds in water using Mg-aminoclay coated nanoscale zero valent iron. *Chemical Engineering Journal*, 262, 133–139. doi: 10.1016/j.cej.2014.09.079.

Bao, Y., Huang, J., Cagnetta, G., & Yu, G. (2019). Removal of F–53B as PFOS alternative in chrome plating wastewater by UV/Sulfite reduction. *Water Research*, 163, 114907. doi: 10.1016/j.watres.2019.114907.

Barisci, S., & Suri, R. (2020). Electrooxidation of short and long chain perfluorocarboxylic acids using boron doped diamond electrodes. *Chemosphere*, 243, 125349. doi: 10.1016/j.chemosphere.2019.125349.

Bentel, M. J., Yu, Y., Xu, L., Li, Z., Wong, B. M., Men, Y., & Liu, J. (2019). Defluorination of Per- and Polyfluoroalkyl Substances (PFASs) with Hydrated Electrons: Structural Dependence and Implications to PFAS Remediation and Management. *Environmental Science and Technology*, 53(7), 3718–3728. doi: 10.1021/acs.est.8b06648.

Blackbeard, J., Lloyd, J., Magyar, M., Mieog, J., Linden, K. G., & Lester, Y. (2016). Demonstrating organic contaminant removal in an ozone-based water reuse process at full scale. *Environmental Science: Water Research and Technology*, 2(1), 213–222. doi: 10.1039/c5ew00186b.

Bruton, T. A., & Sedlak, D. L. (2017). Treatment of aqueous film-forming foam by heat-activated persulfate under conditions representative of in situ chemical oxidation. *Environmental Science and Technology*, 51(23), 13878–13885. doi: 10.1021/acs.est.7b03969.

Buxton, G. V., Greenstock, C. L., Helman, W. P., & Ross, A. B. (1988). Critical Review of rate constants for reactions of hydrated electrons, hydrogen atoms and hydroxyl radicals (OH/·O–in Aqueous Solution). *Journal of Physical and Chemical Reference Data*, 17(2), 513–886. doi: 10.1063/1.555805.

Chen, S. H., Živný, O., Mašláni, A., & Chau, S. W. (2019). Abatement of fluorinated compounds in thermal plasma flow. *Journal of Fluorine Chemistry*, 217(October 2018), 41–49. doi: 10.1016/j.jfluchem.2018.10.004.

Chen, S., Zhou, Y., Meng, J., & Wang, T. (2018). Seasonal and annual variations in removal efficiency of perfluoroalkyl substances by different wastewater treatment processes. *Environmental Pollution*, 242, 2059–2067. doi: 10.1016/j.envpol.2018.06.078.

Coggan, T. L., Moodie, D., Kolobaric, A., Szabo, D., Shimeta, J., Crosbie, N. D., … Clarke, B. O. (2019). An investigation into per- and polyfluoroalkyl substances (PFAS) in nineteen Australian wastewater treatment plants (WWTPs). *Heliyon*, 5(8), e02316. doi: 10.1016/j.heliyon.2019.e02316.

Crone, B. C., Speth, T. F., Wahman, D. G., Smith, S. J., Abulikemu, G., Kleiner, E. J., & Pressman, J. G. (2019). Occurrence of per- and polyfluoroalkyl substances (PFAS) in source water and their treatment in drinking water. *Critical Reviews in Environmental Science and Technology*, 49(24), 2359–2396. doi: 10.1080/10643389.2019.1614848.

Cui, J., Gao, P., & Deng, Y. (2020). Destruction of per- A nd polyfluoroalkyl substances (PFAS) with advanced reduction processes (ARPs): A critical review. *Environmental Science and Technology*, 54(7), 3752–3766. doi: 10.1021/acs.est.9b05565.

Emmett, E. A., Shofer, F. S., Zhang, H., Freeman, D., Desai, C., & Shaw, L. M. (2006). Community exposure to perfluorooctanoate: Relationships between serum concentrations and exposure sources. *Journal of Occupational and Environmental Medicine*, 48(8), 759–770. doi: 10.1097/01.jom.0000232486.07658.74.

Furuta, M., Yamaguchi, M., Tsukamoto, T., Yim, B., Stavarache, C. E., Hasiba, K., & Maeda, Y. (2004). Inactivation of Escherichia coli by ultrasonic irradiation. *Ultrasonics Sonochemistry*, 11(2), 57–60. doi: 10.1016/S1350-4177(03)00136-6.

Glover, C. M., Quiñones, O., & Dickenson, E. R. V. (2018). Removal of perfluoroalkyl and polyfluoroalkyl substances in potable reuse systems. *Water Research*, 144, 454–461. doi: 10.1016/j.watres.2018.07.018.

Gole, V. L., Sierra-Alvarez, R., Peng, H., Giesy, J. P., Deymier, P., & Keswani, M. (2018). Sono-chemical treatment of per- and poly-fluoroalkyl compounds in aqueous film-forming foams by use of a large-scale multi-transducer dual-frequency based acoustic reactor. *Ultrasonics Sonochemistry*, 45(January), 213–222. doi: 10.1016/j.ultsonch.2018.02.014.

Gomez-Ruiz, B., Gómez-Lavín, S., Diban, N., Boiteux, V., Colin, A., Dauchy, X., & Urtiaga, A. (2017). Efficient electrochemical degradation of poly- and perfluoroalkyl substances (PFASs) from the effluents of an industrial wastewater treatment plant. *Chemical Engineering Journal*, 322, 196–204. doi: 10.1016/j.cej.2017.04.040.

Gu, Y., Dong, W., Luo, C., & Liu, T. (2016). Efficient reductive decomposition of perfluorooctanesulfonate in a high photon flux UV/sulfite system. *Environmental Science and Technology*, 50(19), 10554–10561. doi: 10.1021/acs.est.6b03261.

Gu, Y., Liu, T., Wang, H., Han, H., & Dong, W. (2017). Hydrated electron based decomposition of perfluorooctane sulfonate (PFOS) in the VUV/sulfite system. *Science of the Total Environment*, 607–608, 541–548. doi: 10.1016/j.scitotenv.2017.06.197.

Guo, C., Zhang, C., Sun, Z., Zhao, X., Zhou, Q., & Hoffmann, M. R. (2019). Synergistic impact of humic acid on the photo-reductive decomposition of perfluorooctanoic acid. *Chemical Engineering Journal*, 360(October 2018), 1101–1110. doi: 10.1016/j.cej.2018.10.204.

Guo, Y., Lou, X., Fang, C., Xiao, D., Wang, Z., & Liu, J. (2013). Novel Photo-Sulfite System: Toward Simultaneous Transformations of Inorganic and Organic Pollutants. *Environmental Science and Technology*, doi: 10.1021/es403199p.

Hoff, P. (2009). PFCs in Minnesota's Ambient Environment: 2008 Progress Report. 1–91. Available at: https://www.pca.state.mn.us/sites/default/files/c-pfc1-02.pdf (Accessed August 29, 2020).

Hori, H., Hayakawa, E., Einaga, H., Kutsuna, S., Koike, K., Ibusuki, T., … Arakawa, R. (2004). Decomposition of environmentally persistent perfluorooctanoic acid in water by photochemical approaches. *Environmental Science and Technology*, 38(22), 6118–6124. doi: 10.1021/es049719n.

Hori, H., Murayama, M., Inoue, N., Ishida, K., & Kutsuna, S. (2010). Efficient mineralization of hydroperfluorocarboxylic acids with persulfate in hot water. *Catalysis Today*, 151(1–2), 131–136. doi: 10.1016/j.cattod.2010.02.023.

Hori, H., Nagaoka, Y., Yamamoto, A., Sano, T., Yamashita, N., Taniyasu, S., … Arakawa, R. (2006). Efficient decomposition of environmentally persistent perfluorooctanesulfonate and related fluorochemicals using zerovalent iron in subcritical water. *Environmental Science and Technology*, 40(3), 1049–1054. doi: 10.1021/es0517419.

Hu, Y., Lo, S. L., Li, Y. F., Lee, Y. C., Chen, M. J., & Lin, J. C. (2018). Autocatalytic degradation of perfluorooctanoic acid in a permanganate-ultrasonic system. *Water Research*, 140, 148–157. doi: 10.1016/j.watres.2018.04.044.

Kempisty, D. M., Xing, Yun, & Racz, LeeAnn. (2018). Perfluoroalkyl Substances in the Environment. *In Perfluoroalkyl Substances in the Environment*. doi: 10.1201/9780429487125.

Le, T. X. H., Haflich, H., Shah, A. D., & Chaplin, B. P. (2019). Energy-Efficient Electrochemical Oxidation of Perfluoroalkyl Substances Using a Ti4O7 Reactive Electrochemical Membrane Anode. *Environmental Science and Technology Letters*, 6(8), 504–510. doi: 10.1021/acs.estlett.9b00397.

Lee, Y., Lo, S., Kuo, J., & Hsieh, C. (2012). Decomposition of perfluorooctanoic acid by microwaveactivated persulfate: Effects of temperature, pH, and chloride ions. *Frontiers of Environmental Science and Engineering in China*, 6(1), 17–25. doi: 10.1007/s11783-011-0371-x.

Liang, S., David, Pierce R., Lin, H., Chiang, S. Y. D., & Huang, Q. Jack. (2018). Electrochemical oxidation of PFOA and PFOS in concentrated waste streams. *Remediation*, 28(2), 127–134. doi: 10.1002/rem.21554.

Liu, Y. Y., Ptacek, C. J., Baldwin, R. J., Cooper, J. M., & Blowes, D. W. (2020). Application of zero-valent iron coupled with biochar for removal of perfluoroalkyl carboxylic and sulfonic acids from water under ambient environmental conditions. *Science of the Total Environment*, 719, 137372. doi: 10.1016/j.scitotenv.2020.137372.

Lutze, H. V., Brekenfeld, J., Naumov, S., von Sonntag, C., & Schmidt, T. C. (2018). Degradation of perfluorinated compounds by sulfate radicals – New mechanistic aspects and economical considerations. *Water Research*, 129, 509–519. doi: 10.1016/j.watres.2017.10.067.

Mahyar, A., Miessner, H., Mueller, S., Hama Aziz, K. H., Kalass, D., Moeller, D., … Noack, J. (2019). Development and Application of Different Non-thermal Plasma Reactors for the Removal of Perfluorosurfactants in Water: A Comparative Study. *Plasma Chemistry and Plasma Processing*, 39(3), 531–544. doi: 10.1007/s11090-019-09977-6.

Merino, N., Qu, Y., Deeb, R. A., Hawley, E. L., Hoffmann, M. R., & Mahendra, S. (2016). Degradation and Removal Methods for Perfluoroalkyl and Polyfluoroalkyl Substances in Water. *Environmental Engineering Science*, 33(9), 615–649. doi: 10.1089/ees.2016.0233.

Mohr, T. K. G. (2010). *Environmental Investigation and Remediation: 1,4-Dioxane and other Solvent Stabilizers*. CRC Press: Boca Roca, FL.

Mueller, R., & Yingling, V. (2018). Regulations, Guidance, and Advisories for Per- and Polyfluoroalkyl Substances (PFAS) [fact sheet]. *Interstate Technology Regulatory Council Sheets*, 1–6. Retrieved from https://pfas-1.itrcweb.org/wp-content/uploads/2018/01/pfas_fact_sheet_regulations__1_4_18.pdf.

NY Department of Health. (2019). *Westhampton Beach and Quogue Area PFAS Blood Testing : Group-Level Results*. Available at: https://health.ny.gov/environmental/investigations/drinkingwaterresponse/docs/westhampton_quogue_group_level_blood_testing (Accessed August 29, 2020).

Nzeribe, B. N., Crimi, M., Mededovic Thagard, S., & Holsen, T. M. (2019). Physico-Chemical Processes for the Treatment of Per- And Polyfluoroalkyl Substances (PFAS): A review. *Critical Reviews in Environmental Science and Technology*, 49(10), 866–915. doi: 10.1080/10643389.2018.1542916.

Pan, C. G., Liu, Y. S., & Ying, G. G. (2016). Perfluoroalkyl substances (PFASs) in wastewater treatment plants and drinking water treatment plants: Removal efficiency and exposure risk. *Water Research*, 106, 562–570. doi: 10.1016/j.watres.2016.10.045.

Park, H., Vecitis, C. D., Cheng, J., Choi, W., Mader, B. T., & Hoffmann, M. R. (2009). Reductive defluorination of aqueous perfluorinated alkyl surfactants: Effects of ionic headgroup and chain length. *Journal of Physical Chemistry A*, 113(4), 690–696. doi: 10.1021/jp807116q.

Phillips, D. H., Van Nooten, T., Bastiaens, L., Russell, M. I., Dickson, K., Plant, S., … Kalin, R. M. (2010). Ten year performance evaluation of a field-scale zero-valent iron permeable reactive barrier installed to remediate trichloroethene contaminated groundwater. *Environmental Science and Technology*, 44(10), 3861–3869. doi: 10.1021/es902737t.

Prabhu, A. V., Gogate, P. R., & Pandit, A. B. (2004). Optimization of multiple-frequency sonochemical reactors. *Chemical Engineering Science*, 59(22–23), 4991–4998. doi: 10.1016/j.ces.2004.09.033.

Qu, Y., Zhang, C., Li, F., Chen, J., & Zhou, Q. (2010). Photo-reductive defluorination of perfluorooctanoic acid in water. *Water Research*, 44(9), 2939–2947. doi: 10.1016/j.watres.2010.02.019.

Radjenovic, J., & Sedlak, D. L. (2015). Challenges and Opportunities for Electrochemical Processes as Next-Generation Technologies for the Treatment of Contaminated Water. *Environmental Science and Technology*, 49(19), 11292–11302. doi: 10.1021/acs.est.5b02414.

Rahman, M. F., Peldszus, S., & Anderson, W. B. (2014). Behaviour and fate of perfluoroalkyl and polyfluoroalkyl substances (PFASs) in drinking water treatment: A review. *Water Research*, 50, 318–340. doi: 10.1016/j.watres.2013.10.045.

Rayaroth, M. P., Aravind, U. K., & Aravindakumar, C. T. (2016). Degradation of pharmaceuticals by ultrasound-based advanced oxidation process. *Environmental Chemistry Letters*, 14(3), 259–290. doi: 10.1007/s10311-016-0568-0.

Saito, N., Sasaki, K., Nakatome, K., Harada, K., Yoshinaga, T., & Koizumi, A. (2003). Perfluorooctane sulfonate concentrations in surface water in Japan. *Archives of Environmental Contamination and Toxicology*, 45(2), 149–158. doi: 10.1007/s00244-003-0163-9.

Saleem, M., Biondo, O., Sretenović, G., Tomei, G., Magarotto, M., Pavarin, D., Paradisi, C. (2020). Comparative performance assessment of plasma reactors for the treatment of PFOA; reactor design, kinetics, mineralization and energy yield. *Chemical Engineering Journal*, 382(September 2019), 123031. doi: 10.1016/j.cej.2019.123031.

Schaefer, C. E., Andaya, C., Urtiaga, A., McKenzie, E. R., & Higgins, C. P. (2015). Electrochemical treatment of perfluorooctanoic acid (PFOA) and perfluorooctane sulfonic acid (PFOS) in groundwater impacted by aqueous film forming foams (AFFFs). *Journal of Hazardous Materials*, 295, 170–175. doi: 10.1016/j.jhazmat.2015.04.024.

Sekiguchi, K., Kudo, T., & Sankoda, K. (2017). Combined sonochemical and short-wavelength UV degradation of hydrophobic perfluorinated compounds. *Ultrasonics Sonochemistry*, 39(January), 87–92. doi: 10.1016/j.ultsonch.2017.04.002.

Singh, R. K., Fernando, S., Baygi, S. F., Multari, N., Thagard, S. M., & Holsen, T. M. (2019a). Breakdown Products from Perfluorinated Alkyl Substances (PFAS) Degradation in a Plasma-Based Water Treatment Process. *Environmental Science and Technology*, 53(5), 2731–2738. doi: 10.1021/acs.est.8b07031.

Singh, R. K., Multari, N., Nau-Hix, C., Anderson, R. H., Richardson, S. D., Holsen, T. M., & Mededovic Thagard, S. (2019b). Rapid Removal of Poly- and Perfluorinated Compounds from Investigation-Derived Waste (IDW) in a Pilot-Scale Plasma Reactor [Research-article]. *Environmental Science and Technology*, 53, 11375–11382. doi: 10.1021/acs.est.9b02964.

Song, Z., Tang, H., Wang, N., & Zhu, L. (2013). Reductive defluorination of perfluorooctanoic acid by hydrated electrons in a sulfite-mediated UV photochemical system. *Journal of Hazardous Materials*, 262, 332–338. doi: 10.1016/j.jhazmat.2013.08.059.

Soriano, Á., Gorri, D., & Urtiaga, A. (2017). Efficient treatment of perfluorohexanoic acid by nanofiltration followed by electrochemical degradation of the NF concentrate. *Water Research*, 112, 147–156. doi: 10.1016/j.watres.2017.01.043.

Sun, M., Zhou, H., Xu, B., & Bao, J. (2018a). Distribution of perfluorinated compounds in drinking water treatment plant and reductive degradation by UV/SO32− process. *Environmental Science and Pollution Research*, 25(8), 7443–7453. doi: 10.1007/s11356-017-1024-9.

Sun, Z., Zhang, C., Chen, P., Zhou, Q., & Hoffmann, M. R. (2017). Impact of humic acid on the photoreductive degradation of perfluorooctane sulfonate (PFOS) by UV/Iodide process. *Water Research*, 127, 50–58. doi: 10.1016/j.watres.2017.10.010.

Sun, Z., Zhang, C., Xing, L., Zhou, Q., Dong, W., & Hoffmann, M. R. (2018b). UV/Nitrilotriacetic Acid Process as a Novel Strategy for Efficient Photoreductive Degradation of Perfluorooctanesulfonate. *Environmental Science and Technology*, 52(5), 2953–2962. doi: 10.1021/acs.est.7b05912.

Trojanowicz, M., Bartosiewicz, I., Bojanowska-Czajka, A., Kulisa, K., Szreder, T., Bobrowski, K., … Kisała, J. (2019). Application of ionizing radiation in decomposition of perfluorooctanoate (PFOA) in waters. *Chemical Engineering Journal*, 357(July 2018), 698–714. doi: 10.1016/j.cej.2018.09.065.

Trojanowicz, M., Bartosiewicz, I., Bojanowska-Czajka, A., Szreder, T., Bobrowski, K., Nałęcz-Jawecki, G., … Nichipor, H. (2020). Application of ionizing radiation in decomposition of perfluorooctane sulfonate (PFOS) in aqueous solutions. *Chemical Engineering Journal*, 379(April 2019), 122303. doi: 10.1016/j.cej.2019.122303.

Truex, M. J., MacBeth, T. W., Vermeul, V. R., Fritz, B. G., Mendoza, D. P., MacKley, R. D., … Oostrom, M. (2011). Demonstration of combined zero-valent iron and electrical resistance heating for in situ trichloroethene remediation. *Environmental Science and Technology*, 45(12), 5346–5351. doi: 10.1021/es104266a.

U.S. Department of Health and Human Services. (2019). Fourth National Report on Human Exposure to Environmental Chemicals. Updated Tables, January 2019, Volume One. *Fourth National Report on Human Exposure to Environmental Chemicals*, (January), 1–529. Available at: https://www.cdc.gov/exposurereport/ (Accessed August 29 2020).

Vellanki, B. P., Batchelor, B., & Abdel-Wahab, A. (2013). Advanced Reduction Processes: A New Class of Treatment Processes. *Environmental Engineering Science*, 30(5), 264–271. doi: 10.1089/ees.2012.0273.

Wang, L., Lu, J., Li, L., Wang, Y., & Huang, Q. (2020a). Effects of chloride on electrochemical degradation of perfluorooctanesulfonate by Magnéli phase Ti4O7 and boron doped diamond anodes. *Water Research*, 170, 115254. doi: 10.1016/j.watres.2019.115254.

Wang, N., Lv, H., Zhou, Y., Zhu, L., Hu, Y., Majima, T., & Tang, H. (2019). Complete defluorination and mineralization of perfluorooctanoic acid by a mechanochemical method using alumina and persulfate. *Environmental Science and Technology*, 53(14), 8302–8313. doi: 10.1021/acs.est.9b00486.

Wang, Y., Pierce R. David, Shi, H., Li, C., & Huang, Q. (2020b). Electrochemical degradation of perfluoroalkyl acids by titanium suboxide anodes. *Environmental Science: Water Research and Technology*, 6(1), 144–152. doi: 10.1039/c9ew00759h.

Wilhelm, M., Angerer, J., Fromme, H., & Hölzer, J. (2009). Contribution to the evaluation of reference values for PFOA and PFOS in plasma of children and adults from Germany. *International Journal of Hygiene and Environmental Health*, 212(1), 56–60. doi: 10.1016/j.ijheh.2007.11.002.

Wu, B., Hao, S., Choi, Y., Higgins, C. P., Deeb, R., & Strathmann, T. J. (2019). Rapid Destruction and Defluorination of Perfluorooctanesulfonate by Alkaline Hydrothermal Reaction. *Environmental Science and Technology Letters*, 6(10), 630–636. doi: 10.1021/acs.estlett.9b00506.

Yin, P., Hu, Z., Song, X., Liu, J., & Lin, N. (2016). Activated persulfate oxidation of perfluorooctanoic acid (PFOA) in groundwater under acidic conditions. *International Journal of Environmental Research and Public Health*, 13(6). doi: 10.3390/ijerph13060602.

Yu, K., Li, X., Chen, L., Fang, J., Chen, H., Li, Q., … Ma, J. (2018). Mechanism and efficiency of contaminant reduction by hydrated electron in the sulfite/iodide/UV process. *Water Research*, 129, 357–364. doi: 10.1016/j.watres.2017.11.030.

14 Novel Cyclodextrin Polymer Adsorbents for PFAS Removal

Yuhan Ling, Gokhan Barin, Shan Li, and Matthew J. Notter
Cyclopure Inc., USA

CONTENTS

14.1 Properties of DEXSORB+® - A Novel Mesoporous β-Cyclodextrin Adsorbent....292
- 14.1.1 Structure....292
- 14.1.2 Surface Area, Porosity, Density and Particle Size....293
- 14.1.3 Swelling....294
- 14.1.4 Material Stability....294
- 14.1.5 Regeneration....294

14.2 PFAS Adsorption Mechanisms....295
- 14.2.1 Kinetics....295
- 14.2.2 Thermodynamics....296
- 14.2.3 Hydrophobic and Electrostatic Interactions....298

14.3 PFAS Uptake in Real Water Matrices....298
- 14.3.1 Batch Uptake....299
- 14.3.2 Packed-Bed Filtration....300
  - 14.3.2.1 Rapid Small-Scale Column Tests....300
  - 14.3.2.2 Pilot-Scale Column Tests....302

14.4 Potential Effects of Water Matrices....303
- 14.4.1 Suspended Solids....303
- 14.4.2 Ionic Strength....304
- 14.4.3 Natural Organic Matter....304
- 14.4.4 pH....305

14.5 Cyclodextrin Polymer Operation Cycles....305
- 14.5.1 Loading of Fresh Adsorbents....306
- 14.5.2 Regeneration of Exhausted Adsorbents....307
- 14.5.3 Reuse of Regenerated Adsorbents....307
- 14.5.4 Destruction of PFAS Concentrate....308

14.6 Adsorption Treatment Applications....308
- 14.6.1 Batch Adsorption Followed by Ultrafiltration....308

14.6.1.1 Preparation of Powdered Adsorbent 308
14.6.1.2 Hydraulic Residence Time 309
14.6.1.3 Filtration of Adsorbent 309
14.6.2 Packed-bed Filtration 309
14.6.2.1 Preparation of Granular Adsorbent 309
14.6.2.2 Pretreatment Considerations 309
14.6.2.3 Hydraulic Conditions 310
14.6.2.4 Fixed or Backwashable Bed 310
14.6.2.5 Regeneration and Concentrate Destruction 310
14.6.3 Others 310
14.6.3.1 Batch Adsorption Followed by Coagulation and Flocculation 310
14.6.3.2 Small Drinking Water Systems 311
14.7 Summary 311
References 312

14.1 PROPERTIES OF DEXSORB+® - A NOVEL MESOPOROUS β-CYCLODEXTRIN ADSORBENT

Per- and polyfluorinated alkyl substances (PFAS) have emerged as a pervasive and persistent threat to drinking water supplies throughout the world (Herkert et al. 2020; Seo et al. 2018). Favored for their chemical/thermal stability and hydrophilic-lipophilic nature, PFAS have been ubiquitous in industrial and consumer products. Environmental transport from emission sites (e.g., direct release from manufacturing processes, firefighting foams, wastewater discharge, landfill leachate, etc.) to water sources is facilitated by their carbon–fluorine bonds that are highly resistant to biodegradation and highly soluble in water (Wang et al. 2017).

The class of synthetically made PFAS consists of over 4,000 chemicals, characterized by complex and diverse compositions such as functional group (e.g., carboxylic, sulfonic acids, etc.), fluorocarbon chain length, and isomer structure (e.g., branched or linear). As demonstrated in recent publications (Wang et al. 2017), the complex compositions and special physicochemical properties of PFAS render their removal during water and wastewater treatment a significant challenge. Among all available treatment options, adsorption is regarded as a highly effective technology for PFAS treatment, though the adsorption mechanisms of a sorbent can greatly affect its efficacy. Positively charged, mesoporous β-cyclodextrin adsorbents, known by the tradename DEXSORB+®, featuring uniform adsorption sites (i.e., β-cyclodextrin cavities) and dual adsorption mechanisms with hydrophobic and electrostatic interaction have been developed as a novel strategy for efficient and effective removal of broad-spectrum PFAS.

14.1.1 STRUCTURE

β-cyclodextrin is an inexpensive and renewable macrocycle comprised of seven glucose molecules which form well-defined hydrophobic cavities that are ideally suited to binding organic micropollutants (Crini 2014). Thousands of micropollutants

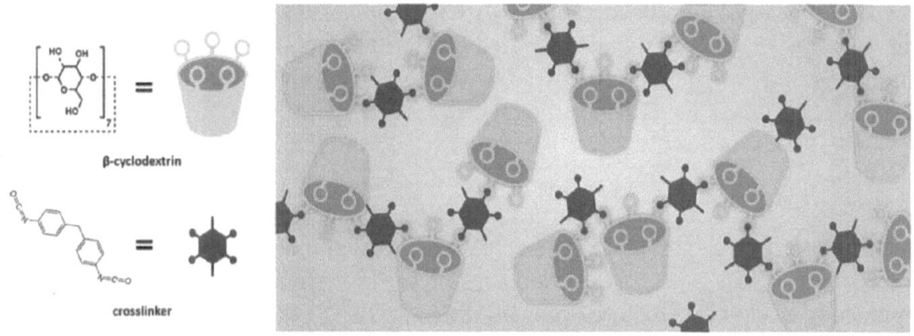

FIGURE 14.1 Cyclodextrin polymers formed in a single-step reaction by means of nucleophilic aromatic substitution with co-monomer.

including PFAS can form host-guest complexes with β-cyclodextrin in its uniform 0.78 nm interior cavity (Crini 2014). However, because β-cyclodextrin is soluble in water, it first needs to be polymerized to render an insoluble polymer form before its use as an adsorbent. Synthesis of a high surface area, porous, insoluble polymer containing β-cyclodextrin was first reported in *Nature* in January 2016 (Alsbaiee et al. 2015).

Porous β-cyclodextrin polymer networks are formed in a single-step reaction by means of nucleophilic aromatic substitution between β-cyclodextrin and a co-monomer (Alsbaiee et al. 2015) as illustrated in Figure 14.1. Notably, one gram of the β-cyclodextrin polymer adsorbent contains approximately $3x10^{20}$ cyclodextrin cups as active sites for adsorption. As discussed further below, the 0.78 nm interior cavity of β-cyclodextrin confers size-selective adsorption of micropollutants and provides size-exclusion resistance to fouling caused by natural organic matter (NOM) (Ling et al., 2020, 2017). This β-cyclodextrin polymer, known by the tradename DEXSORB®, carries negative charge. With no further modification, the original DEXSORB polymer favors positively charged compounds, and exhibits weak affinity for negatively charged PFOA and PFOS. Preserving the hydrophobic and mesoporous features of DEXSORB, the media charge can be modulated through the incorporation of positively charged units. The resulting material, DEXSORB+, is a positively charged β-cyclodextrin polymer that involves two types of adsorption sites: hydrophobic β-cyclodextrin cavities and cationic quaternary ammonium sites.

14.1.2 Surface Area, Porosity, Density and Particle Size

The as-synthesized cyclodextrin polymers are pale yellow powders with an average particle size of 0.3 mm, featuring a surface area of up to 450 m^2/g, uniform mesoporous structures, and an average apparent density of 0.56 g/mL in its dry form. Benefitting from its compatibility with industrial milling and granulation techniques, the particle size of cyclodextrin polymers is easily tunable and can be further prepared in both powder and granular form.

With industrial milling, DEXSORB+ can be sized down to 30 μm for use in powdered adsorbent treatment processes (e.g., batch adsorption followed by ultrafiltration

(BA-UF) or coagulation/flocculation (BA-C/F)), filter paper (e.g., passive sampling with solid phase extraction), and extruded block (i.e., household point-of-use filters). Through granulation, DEXSORB+ can be sized up to 2 mm to meet the requirements of applications like packed-bed filtration processes (PBF).

14.1.3 Swelling

Swelling of granular DEXSORB+ occurs when dry granules come in contact with aqueous phase. An average expansion factor of 75% should be considered in the engineering design and material preparations for DEXSORB+ granules. As a result of swelling, the apparent density of DEXSORB+ granules decreases from 0.56 g/mL in its dry form, to less than 0.40 g/mL in its wet form. The mechanical strength and rigid porous structure of DEXSORB+ granules are not affected by swelling. The skeleton density of post-swelling DEXSORB+ remains higher than the density of water. As a result, the terminal velocity for swelled DEXSORB+ granules with an average particle size of 0.85 mm is generally more than 2 meters per minute. Due to these features, DEXSORB+ granules are compatible with all PBF and fluidized systems.

14.1.4 Material Stability

Cyclodextrin polymers are highly stable in multiple aspects of their use. Based on the results from thermogravimetric analysis (TGA), cyclodextrin polymers are thermally stable and can operate at temperatures as high as 300°C, allowing for adaptive uses such as block extrusion. With a rigid polymeric network, cyclodextrin polymers are not subject to aerobic or anaerobic degradation and are not affected by disinfection processes like chlorination. The single step synthesis provides a clean, insoluble, fully polymerized structure that shows no leaching in aqueous environments. Contaminants that are adsorbed onto the media are held in bond without displacement by other contaminants until desorption is triggered by regeneration processes. It is important to note that the variance among general water quality (e.g., pH, total dissolved solids (TDS), Total Organic Carbon (TOC), etc.) has very limited impacts on the adsorption processes of cyclodextrin polymers for PFAS uptake. This superior feature and how it contributes to the consistent performance of cyclodextrin polymers across different water matrices is discussed in detail in the following sections. DEXSORB+ is now certified by NSF International under NSF/ANSI 61 for its compatibility with municipal drinking water treatment.

14.1.5 Regeneration

The adsorption of PFAS and other organic contaminants on DEXSORB+ relies on the formation of host-guest complexes in the interior cavity of β-cyclodextrin and interaction with its positively charged surface (Ling et al. 2020), both of which are reversible under specific regeneration conditions. Desorption of PFAS from cyclodextrin polymers is accomplished by a quick wash of the polymer with a small amount of regeneration solution at ambient temperature. In general, a volume ratio of 5:1 of regeneration solution to spent cyclodextrin polymer is sufficient to provide complete desorption of PFAS. To date, five consecutive cycles of adsorption/desorption have been performed for cyclodextrin polymers with minor effect to performance compared to the as-synthesized adsorbent (Alsbaiee et al., 2015; Xiao et al., 2017).

Easy and multicycle regeneration makes cyclodextrin polymers a cost-effective, energy-efficient, and sustainable treatment technology for PFAS removal from contaminated water. Of further importance, the regeneration solution can be treated with evaporation/crystallization or membrane filtration to concentrate the desorbed PFAS into a "neat" waste product for cost-effective handling and disposal. The concentrated PFAS waste product can then be destroyed with non-combustion mineralization such as mechano-chemical, plasma, hydro-thermal, and electrochemical processes (Zhang et al. 2013; Merino et al. 2016; Wu et al. 2019). Regeneration of cyclodextrin polymer granules is discussed further in this chapter.

14.2 PFAS ADSORPTION MECHANISMS

The characterization of a novel adsorbent starts with investigation of the adsorption kinetics and thermodynamics for targeted contaminants at environmentally relevant concentrations in batch experiments. Kinetic and thermodynamic studies provide insights on the adsorption mechanisms and the microstructures of an adsorbent. Adsorption kinetics and equilibrium removal of PFAS are widely viewed as major determinants of the feasibility of an adsorbent because of their correlation with treatment efficiency and capacity (Ling et al. 2017; Xiao et al. 2017). This section discusses adsorption kinetics, isotherms, and major mechanisms of PFAS adsorption on DEXSORB+.

14.2.1 Kinetics

Adsorption kinetics describes the rate at which equilibrium is achieved between adsorbates and adsorbents. It is a measure of the uptake speed of targeted contaminants on adsorbents (Ling et al., 2017). Benefitting from uniform mesoporous structures and selective adsorption, DEXSORB+ exhibits consistently rapid adsorption kinetics for PFAS uptake across clean water matrices (e.g. distilled deionized DDI water) and complex water matrices (e.g. landfill leachate). The mesoporous structures significantly reduce the time required for intraparticle diffusion, while the selective adsorption resulting from size exclusion provides ready access for PFAS. Importantly, fast kinetics are equally observed across varying particle sizes in both powder and granular forms. The percent removal of 7 PFAS from a BA-C/F process simulation in surface water is presented in Figure 14.2 to exemplify the rapid adsorption kinetics of DEXSORB+. As shown, the differences among 5-min, 10-min, and 70-min removal are insignificant for each PFAS, indicating that a contact time as short as 5 minutes in powdered treatment processes is sufficient for the effective removal of PFAS with DEXSORB+. In water treatment processes, rapid adsorption kinetics translate to low adsorbent dosing and short contact times.

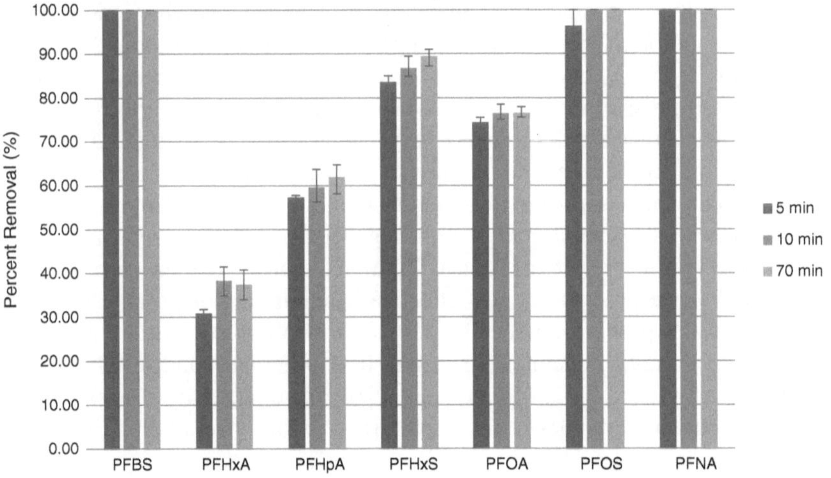

FIGURE 14.2 Percent uptake of seven detected PFAS at 20 mg/L doses of DEXSORB+. Error bars, minimum and maximum uptake.

14.2.2 Thermodynamics

Insights into both the maximum capacity and adsorption affinity of PFAS on DEXSORB+ are obtained from isotherms for three PFAS. Batch isotherm experiments for PFOA, PFOS, and HFPO-DA (GenX) were performed in DDI water to provide equilibrium data at ambient temperature. The resulting isotherm data was applied for the fitting of Langmuir and Freundlich models, which are presented in Table 14.1.

Based on Langmuir model fitting output, high adsorption affinity and capacity of DEXSORB+ for PFOA, PFOS and GenX is demonstrated. Figure 14.3 details the complete isotherm for PFOA, PFOS, and GenX with Langmuir and Freundlich curve fits. The q_{max} for PFOA and PFOS are well above 180 mg/g, which outperforms most commercially available adsorbents targeting the removal of PFAS from drinking water and wastewater. Additionally, the DEXSORB+ q_{max} for GenX is also among the highest of all tested adsorbents, indicating its potential for the removal of short-chain PFAS and PFAS alternatives of emerging concern.

TABLE 14.1
The isotherm and model fitting output of PFOA, PFOS, and GenX for DEXSORB+.

	Langmuir Isotherm			Freundlich Isotherm		
Compound	K_L (L/g)	q_{max}(mg/g)	R^2	K_F (L/g)	n	R^2
Perfluorooctanoic acid (PFOA)	1.11	240.9	0.99	129.2	3.88	0.97
Perfluorooctanesulfonic acid (PFOS)	2.35	187.9	0.98	126.0	5.56	0.94
HFPO-DA (GenX)	1.29	93.9	0.96	66.1	8.17	0.96

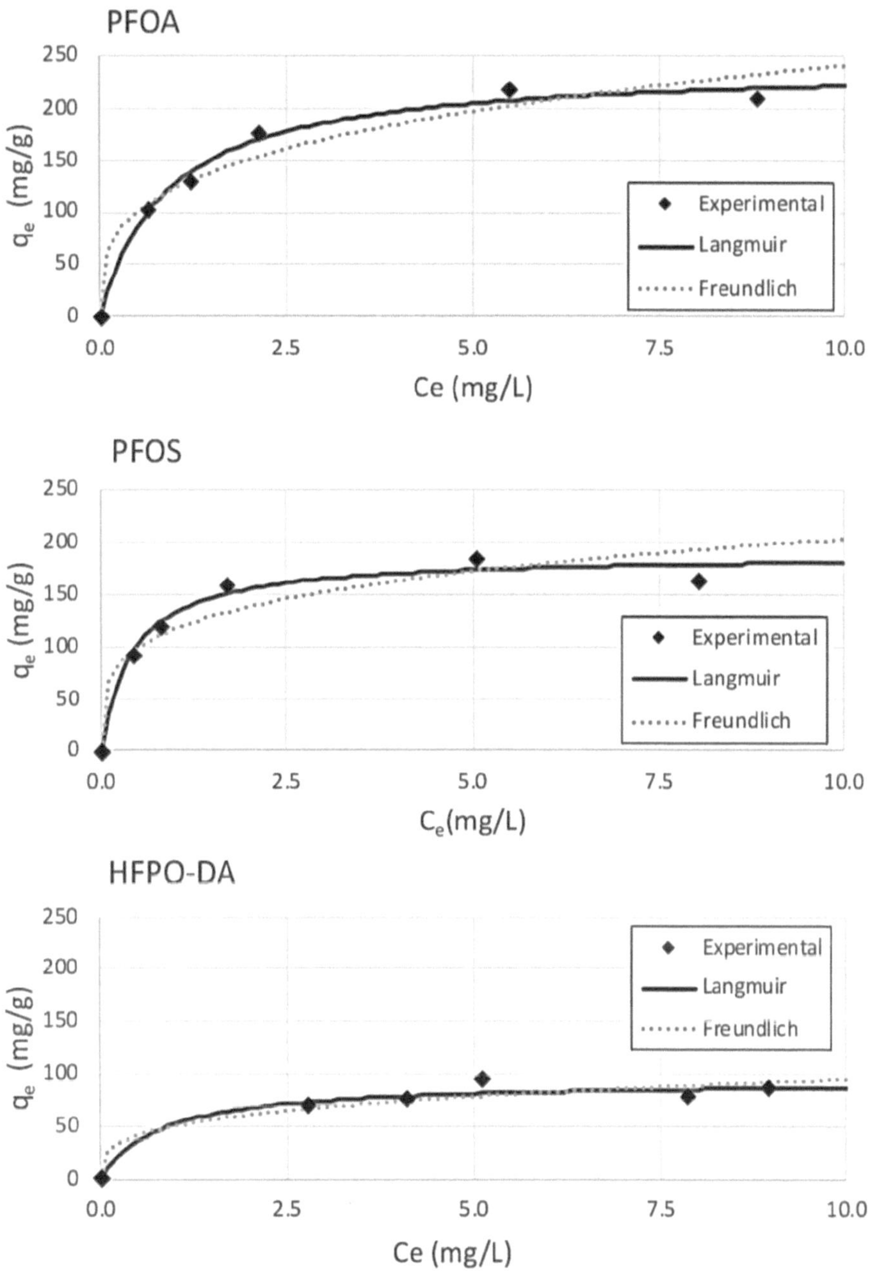

FIGURE 14.3 The isotherm of PFOA, PFOS and HFPO-DA (GenX) for DEXSORB+ with Langmuir and Freundlich fitting curves.

PFAS and other micropollutants generally occur in the environment at low concentrations (i.e., ng/L to low µg/L). Using values of q_{max} alone, limited insights can be gained into the efficacy of an adsorbent under real circumstances. Figure 14.3 also shows that DEXSORB+ exhibits high Langmuir adsorption affinity (i.e., K_L for each PFAS) at low concentrations, demonstrating the feasibility of DEXSORB+ in actual drinking water and wastewater treatment applications. Further discussion of the performance DEXSORB+ in real water matrices at environmentally relevant concentrations can be found in Section 14.3.

14.2.3 Hydrophobic and Electrostatic Interactions

As demonstrated in recent publications (Raza et al. 2019; Wang et al. 2017), the treatment of diverse PFAS compounds remains a great challenge due to their complex structures and unique physicochemical properties. Their surfactant nature derives from the basic structural elements of a hydrophobic tail (i.e., per- and polyfluorinated carbon chain) and a hydrophilic head group (e.g., carboxylic acid like PFOA or sulfonic acid like PFOS). Because of the co-presence of a hydrophobic part and an ionized head group, most PFAS can interact with adsorption materials through both hydrophobic and electrostatic mechanisms.

As described in previous sections, DEXSORB+ has been designed to combine *both* hydrophobic and electrostatic adsorption through the use of β-cyclodextrin cups and cationic quaternary ammonium sites. The 0.78 nm interior cavity of β-cyclodextrin is ideally suited to form host-guest complexes with the hydrophobic tails of PFAS, accommodating both linear and branched structures. The uniform cavities of DEXSORB+ prevent fouling by excluding larger dissolved organic matter (DOM) from adsorption sites, offering equal access to both short- and long-chain PFAS. The positive surface charge enhances the adsorption of PFAS with short carbon chains (i.e., more hydrophilic) on DEXSORB+ by providing active sites for electrostatic interactions.

Dual binding by hydrophobic and electrostatic interactions allows DEXSORB+ to uniquely address the structural variances of chain length, functional groups, and isomer structures among PFAS. These dual mechanisms of adsorption, coupled with uniform, mesoporous structures and well-defined β-cyclodextrin cavities, provides DEXSORB+ with rapid kinetics, high treatment capacity for PFAS removal, and complimentary resistance to fouling caused by DOM in complex water matrices.

14.3 PFAS UPTAKE IN REAL WATER MATRICES

The performance of DEXSORB+ for PFAS uptake under environmentally relevant scenarios has been systematically studied in a variety of water matrices, including tap water, groundwater, surface water, wastewater effluent, and landfill leachate. The treatment capacity of DEXSORB+ for diverse groups of PFAS has been investigated through batch uptake experiments (i.e., a simulation of powdered adsorbent treatment processes) and flow-through column testing (i.e., a simulation of PBF processes).

14.3.1 Batch Uptake

Batch uptake experiments provide a reliable and fast way to evaluate DEXSORB+ as a potential candidate for PFAS treatment in water and wastewater treatment processes. With proper design, batch experiments can simulate powdered adsorbent treatment processes (e.g., BA-UF or BA-C/F) to provide reliable estimations of DEXSORB+ performance in full-scale systems.

Batch uptake experiments were performed in a consistent setup that is compatible with and recommended for PFAS (i.e., polypropylene containers with tumbler mixing). All water samples were systematically characterized for background PFAS concentrations and general water quality parameters (e.g., pH, ionic strength, TOC, DOM, inorganic ions, etc.). All real-water batch experiments were performed at the original conditions of the water samples without pretreatment or adjustment. The powdered DEXSORB+ used in batch uptake experiments has a size range of 40 to 80 microns. DEXSORB+ and benchmark materials were first wetted to create a well-mixed suspension and then spiked into water samples to achieve the desired sorbent doses. The typical sorbent doses were 10 mg/L, 20 mg/L, 50 mg/L, and 100 mg/L. Samples were collected at a series of predetermined time points to investigate PFAS removal as a function of contact time. The quantification of analytes from batch experiments was performed by means of HPLC-MS/MS (QExactive, ThermoFisher Scientific) with reference to EPA methods 533 & 537 and ASTM 7979.

Across water matrices with diverse background compositions, DEXSORB+ exhibits consistent removal rate and rapid adsorption kinetics for each PFAS where the concentration of DOM is between 0.5 mg/L and 10 mg/L. The removal of PFOA and PFOS after 30-min and 24-hour contact times for groundwater, surface water, and wastewater effluent are presented in Figure 14.4.

As described above, DEXSORB+ exhibits consistently rapid adsorption kinetics for PFAS uptake in clean water matrices (e.g. DDI water and tap water). In Figure 14.4, the similar removal of PFOA and PFOS after 30-min and 24-hour contact times in each water matrix indicates that the rapid adsorption kinetics of DEXSORB+ are preserved from simple water matrices (e.g., groundwater and surface water) to

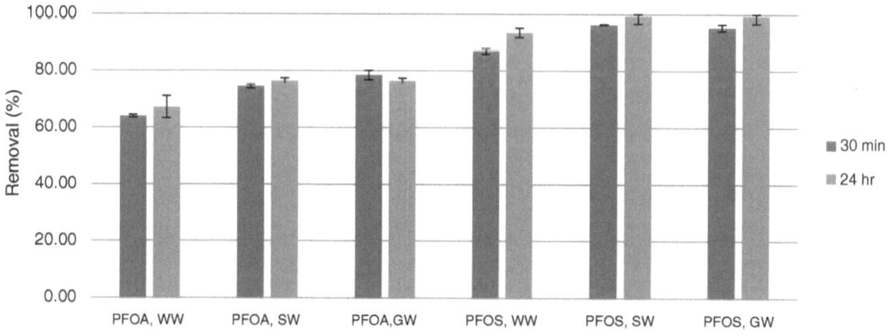

FIGURE 14.4 PFOA and PFOS removal by DEXSORB+ in wastewater (WW), surface water (SW), and groundwater (GW) at 30 minutes and 24 hours.

TABLE 14.2
Water Quality for PFAS batch uptake tests with DEXSORB+.

Parameter	Groundwater	Surface Water	Wastewater	RL[a]
Calcium [Ca^{2+}] (mg/L)	37.1	52.4	90.0	1.0
Magnesium [Mg^{2+}] (mg/L)	18.2	11.1	38.0	1.0
Potassium [K^{+}] (mg/L)	8.6	1.7	18.4	1.0
Sodium [Na^{+}] (mg/L)	88.2	68.8	158	1.0
Chloride [Cl^{-}] (mg/L)	112	159	199	1.0
Sulfate [SO_4^{2-}] (mg/L)	101	26.4	320	1.0
Nitrate/Nitrite [NO_3^-/ NO_2^-] (mg/L)	0.025[b]	1.2	13.4	1.0
DOM (mg/L)	1.2	2.5	5.4	1.0
PFOA (ppt, ng/L)	16.4	35.5	537.8	1.0
PFOS (ppt, ng/L)	26.3	9.5	483.7	1.0

[a] This column provides the reporting limit of each parameter.
[b] This indicates that the estimated concentration is above the adjusted method detection limit, but below the adjusted reporting limit.

complex water matrices (e.g., wastewater). Importantly, the variance among background matrix compositions shows no impact on the PFAS adsorption kinetics of DEXSORB+.

Though the compositions of inorganic ions and the levels of DOM are significantly different, as shown in Table 14.2, the removal of PFOA and PFOS by DEXSORB+ remains constant in these three water matrices. This observation is consistent with previous publications about cyclodextrin polymers, demonstrating the resistance of cyclodextrin polymers to fouling caused by DOM and inorganic ions (Ling et al., 2020). This feature, coupled with the high capacity of DEXSORB+ for PFAS uptake, can be applied in batch processes to reduce hydraulic retention time in contactors and sorbent dosing while achieving high PFAS removal.

14.3.2 Packed-Bed Filtration

Though results from batch experiments can provide insight into the performance of cyclodextrin polymers in powdered adsorbent treatment processes, it is necessary to perform column-type experiments to study the relationship between PFAS breakthrough behaviors and hydraulic conditions in packed bed configurations.

14.3.2.1 Rapid Small-Scale Column Tests

A Rapid Small-Scale Column Test (RSSCT) is a time-efficient and cost-effective bench-scale simulation of pilot- and full-scale PBF systems. Compared with pilot-scale column tests, the advantages of RSSCT include: (i) simple setup; (ii) small column dimensions; (iii) short duration; and (iv) reduced water sample volume. With proper design, RSSCT can simulate PBF treatment processes (e.g., pressure filtration and gravity filtration) to provide reliable estimations of the breakthrough behaviors of PFAS on cyclodextrin polymers in full-scale systems.

TABLE 14.3
Design parameters for RSSCT compared to the pilot-scale column being simulated.

Column parameters	Pilot-scale column	RSSCT
Particle diameter (mm)	0.4 to 2	0.05 to 0.25
EBCT (min)	2 to 10	0.05 to 0.5
Duration of test (hr)	1500 to 25000	40 to 1250
Reynolds number	>1	0.1 – 1
Throughput (bed volume)	50000 to 150000	50000 to 150000

The configurations of RSSCT columns and operational parameters are derived from previous studies (Knappe et al. 1997; Ling et al. 2020) and the following scaling function (*x = 2 constant diffusivity, CD; x = 1 proportional diffusivity, PD*):

$$\frac{EBCT_{sc}}{EBCT_{lc}} = \left(\frac{d_{sc}}{d_{lc}}\right)^{x} = \frac{t_{sc}}{t_{lc}} \tag{14.1}$$

where $EBCT_{sc}$, d_{sc}, and t_{sc} are the empty bed contact time (EBCT), particle diameter, and operation time for small columns, respectively; and $EBCT_{lc}$, d_{lc}, and t_{lc} are the EBCT, particle diameter, and operation time for simulated large columns, respectively.

The CD assumption has been demonstrated to work well in prior studies to simulate PFAS breakthrough for GAC and cyclodextrin polymer adsorbents in natural waters (Ling et al. 2020). The design of the RSSCT requires a pilot-scale or full-scale column with known operational parameters to simulate. Table 14.3 shows the recommended RSSCT parameters for cyclodextrin polymers together with the respective operating parameters of the simulated pilot-scale columns:

The breakthrough behaviors of PFAS on DEXSORB+ have been extensively studied through RSSCT in PFAS-contaminated water samples. The treatment capacity of DEXSORB+ for four PFAS, evaluated by the metric of specific throughput, is shown below in Figure 14.5.

Specific throughput, generally represented by the unit of L/Kg, provides the critical information: liters of treated water per Kg of an adsorbent. This enables performance comparisons among: (a) columns packed with the same material but with different bed densities; or (b) columns packed with different materials. The general water quality parameters and PFAS contamination levels for the groundwater sample used in RSSCT are presented in Table 14.4.

The high treatment capacity for both short- and long-chain PFAS are consistent with the data provided in previous sections, indicating the resistance of cyclodextrin polymers to water matrix effects compared to other adsorption materials. The results of RSSCT in groundwater show that DEXSORB+ maintains its rapid kinetics and high treatment capacity for PFAS in real water matrices for PBF processes. DEXSORB+ was able to treat all four PFAS of concern for a throughput of 100,000 L/kg before 30% breakthrough, translating to a use rate of less than 10 mg

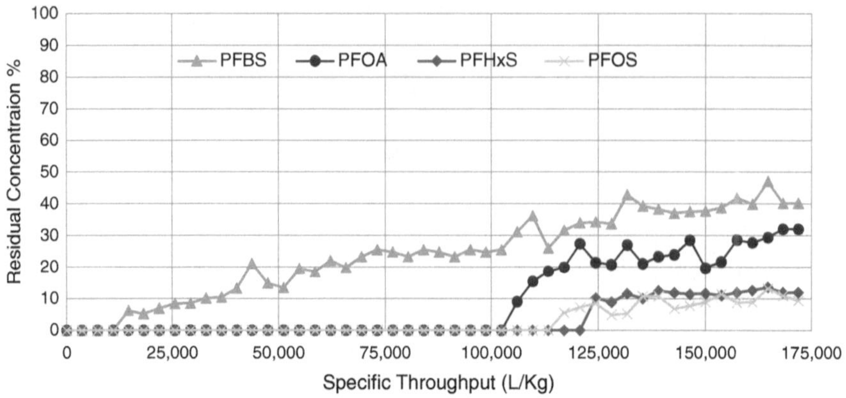

FIGURE 14.5 Breakthrough curve of four PFAS for DEXSORB+ with an EBCT of 3.2 minutes.

TABLE 14.4
Ion, DOM, and PFAS concentration in groundwater used for a 3.2-minute-EBCT RSSCT for DEXSORB+.

Parameter	Groundwater	Reporting Limit
Calcium [Ca^{2+}] (mg/L)	37.1	1.0
Magnesium [Mg^{2+}] (mg/L)	18.2	1.0
Potassium [K^{+}] (mg/L)	8.6	1.0
Sodium [Na^{+}] (mg/L)	88.2	1.0
Chloride [Cl^{-}] (mg/L)	112	1.0
Sulfate [SO_4^{2-}] (mg/L)	101	1.0
Nitrate/Nitrite [NO_3^{-}/ NO_2^{-}] (mg/L)	0.025[a]	1.0
DOM (mg/L)	1.2	1.0
PFBS (ppt, ng/L)	22.2	2.5
PFOA (ppt, ng/L)	18.1	2.5
PFHxS (ppt, ng/L)	11.7	2.5
PFOS (ppt, ng/L)	21.8	2.5

[a] Indicates that the estimated concentration is above the adjusted method detection limit, but below the adjusted reporting limit.

of DEXSORB+ per liter of water treated. The high performance of DEXSORB+ was consistent among both short- and long-chain PFAS, as shown by its high capacity for both PFBS and PFOS. The short EBCT required by cyclodextrin polymers to remove PFAS can be leveraged to reduce the footprint of pressure or gravity filtration systems and increase the throughput of water treatment.

14.3.2.2 Pilot-Scale Column Tests

Though RSSCT can provide estimates of the performance of an adsorbent, it is still necessary to perform pilot-scale column tests to: (i) confirm the results of RSSCT;

(ii) establish the relationship between particle size, hydraulic loading, and pressure drop; (iii) evaluate the material stability and mechanical strength in long-term operation; and (iv) support the design of full-scale systems that are compatible with the physicochemical properties of the adsorbent. Pilot scale results are also useful to provide when obtaining NSF and state regulatory approval for use in drinking water treatment systems.

At the time of this writing, a pilot-scale field study for DEXSORB+ was in progress with a column ID of 2 inches, a bed height of 29 inches, and an EBCT of 5 minutes. The water matrix for this pilot study was PFAS-contaminated groundwater from Orange County, CA. Other pilot-scale studies in the Netherlands, Michigan, and Massachusetts with larger configurations are under design for cyclodextrin polymers.

14.4 POTENTIAL EFFECTS OF WATER MATRICES

The fouling of traditional adsorbents by water matrix constituents is one of the major concerns for PFAS removal in both batch and PBF processes. For adsorbents that rely primarily on hydrophobic interactions, like AC, DOM is regarded as a significant inhibitor of treatment capacity for PFAS uptake. As AC adsorption is non-selective, both PFAS and DOM, which is generally present in waters at concentrations in the low mg/L range, are adsorbed by means of hydrophobic interactions. As a result, adsorption of PFAS on AC can be inhibited through: (i) direct site competition with DOM that is similar in size to PFAS molecules; and (ii) pore blockage caused by large DOM that can block the porous structures, thereby reducing the effective surface area of AC (Ding et al. 2008; Newcombe et al. 1997).

For adsorbents that rely primarily on electrostatic interaction like ion exchange resins, the ionic strength and the composition of inorganic ions are regarded as the major factors that determine their performance for PFAS uptake. The presence of inorganic ions in bulk water can diminish the surface charge of ion exchange resins, thereby limiting their electrostatic interactions with PFAS. In addition, DOM generally take negative charges, which can compete with PFAS through direct site competition (Merino et al. 2016).

The adsorption of PFAS and other micropollutants on cyclodextrin polymers relies on the formation of host-guest complexes in the 0.78 nm interior cavity of the β-cyclodextrin monomer. The uniform cavity prevents the binding of oversized organic matter through size exclusion. Secondary attraction to the positive surface charge on DEXSORB+ further enhances the adsorption of anionic molecules like most PFAS. Though the resilience of cyclodextrin polymers in complex water matrices has been demonstrated in previous studies (Ling et al. 2020; Wang et al. 2020), certain water quality parameters should be considered as part of the treatment process.

14.4.1 Suspended Solids

Suspended solids are pervasively present in all waters, from either natural or anthropogenic sources, and generally do not interfere with the adsorption kinetics or treatment capacity of cyclodextrin polymers in the removal of PFAS. As a result, for powdered adsorbent treatment processes like BA-UF and BA-C/F, pretreatments to remove suspended solids are not required for cyclodextrin polymers applications.

PBF processes like gravity filtration and pressure filtration require pretreatment of influent water to reduce suspended solids. Removal of suspended solids will reduce pressure drop and prevent channeling and bed heterogeneity for any adsorption material. Recommended pretreatments include coagulation/flocculation, small-pore filtration, deep-bed sand filtration, and multimedia filtration. Although suspended solids do not interfere with PFAS adsorption on cyclodextrin polymers, they can accumulate in a packed bed resulting in increased backpressure and clogging. Incorporating pretreatment to remove suspended solids will prolong the life of PBF filters and reduce energy consumption from water plumbing. In addition, pretreatments will also benefit backwash and regeneration cycles.

14.4.2 Ionic Strength

Ionic strength is generally evaluated by the metrics of TDS and conductivity. While ionic strength varies considerably among different water matrices, no significant effects have been observed on the treatment capacity of cyclodextrin polymers for PFAS from the level of ionic strength in water. As noted in previous sections, PFAS adsorption on DEXSORB+ relies on the formation of host-guest complexes in the 0.78 nm interior cavity of β-cyclodextrin, and interactions with positive surface charges to further increase adsorption affinity and selectivity.

The high adsorption affinity and capacity of cyclodextrin polymers for short-chain PFAS like PFBA observed in DDI water are impacted lightly in real water matrices with elevated ionic strength. Due to its small molecular size and short carbon chain length, PFBA is more hydrophilic than other PFAS. This leads to increased reliance on electrostatic interactions instead of hydrophobic interactions for PFBA adsorption on cyclodextrin polymers. An elevated ionic strength in aqueous phase can diminish the surface charge of adsorbents by screening local ionizable functional groups, limiting their electrostatic interactions with PFBA. In general, the impacts of ionic strength on PFAS treatment are limited for cyclodextrin polymers applications, eliminating the need for pretreatment to prevent ionic effects.

14.4.3 Natural Organic Matter

NOM is a collective definition of the complex mixture of organic substances in natural waters. Resulting from decomposition of organisms like plants and animals and their waste products, NOM refers to a heterogeneous mixture including insoluble, colloidal, and soluble organic substances. The impacts of insoluble and colloidal fractions were introduced earlier in this chapter.

The soluble fraction of NOM is commonly referred to as DOM, of which the major constituents include humic acids and fulvic acids. Humic acids are large and complex organic acids with a typical molecular weight range of 1 to 100 kDa. Because of their large molecular size, humic acids cannot bind with the interior cavity of β-cyclodextrin due to the size-exclusion mechanism, having no effect on PFAS adsorption on cyclodextrin polymers. Compared to humic acids, fulvic acids are much smaller with a typical molecular weight range of 0.1 to 10 kDa.

The presence of DOM in aqueous phase can greatly interfere with adsorption processes on traditional activated carbon adsorbents that rely primarily on hydrophobic interactions, inhibiting their treatment capacity for target contaminants (Ding et al. 2008, Newcombe et al. 1997). In both batch and column studies, cyclodextrin polymers demonstrate consistent treatment capacity for PFAS in different water matrices with varying levels of TOC and DOM. This observation is not unexpected; as introduced earlier, cyclodextrin polymers have uniform mesoporous structures and well-defined binding sites of β-cyclodextrin contained in 0.78 nm diameter interior cavities (Ding et al. 2008; Newcombe et al. 1997). Due to the size-exclusion mechanism, large organic matter cannot bind well with cyclodextrin polymers. It has been reported that DOM with molecular weight greater than 600 Da has no impact on the treatment capacity of cyclodextrin polymers, whereas small fulvic acids with molecular weight less than 600 Da could compete with PFAS for adsorption sites (Ling et al. 2020). Overriding this potential competition, PFAS removal by cyclodextrin polymers remains consistent across diverse water matrices with varying compositions of DOM, indicating limited fouling by DOM (Ling et al. 2020).

14.4.4 pH

Generally, the pH variance in water and wastewater treatment is in the range of 5 to 9 pH units. For this range, treatment data shows no increased or decreased influence by pH on the treatment capacity of cyclodextrin polymers for PFAS. As introduced in previous sections, DEXSORB+ features the incorporation of positively charged units in combination with its hydrophobic and mesoporous properties. The surface charge of DEXSORB+ remains positive across pH variances found in water and wastewater matrices. Accordingly, there is no observed or expected influence on the removal of PFAS by DEXSORB+ for pH variances from 5 to 9 pH units.

14.5 CYCLODEXTRIN POLYMER OPERATION CYCLES

Cyclodextrin polymers feature rapid adsorption kinetics, high treatment capacity at high and low pollutant concentrations, resistance to complex matrix effects, and easy regeneration. Each of these features can be leveraged in the operation cycles of cyclodextrin polymers for PFAS treatment to maximize water treatment throughput and minimize the cost. For batch processes like BA-UF and BA-C/F, these features enable reduced hydraulic retention time, small contactors, and low adsorbent use rates. For PBF processes like gravity and pressure filtration, these features permit short EBCT, compact vessels, and long lifecycles.

Previous sections have noted that adsorption on cyclodextrin polymers relies on the formation of host-guest complexes in the interior cavity of β-cyclodextrin. The host-guest complexation can be easily reversed (i.e., desorbed) by using a small amount of organic phase amended with salts as a regeneration solution. Accounting for operability and cost, greater efficiency is likely with single use of powdered cyclodextrin polymers in batch processes, and multiple cycles of reuse by off-site

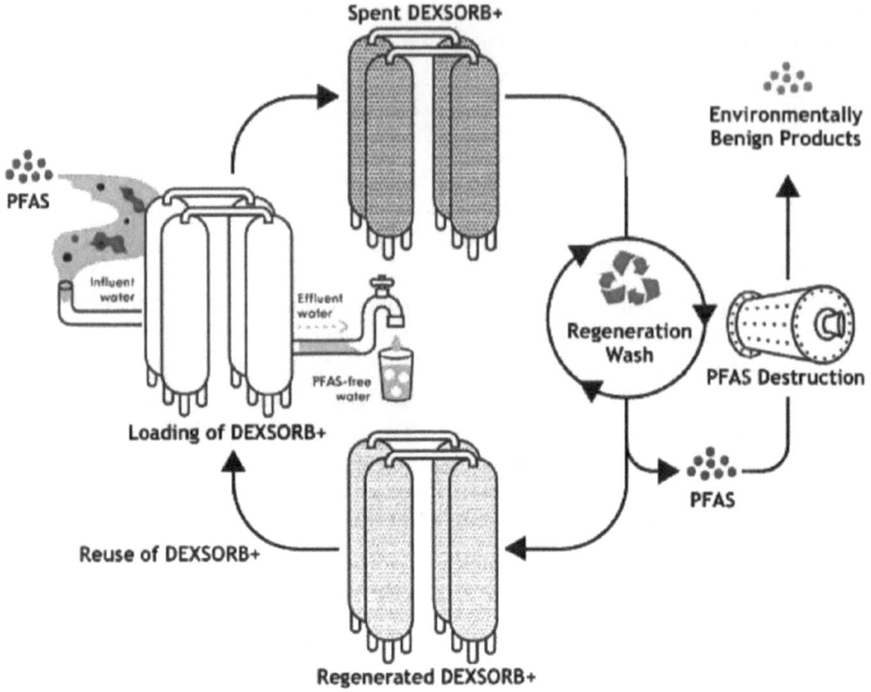

FIGURE 14.6 Treatment train for off-site regeneration of compact vessels with DEXSORB+ and PFAS destruction.

regeneration for granular cyclodextrin polymers in PBF systems. This section discusses the four operation cycles of regenerable granular cyclodextrin polymers: Loading, Regeneration, Reuse, and PFAS Disposal. Figure 14.6 presents a diagram to illustrate the operation cycles.

14.5.1 Loading of Fresh Adsorbents

For granular cyclodextrin polymers in PBF systems, the pressure drop is similar to that of ion exchange resins at the same hydraulic loading. Pretreatment is recommended in certain instances to remove suspended solids to reduce pressure drop and prevent channeling and bed heterogeneity. For the polishing of drinking water, the existing treatments for suspended solids, like sand filtration and coagulation/flocculation, should be sufficient to ensure the smooth operation of PBF systems loaded with granular cyclodextrin polymers. For wastewater treatment, groundwater remediation, and landfill leachate treatment, prefiltration such as small-pore filtration, sand filtration, or multimedia filtration is advisable to reduce the amount of suspended solids.

Contaminated water can be directly loaded to PBF systems with no requirement of additional treatments for pH adjustment, inorganic ion removal, or DOM reduction. Due to the adsorption features of cyclodextrin polymers, an EBCT design of

3 to 5 minutes is recommended under most circumstances. For the treatment of highly complex matrices like landfill leachate, an extended EBCT of 5 to 10 minutes is recommended to exhaust the full capacity of cyclodextrin polymers.

The operation life of PBF systems loaded with cyclodextrin polymers is consistent across different water matrices when PFAS contamination levels do not differ significantly. As a result, the adsorbent exhaustion limit can be determined from a universal cyclodextrin polymer use rate accounting for: (i) PFAS contamination level (e.g., a material use rate of ~10 mg per liter of water is generally required for AFFF-impacted waters with high PFAS levels); (ii) vessel configuration (e.g., a lead/lag configuration will allow for high water throughput per vessel without the risk of contaminant breakthrough); and (iii) hydraulic conditions like EBCT and hydraulic loading.

14.5.2 Regeneration of Exhausted Adsorbents

It is desirable to regenerate exhausted cyclodextrin polymers in PBF systems to increase sustainability and decrease capital cost. Regeneration and disposal of PFAS waste are expected to be performed at centralized off-site facilities to reduce the capital cost and complications of on-site PBF systems.

Due to the advanced features of cyclodextrin polymers, small compact vessels can be employed for PFAS treatment in most circumstances. Compact vessels provide the benefit of a flexible configuration, where vessels can be installed and uninstalled in a rotating fashion, greatly facilitating exchange, transport, and regeneration of spent vessels. Exhausted vessels can be regenerated at centralized facilities and inspected for PFAS residuals and other organics before redeployment.

Before regeneration, spent vessels are first backwashed: (i) to remove suspended solids and debris that accumulate in the packed bed during filtration, reducing pressure drop in future operation cycles; and (ii) to fluidize the bed to prevent the development of bed heterogeneity and channeling, increasing the use life in future operation cycles. Note that regeneration occurs through desorption of PFAS and other organic contaminants from spent adsorbent by a quick wash with regeneration solution at ambient temperature. Regeneration can be performed with spent vessels arranged in series, increasing efficiency by maximizing the concentration of PFAS in regeneration solution. Typically, a total of 5 bed volumes of regeneration solution (i.e., less than 0.01% of treated water throughput) is sufficient for the complete regeneration of spent vessels.

The regenerate stream can be further treated with evaporation/crystallization or membrane filtration to concentrate the PFAS waste stream and allow for recycle of the organic phase in regeneration treatments.

14.5.3 Reuse of Regenerated Adsorbents

Following inspection for any residual PFAS and other organics, regenerated vessels can be re-installed for another use cycle at a municipal facility. Vessel exhaustion limits will be estimated and provided to customers to guide the use of regenerated vessels. It is highly recommended to monitor PFAS breakthrough at 25%, 50%, and

75% of the estimated operation life to confirm use cycle expectations. Five consecutive regeneration cycles have been performed on cyclodextrin polymers with minor effect to performance (Alsbaiee et al. 2015; Xiao et al. 2017). Under most circumstances, at least 5 to 10 regeneration cycles should be expected for DEXSORB+ vessels.

14.5.4 Destruction of PFAS Concentrate

The PFAS concentrate isolated from regeneration solution can be destroyed through mechano-chemical, electrochemical, plasma, or hydro-thermal processes. Though high temperature incineration can also degrade PFAS to some extent, it is subject to increasing environmental concern due to: (i) inefficient energy consumption; (ii) incomplete mineralization of some PFAS, which have been detected in airborne exhaust streams; and (iii) the presence of hazardous by-products like hydrofluoric acid (HF), which is toxic and corrosive to refractories.

14.6 ADSORPTION TREATMENT APPLICATIONS

Cyclodextrin polymers are compatible with existing adsorption treatment processes because of their controllable particle size, high mechanical strength, and high chemical, thermal, and biological stability. Due to rapid kinetics, high treatment capacity, resistance to fouling, and easy regeneration of cyclodextrin polymers, the design parameters of existing processes can be optimized to enhance both treatment efficiency and sustainability, reducing capital costs and energy consumption. This section discusses the use of cyclodextrin polymers in major adsorption treatment processes.

14.6.1 Batch Adsorption Followed by Ultrafiltration

BA-UF is an emerging treatment process to address rising concerns about anthropogenic compounds like PFAS. In this process, powdered adsorbents like PAC are typically applied in contactors to adsorb targeted contaminants such as PFAS, followed by a polishing step of membrane filtration to remove adsorbents from aqueous phase. Full-scale BA-UF trials have demonstrated efficient removal of a broad spectrum of anthropogenic compounds, and reduction in effluent toxicity (Bonvin et al. 2015). However, due to the slow kinetics and easy fouling of PAC, this process requires significant infrastructure upgrade (i.e., large contactors) and a high dosage of PAC (Bonvin et al. 2015). The unique adsorption features of cyclodextrin polymers make them well-suited for BA-UF applications.

14.6.1.1 Preparation of Powdered Adsorbent

Preparation of powdered cyclodextrin polymers for BA-UF requires preliminary wetting and mixing of adsorbents. The adsorbent suspension can then be directly introduced into batch contactors for PFAS removal. Small particle sizes are preferred in BA-UF processes, as they provide rapid adsorption kinetics. Particle size selection will need to consider and balance fast adsorption kinetics with the effect that overly small particle sizes have on membrane longevity.

14.6.1.2 Hydraulic Residence Time

As described in previous sections, the adsorption kinetics of cyclodextrin polymer for PFAS uptake are consistently fast across diverse water matrices. As a result, a hydraulic residence time of 5 to 10 minutes is sufficient for cyclodextrin polymers to achieve complete PFAS uptake. Short hydraulic residence time translates to: (i) small contactor footprints; and (ii) high water throughput.

14.6.1.3 Filtration of Adsorbent

The ultrafiltration process is necessary to separate saturated cyclodextrin polymers from treated water. Reduced sorbent dosing for cyclodextrin polymers will reduce membrane replacement frequency.

14.6.2 Packed-bed Filtration

PBF is the most common adsorption treatment process and is generally employed as a polishing step for water purification or for emergency remediation of contaminated sites. In this process, granular adsorbents like GAC and IX resins have been applied in PBF vessels to adsorb target contaminants such as PFAS. Though GAC and ion exchange resins have shown effectiveness in PFAS treatment (Merino et al. 2016), PFAS breakthrough occurs quickly on GAC, resulting in high material use rates, and IX resins are either single use, or require a large amount of regeneration solution (i.e., over 2% of treated water). Increasingly, drinking water regulations are adding short-chain PFAS to their maximum contaminant level (MCL) lists alongside C-8 PFOA and PFOS. The advanced features of DEXSORB+, demonstrated in previous sections, uniquely positions it to address expanding regulations that require the removal of both short- and long-chain PFAS in household, municipal, and industrial PBF applications.

14.6.2.1 Preparation of Granular Adsorbent

For optimal performance of DEXSORB+, preliminary activation is required as an initial step prior to use in PBF systems. Since dry DEXSORB+ granules swell when exposed to aqueous phase, extra space must be provided based on the expansion rate of DEXSORB+ to allow for free expansion of granules in compact vessels. Activation can be accomplished by: (i) performing wet loading of DEXSORB+ (i.e., adding dry material into water-loaded vessels); or (ii) pre-wetting the materials before addition into the compact vessels.

14.6.2.2 Pretreatment Considerations

Cyclodextrin polymers are distinguished by their high physical, chemical, and biological stability along with resilience in complex water matrix compositions. Therefore, no additional water pretreatment is required for the optimal performance of cyclodextrin polymers, though pretreatments that remove suspended solids could be required for some applications such as the treatment of leachate.

Recall that similar to PBF processes with GAC and IX resins, pretreatments like sand filtration, coagulation/flocculation, small-pore filtration, or multimedia filtration may be required in complex matrices. When applied, these pretreatments will remove suspended solids to reduce pressure drop, and prevent channeling and bed heterogeneity in PBF systems.

14.6.2.3 Hydraulic Conditions

Rapid adsorption kinetics in diverse water matrices allow the adoption of short EBCTs in PBF systems. An EBCT design between 3 to 5 minutes is recommended for most circumstances, including drinking water polishing, wastewater treatment, and groundwater remediation. For the treatment of highly complex matrices such as landfill leachate, an extended EBCT design between 5 to 10 minutes may be necessary to exhaust the capacity of cyclodextrin polymers. Hydraulic loading similar to PBF-GAC and PBF-IX is appropriate for cyclodextrin polymers.

14.6.2.4 Fixed or Backwashable Bed

For regeneration purposes, it is recommended to use backwashable vessels for cyclodextrin polymers. Initial treatment configurations envision the return of spent transportable vessels to a centralized regeneration facility for backwashing and regeneration to extend use life.

Treatment facilities that have already installed large vessels for PFAS treatment can be repurposed for use with cyclodextrin polymers. Full loading of large vessels with cyclodextrin polymers will yield a significant increase in capacity and breakthrough volume compared to similar loading with GAC. Spent adsorbent can be removed and transported to a centralized facility for regeneration in large volume containers for re-use. Because of the kinetics and capacity of cyclodextrin polymers, transportable vessels that can be flexibly replaced and rotated provide significant cost advantages for long-term operation.

14.6.2.5 Regeneration and Concentrate Destruction

Following backwash, regeneration solution is then passed through spent vessels to regenerate (i.e., desorb) the exhausted adsorbents. Spent vessels can be treated in a sequential line assembly, providing a scalable and efficient use of regeneration solution. In general, a limited amount of regeneration solution (i.e., less than 0.01% of treated water throughput) is required to desorb PFAS from spent adsorbent. The PFAS can then be concentrated and disposed of as detailed in Section 14.5.

14.6.3 Others

14.6.3.1 Batch Adsorption Followed by Coagulation and Flocculation

BA-C/F is another emerging treatment process that can address PFAS contamination and relieve the contaminant loading on downstream treatment processes without investment in additional infrastructure. By employing powdered adsorbents directly to the influent of raw waters, the sorbent maintains contact with the aqueous phase to adsorb target contaminants until the powder is removed by coagulation and flocculation. Despite the similar concept and use of powdered adsorbents, BA-C/F is distinguished from BA-UF with respect to infrastructures, hydraulic conditions, mixing patterns, and separation technologies.

In contrast to BA-UF, BA-C/F processes have no requirement for significant infrastructure investment. It utilizes existing C/F treatment instead of membrane filtration as the separation technology, leading to significantly reduced capital cost without requirement of additional adsorption contactors. Due to the short hydraulic residence time and the lack of controlled mixing, dosing requirements of non-selective adsorbents like AC would be significantly higher. The adsorption features of cyclodextrin polymers allow for hydraulic residence times as short as 5 minutes and a low adsorbent dose of 10 to 20 mg/L for sufficient PFAS removal from influent in most applications.

14.6.3.2 Small Drinking Water Systems

Small drinking water systems provide drinking water to communities with fewer than 10,000 people, removing pathogens and contaminants of concern like PFAS from water sources (EPA 2017). Small community treatment systems must be highly efficient to keep costs down. Per capita considerations are magnified when costs are shared among smaller populations compared to large municipal systems. In addition, small drinking water systems must be compatible with complementary treatment processes and easily adaptable across varying conditions of water demand and source-water quality. Processes like chlorination and UV treatment are generally applied to disinfect water and integrated with treatments such as filtration to address community-specific contamination concerns. Due to their chemical/thermal stability and surfactant nature, PFAS remain a significant challenge to traditional small community water treatment systems. Among available treatment options, adsorption through PBF is regarded as a highly effective method to address PFAS contamination in small drinking water systems.

As introduced in previous sections, the advanced features of cyclodextrin polymers allow for application in small vessels, which are ideally suited for use in small drinking water systems. Due to rapid adsorption kinetics, high treatment capacity, and resistance to matrix constituents, cyclodextrin polymers in small vessel PBF systems are able to ensure consistent water quality regardless of fluctuations in water demand and source-water quality. Multicycle regeneration of cyclodextrin adsorbents allows for cost-effective utilization, providing affordable solution alternatives to address the per capita cost sensitivities of small communities. Small vessel PBF systems provide flexibility to incorporate the number of vessels required to meet the specific water demands of a community. As described in Section 14.5, exhausted small vessels can be transported to a centralized facility for backwashing and regeneration to allow for reuse. Centralized regeneration facilities eliminate the requirement of dedicated backwash equipment and infrastructure for small community systems. For small communities impacted by PFAS contamination, cyclodextrin adsorbents in small vessels offer the opportunity of municipal quality treatment and performance in a scalable, cost-effective solutions.

14.7 SUMMARY

DEXSORB+® represents a breakthrough solution for the safe, rapid, and cost-effective removal of diverse PFAS, allowing for the delivery of affordable, PFAS-free drinking water to households. With dual-adsorption mechanisms of hydrophobic and electrostatic binding to well-defined β-cyclodextrin cavities, DEXSORB+ features efficient removal of both short- and long-chain PFAS, including carboxylic acids (e.g., PFBA, GenX, PFOA) and sulfonic acids (e.g., PFBS, PFHxS, PFOS) from contaminated water.

The adsorption of PFAS on DEXSORB+ derives from the formation of host–guest complexes in the interior cavities of β-cyclodextrins and interactions with its positive surface charge. The well-defined 0.78-nm interior cavity of ß-cyclodextrin is ideally suited to binding PFAS, and at the same time resists adsorption of oversized NOM by *size exclusion*. This results in consistently high PFAS uptake on DEXSORB+ across water sources with diverse matrix compositions (e.g., drinking water, groundwater, surface water, and landfill leachate). Resilience to variances among water quality makes DEXSORB+ a highly flexible treatment solution that can be adapted to the most challenging environments with limited pretreatment, short contact time, low dosing and high capacity.

With the ability to control particle size in powder and granular forms, DEXSORB+ can be flexibly used in a variety of large volume treatment configurations, including packed bed filtration, batch + coagulation/flocculation and batch + ultrafiltration, as well as small community drinking water systems.

Additionally, due to the reversibility of host–guest complexation and electrostatic interactions, spent DEXSORB+ can be easily regenerated through desorption by a quick wash of the polymer with a small amount of regeneration solution at ambient temperature, allowing for multiple cycles without loss in performance. With high treatment capacity across diverse water matrices, DEXSORB+ offers lower operational costs and increased sustainability for water treatment processes. Completing the treatment life cycle, the processed regenerate solution can be further concentrated to isolate the desorbed PFAS into a "neat" waste product for cost-effective handling and mineralization using non-combustion techniques to fully detoxify the PFAS waste.

REFERENCES

Alsbaiee, A., Smith, B.J., Xiao, L., Ling, Y., Helbling, D.E., Dichtel, W.R. Rapid removal of organic micropollutants from water by a porous β-cyclodextrin polymer. *Nature* 2015 529, 190–194. doi: 10.1038/nature16185.

Bonvin, F., Jost, L., Randin, L., Bonvin, E., Kohn, T. Super-fine powdered activated carbon (SPAC) for efficient removal of micropollutants from wastewater treatment plant effluent. *Water Res.* 2015 90, 90–99. doi: 10.1016/j.watres.2015.12.001.

Crini, G. Review: A history of cyclodextrins. *Chem. Rev.* 2014 114, 10940–10975. doi: 10.1021/cr500081p.

Ding, L., Snoeyink, V.L., Mariñas, B.J., Yue, Z., Economy, J. Effects of powdered activated carbon pore size distribution on the competitive adsorption of aqueous atrazine and natural organic matter. *Environ. Sci. Technol.* 2008 42, 1227–1231. doi: 10.1021/es0710555.

EPA, 2017. *Small Drinking Water Systems Research*. Available at:. https://www.epa.gov/water-research/small-drinking-water-systems-research (Accessed August 26, 2020).

Herkert, N.J., Merrill, J., Peters, C., Bollinger, D., Zhang, S., Hoffman, K., Ferguson, P.L., Knappe, D.R.U., Stapleton, H.M. Assessing the Effectiveness of Point-of-Use Residential Drinking Water Filters for Perfluoroalkyl Substances (PFASs). *Environ. Sci. Technol. Lett.* 2020. acs.estlett.0c00004. doi: 10.1021/acs.estlett.0c00004.

Knappe, D.R.U., Snoeyink, V.L., Roche, P., Prados, M.J., Bourbigot, M.M. The effect of preloading on rapid small-scale column test predictions of atrazine removal by GAC adsorbers. *Water Res.* 1997 31, 2899–2909. doi: 10.1016/S0043-1354(97)00148-6.

Ling, Y., Alzate-Sánchez, D.M., Klemes, M.J., Dichtel, W.R., Helbling, D.E. Evaluating the effects of water matrix constituents on micropollutant removal by activated carbon and β-cyclodextrin polymer adsorbents. *Water Res.* 2020 173. doi: 10.1016/j.watres.2020.115551.

Ling, Y., Klemes, M.J.M.J., Xiao, L., Alsbaiee, A., Dichtel, W.R.W.R., Helbling, D.E.D.E. Benchmarking Micropollutant Removal by Activated Carbon and Porous β-Cyclodextrin Polymers under Environmentally Relevant Scenarios. *Environ. Sci. Technol.* 2017 51, 7590–7598. doi: 10.1021/acs.est.7b00906.

Merino, N., Qu, Y., Deeb, R.A., Hawley, E.L., Hoffmann, M.R., Mahendra, S. Degradation and Removal Methods for Perfluoroalkyl and Polyfluoroalkyl Substances in Water. *Environ. Eng. Sci.* 2016 33, 615–649. doi: 10.1089/ees.2016.0233.

Newcombe, G., Drikas, M., Hayes, R. Influence of characterised natural organic material on activated carbon adsorption: II. Effect on pore volume distribution and adsorption of 2-methylisoborneol. *Water Res.* 1997 31, 1065–1073. doi: 10.1016/S0043-1354(96)00325-9.

Raza, A., Bardhan, S., Xu, L., Yamijala, S.S.R.K.C., Lian, C., Kwon, H., Wong, B.M. A Machine Learning Approach for Predicting Defluorination of Per- And Polyfluoroalkyl Substances (PFAS) for Their Efficient Treatment and Removal. *Environ. Sci. Technol. Lett.* 2019. doi: 10.1021/acs.estlett.9b00476.

Seo, S.H., Son, M.H., Choi, S.D., Lee, D.H., Chang, Y.S. Influence of exposure to perfluoroalkyl substances (PFASs) on the Korean general population: 10-year trend and health effects. *Environ. Int.* 2018 113, 149–161. doi: 10.1016/j.envint.2018.01.025.

Wang, R., Ching, C., Dichtel, W.R., Helbling, D.E. Evaluating the Removal of Per- And Polyfluoroalkyl Substances from Contaminated Groundwater with Different Adsorbents Using a Suspect Screening Approach. *Environ. Sci. Technol. Lett.* 2020 7, 954–960. doi: 10.1021/acs.estlett.0c00736.

Wang, Z., Dewitt, J.C., Higgins, C.P., Cousins, I.T. A Never-Ending Story of Per- and Polyfluoroalkyl Substances (PFASs)? *Environ. Sci. Technol.* 2017 51, 2508–2518. doi: 10.1021/acs.est.6b04806.

Wu, B., Hao, S., Choi, Y., Higgins, C.P., Deeb, R., Strathmann, T.J. Rapid Destruction and Defluorination of Perfluorooctanesulfonate by Alkaline Hydrothermal Reaction. *Environ. Sci. Technol. Lett.* 2019 6, 630–636. doi: 10.1021/acs.estlett.9b00506.

Xiao, L., Ling, Y., Alsbaiee, A., Li, C., Helbling, D.E.D.E., Dichtel, W.R.W.R. β-Cyclodextrin Polymer Network Sequesters Perfluorooctanoic Acid at Environmentally Relevant Concentrations. *J. Am. Chem. Soc.* 2017 139, 7689–7692. doi: 10.1021/jacs.7b02381.

Zhang, K., Huang, J., Yu, G., Zhang, Q., Deng, S., Wang, B. Destruction of perfluorooctane sulfonate (PFOS) and perfluorooctanoic acid (PFOA) by ball milling. *Environ. Sci. Technol.* 2013 47, 6471–6477. doi: 10.1021/es400346n.

15 Modeling Water Treatment Performance and Costs for Removal of PFAS from Drinking Water

Jonathan B. Burkhardt
US Environmental Protection Agency, USA

Richard H. Anderson
US Air Force Civil Engineer Center, USA

Rajiv Khera, Levi M. Haupert
US Environmental Protection Agency, USA

Patrick Ransom
Abt Associates, USA

David G. Wahman, Page Jordan, Jonathan G. Pressman, Marc A. Mills and Thomas F. Speth
US Environmental Protection Agency, USA

CONTENTS

15.1 Introduction....316
15.1.1 High-Pressure Membrane Systems....318
15.1.2 Anion Exchange Resins....318
15.1.3 Granular Activated Carbon (GAC)....319

15.2 Modeling PFAS Removal with GAC 321
15.2.1 Parameters 321
15.2.2 Parameter Determination 322
15.2.3 Modeling Full-scale Systems 324
15.2.4 Specific Throughput 325
15.2.5 Case Study: Short Bed Contactors 325
15.2.6 Case Study: Assessing Designs for Future Needs 325
15.3 Cost Modeling 330
15.3.1 Work Breakdown Structure Model 330
15.3.2 DoD Costing Example 331
15.4 Conclusions 335
15.5 Software Information 335
Acknowledgments 336
References 336

15.1 INTRODUCTION

Per- and polyfluoroalkyl substances (PFAS) are a large and diverse class of compounds that have received a great deal of attention recently. Due to a combination of increased analytical capabilities and health effects studies, many states have or are anticipated to set regulatory limits for select PFAS at environmentally relevant concentrations. Also, citizen groups have expressed interest in removing PFAS from drinking and environmental waters. This has mobilized states, water utilities, and entities that utilize PFAS products to evaluate options for treating PFAS-contaminated water. USEPA has consolidated a large collection of PFAS treatment data into its Drinking Water Treatability Database (TDB) (USEPA 2020a) [https://tdb.epa.gov]. At the time of writing, data from 118 references for 26 PFAS treated with 22 technologies can be found in the TDB.

Determining the optimal treatment technology and subsequent system design for a given application can be made more difficult by numerous factors. First and foremost is the question of what PFAS need to be removed and to what level. However, as described in more detail below, while some technologies are very effective for some of the thousands of PFAS compounds, none of them are 100% effective over an extended period of time for all PFAS. Therefore, treatment goals need to consider what PFAS are in the source water, which ones have health impacts at those concentrations, and what finished drinking water concentrations are needed to reduce risk. This approach is the basic construct of the regulatory determination process, and although a complete description of the regulatory determination process at the federal or state level is beyond the scope of this chapter, it is useful to explore: (1) the value data collected during treatment studies; and (2) what treatment goals need to be applied to help direct modeling efforts.

Available analytical chemistry methods are used to characterize the PFAS in a source water, determine their concentrations, and also help determine the treatment effectiveness. Currently, the drinking water PFAS analytical methods can only measure a small subset of known PFAS potentially in the water. However, there are research methods that cover hundreds of PFAS for semiquantitative analysis or occurrence. An example of these methods includes non-targeted methods (Strynar et al. 2015). These approaches are useful for discovering novel PFAS in environmental samples and qualitative assessments of treatment effectiveness; however, they do not offer confirmation of the exact species for the majority of the tentatively identified compounds and cannot be quantified without analytical standards. Therefore, this discussion will refer to methods that provide quantitative analysis of specific PFAS.

A total of 29 PFAS are included in EPA's drinking water methods, where Method 533 (USEPA 2019b, 2020e) includes 25 PFAS and Method 537.1 (USEPA 2019a, 2020e) includes 18 PFAS; 14 compounds are common to both methods, 11 compounds specific to Method 533, and 4 compounds are unique to Method 537.1. It is important to note that use of these methods, as written, can be prescribed under various compliance monitoring activities for PFAS in drinking water. For non-potable water (i.e., water that is not considered drinking water) EPA published SW-846 Method 8327 in 2019 that included 25 PFAS (USEPA 2019). Other research analytical methods have been developed (Sanan and Magnuson 2020, Dasu et al. 2017) and some commercial laboratories often report additional analytes by using related methods or "modifications" of these methods. Care must be taken extending these methods to related water matrices for which they have not been developed, and selection of appropriate methods should consider the data quality objectives of the particular study and be performed in consultation with all concerned parties. Some environmental regulatory or stewardship authorities may have specific requirements for analytical methods and approaches, including how sampling is performed and how samples are prepared for quantification.

Even with available analytical resources for quantification of PFAS concentrations, the consistency of PFAS levels in source waters and how variable they are for that source is not well understood. The natural variability in source water PFAS concentrations are due to site-specific conditions and not related to experimental variability introduced by the analytical approach.

The task of evaluating a range of treatment goals (e.g., which PFAS are to be removed and to what level) gets more difficult when considering health effects levels. Typically, a utility or system has regulated levels set by their state primacy agency or those set by the federal government. Although the USEPA has set health advisory levels (USEPA 2016a, 2016b) for PFOA and PFOS and published draft toxicity assessments for PFBS and GenX (USEPA 2020b), to date, there are no federal drinking water regulations for PFAS under the Safe Drinking Water Act, as amended. Some states have PFAS drinking water regulations that are only for a select few PFAS, and regulated concentrations vary between states. Therefore, the varied regulations can cause difficulties in making PFAS treatment goal decisions for utilities or systems, especially when considering unknowns in the future of PFAS regulation, such as: (1) if the number of regulated PFAS will increase in the future; (2) if the regulated concentrations will change as additional health effects data are obtained; and (3) if improvements in analytical methods will lead to changes in how analytical results are considered within regulations.

These potential uncertainties, particularly as they relate to influent concentrations and treatment goals, point to the value of modeling treatment studies. Modeling can be used to interpolate and extrapolate the results of treatment studies to other conditions, such as variable influent concentrations and changing treatment goals for not only the PFAS being evaluated in the initial treatment study, but also to other PFAS that may be an issue in the future. Because it is cost-prohibitive to run necessary pilot studies, analyze results, and develop models for all technologies, it is important to limit the technologies to only those that can be implemented in terms of treatment and cost effectiveness and permitting. The most commonly implemented technologies for drinking water or minimally impacted groundwater remediation systems are high-pressure membranes (e.g., reverse osmosis), anionic exchange resins (AER), and granular activated carbon (GAC).

15.1.1 High-Pressure Membrane Systems

For high-pressure membrane systems, excellent removal can be achieved for many PFAS (Dickenson and Higgins 2016; Crone et al. 2019). These percent removals may change as membranes foul or develop minor integrity issues; however, the general conclusion remains the same. The choice of membranes is usually limited by the high capital and operating costs associated with high-pressure membrane systems. Also, residual stream management is a concern. High-pressure membrane systems typically discharge approximately 20 percent of the feed flow as concentrated retentate. How this waste stream is managed can increase costs for further treatment or disposal options. With regards to pilot testing, the percent removal is a function of the specific proprietary membrane chosen and the arrangement of pressure vessels with the percent recovery for the array. Because membrane systems behave generally as steady state processes, and because membranes' rejection are generally high and essentially concentration-independent, it is straightforward to model various alternate array scenarios. Although predicting membrane rejection between different membranes can be done, because many factors related to membrane structure, design, and operation impact performance, it is not recommended and results will be different, as will the fouling tendencies. Therefore, because these systems are very membrane-dependent and modeling is straightforward given pilot data for a chosen membrane, the modeling discussion in this chapter concentrates on non-steady state column technologies (ion exchange resins and granular activated carbon).

15.1.2 Anion Exchange Resins

Anionic exchange resins have been shown to be effective for PFAS removal (Chularueangaksorn et al. 2014; McCleaf et al. 2017; Woodard et al. 2017; Ross et al. 2018; Franke et al. 2019; Schaefer et al. 2019; Crone et al. 2019; Zeng et al. 2020). Some resins have been pilot tested and utilized at full-scale facilities. The resins can be in macroporous or gel form. An important aspect of this technology is the impact of background constituents such as other anions than PFAS, including chloride, nitrate, sulfate, and natural organic matter (NOM). Often these parameters are fairly constant at a given utility or system and they are often ignored when planning a pilot-testing effort, likely due to a number of factors, such as the additional cost and complexity to evaluate their impact. The results in such a situation can be used as a direct evaluation of treatment efficiency; however, it hampers the ability to fit the results with models and then extrapolate to other scenarios such as different PFAS concentrations, anion concentrations, flow rates, and filter sizes. This is especially true if selectivity coefficients for the anions for the resin of interest are unknown. There is a paucity of this information in the literature. Few column or pilot studies have been reported for PFAS sorption by strong base anion exchange. Generally, more bed volumes can be treated by ion exchange compared to GAC (see below), though beds are usually smaller; however, due to the general lack of analysis of underlying ion exchange theory, it is not clear if this trend can be expected to hold for all possible conditions.

There are few publicly available models for predicting ion exchange performance. Recently, EPA entered into an agreement with Michigan Technological University (MTU) to make their suite of models available to the public at no cost (see Software Information section below for link, USEPA 2020c). These models were developed by the National Center for Clean Industrial and Treatment Technologies at MTU. They were originally released under the moniker Environmental Technologies Design Option Tool (ETDOT). The model for ion exchange resin is under the name Ion Exchange Design Software (IonExDesignS) Version 1.0.0. The software code is available under USEPA's GitHub site where improvements can be made, such as an update to current operating systems (USEPA 2020d). This program is similar to external/internal diffusion models for GAC discussed below and is directly applicable to macroporous resins. USEPA is developing additional ion exchange models that include detailed mass transport kinetics in both gel-type and macroporous resins. These new models use fundamental theories of ion exchange to describe interference from multiple ionic species commonly found in drinking water, including doubly charged anions such as sulfate. In addition to the column models, the USEPA is working on a set of tools to aid researchers in designing and interpreting batch and column experiments to make the best use of the limited resources and available data.

Ultimately, there have been limited examples available that can be used to demonstrate the application of ion exchange models for PFAS removal, primarily due to the lack of: (1) data on selectivity coefficients for various resins; (2) pilot data that includes influent and effluent anions data; and (3) available models. This is an active area of research at EPA and within the environmental research community and is an area poised for significant progress. Therefore, the remainder of this chapter concentrates on GAC models, the results from GAC pilot systems, and the predictions for those systems for PFAS removal under different conditions to be used in cost modeling for individual sites and for multiple site evaluations.

15.1.3 Granular Activated Carbon (GAC)

Carbon adsorption can be an effective treatment technology for removing PFAS from water (Dickenson and Higgins 2016; Kucharzyk et al. 2017; Crone et al. 2019). GAC can be made of organic materials such as bituminous coal, coconut shells, and wood (Sontheimer et al. 1988). Although the optimal choice of starting material is application-specific, an investigation using a variety of source water qualities suggests that GAC made from bituminous coal is optimal for PFAS treatment (Dickenson and Higgins, 2016). Of the short list of available technologies discussed in this chapter, GAC has been the most studied, perhaps because it can be the least expensive. A host of bench-, pilot-, and full-scale studies are available for review and modeling. However, many of the published studies show varying results even at the bench-scale level with laboratory-pure water. For instance, reported Freundlich isotherm values for PFAS vary widely (Crone et al. 2019), similar to other contaminants. This variability can lead to complications when these values are used in predictive models for design of GAC treatment systems (e.g., bed size) and operating conditions (e.g., flow rates). Probabilistic modeling could be used to capture the impact of uncertainty or variability in these parameters.

GAC has been widely used to treat drinking water for numerous trace contaminants (Sontheimer et al. 1988); however, treating PFAS-contaminated water may require larger contactor beds or longer contact times for PFAS removal compared to many other contaminants. Sizing and design of GAC contactors has been well studied and supported by data from pilot or other smaller column tests (RSSCT: Rapid small-scale column test, or ACT: Accelerated column test). These column tests can also help highlight the impacts of competition with other contaminants and fouling related to NOM for a specific source water.

Column testing provides valuable data for the design process but can be limited to only the configurations that were tested and the specific water's characteristics (e.g., other chemicals or NOM fouling). For example, the expected time to breakthrough for a single species may be approximately doubled if the empty bed contact time (EBCT) is doubled or if the concentration is halved. However, many water utilities are treating to a multi-PFAS objective where the sum of PFAS species is used as the treatment objective. This means that errors in the initial extrapolation from pilot-scale results lead to compounded error when predicting final design, operation and maintenance costs and performance. Running multiple column tests adds costs and time to this stage of the design process. The use of performance models can help provide estimates for numerous conditions based on input parameters rather than direct use of experimental data. This can be done by fitting pilot system data with models that incorporate multiple capacity and kinetic effects that can be used to simulate results based on other water qualities and treatment designs.

Similar to ion exchange, USEPA offers GAC models as developed under the ETDOT program. The Adsorption Design Software (AdDesignS) program is available through the ETDOT suite of programs that is hosted on GitHub (see Software Information section for link). Available models include those that assume instantaneous kinetics and those that consider kinetic factors. Of interest is the pore surface diffusion model (PSDM), which accounts for diffusion through the stagnant external film layer, internal pore diffusion, internal surface diffusion, multi-component adsorption (using the Freundlich isotherm), variable influent concentrations, and NOM preloading. This pseudo-mechanistic model has been used successfully to fit pilot- and full-scale data (Hand et al. 1997; Sontheimer et al. 1988). The added complexity of the PSDM also provides the capability to assess uncertainty (Knappe et al. 1999) within parameters directly, unlike the direct estimation from pilot results that does not. To serve this end, additional models and capabilities (e.g., conversion to Python to incorporate automated parameter fitting routines with statistics and inclusion of prediction capability for multiple treatment objectives) are being developed by the USEPA (see Software Information section for more information).

Simulations can be performed on numerous scenarios like varying influent concentrations, flows, carbon type, system size, or operating configurations. Results from this performance modeling can then be used within a cost model to better assess results on an economic basis. EPA has developed a suite of cost models known as the work breakdown structure (WBS) models (Khera et al. 2013; USEPA 2020d). The WBS models are transparent, spreadsheet-based engineering models for individual treatment technologies linked to a central spreadsheet of component unit costs. The WBS approach involves breaking the technology into discrete components for the purposes of estimating unit costs. The WBS-based approach identifies the components that should be included in a cost analysis, and the models produce a detailed assessment of the capital and operating requirements for a treatment system. Many of the cost models, including the GAC models used herein, have been peer-reviewed by nationally recognized treatment technology experts for completeness, accuracy, and usability and are publicly available (USEPA 2020d).

Although these models were primarily designed for estimating drinking water regulatory compliance costs as part of USEPA's rulemaking process, they may be used by water systems for a contaminated site or in evaluations between sites. One such example of the latter is the recent US Department of Defense (DoD) effort to identify all legacy uncontrolled aqueous film-forming foam (AFFF) releases that have contaminated soil, groundwater, sediment, and/or surface water across the portfolio of installations following the Comprehensive Environmental Response, Compensation, and Liability Act (CERCLA) process (Anderson and Kempisty 2019). During this DoD investigation, potential drinking water exposures are being identified and investigated. There is a need to develop accurate cost predictions to allow for optimal program implementation. Applying treatment data and performance model results in cost models allows for the evaluation of the cost of removing PFAS from contaminated waters.

Therefore, the intent of the remainder of this chapter is to define how GAC treatment performance models can be used to fit and then predict actual GAC treatment performance for a wide range of water quality, design, and flow rate scenarios. WBS cost models will then be used to demonstrate how these treatment predictions can be translated into costs that will allow decision-makers to make informed choices for designs at individual sites and even across sites, allowing the optimization of entire treatment programs. The demonstration across sites will be an initial evaluation of operation and maintenance (O&M) costs for DoD's treatment systems for policy decisions based on performance, cost, and risk management.

15.2 MODELING PFAS REMOVAL WITH GAC

15.2.1 Parameters

The PSDM model requires parameters for the treatment column (or contactor), the adsorbent, general chemical characteristics, and adsorbent-chemical parameters. For the treatment system, information about the bed depth (sometimes reported as column length), diameter, mass of adsorbent and flow rate is needed. For the adsorbent, information about particle radius, porosity, particle density (i.e., density of just the 'spherical' adsorbent particle) and apparent density (i.e., the effective density within a packed bed—mass of adsorbent in volume of bed) are needed. For a chemical, the molecular weight, molar volume, and liquid density are required. Additionally, the model requires film transfer, pore- and surface-diffusion coefficients for each chemical or these values can be estimated from correlations. The adsorbent specific isotherm parameters are also required for each adsorbent-chemical pair—in this model the Freundlich isotherm is used, which requires a K and 1/n (or sometimes just reported as n) parameters (Hand et al. 1997).

15.2.2 Parameter Determination

The isotherm parameters can be determined from laboratory experiments or estimated from pilot data. Ideally, the values for K and 1/n would be determined from controlled laboratory adsorption experiments for ranges of concentrations of interest; however, these are not always available and do not necessarily capture other characteristics like NOM fouling that occur in natural waters over time. Within the model, NOM fouling is simulated by a reduction of available adsorption capacity over time, and this can be applied during fitting of model parameters. Fitting pilot system data with models that incorporate multiple capacity and kinetic effects, in addition to Freundlich values, can be a useful approach to train a model that then can be used to predict results based on other water qualities and treatment designs. The next paragraphs provide a brief explanation of the steps involved in estimating parameters from RSSCT, pilot, or full-scale treatment study data.

Following a test column or in full-scale studies with an adsorbent bed replacement, the time series related to both influent and effluent concentrations can be used to estimate adsorbent capacity (see Figure 15.1 for examples). An estimate of adsorbed mass can be calculated from the area between the influent and effluent curves that precede breakthrough. Complete breakthrough is preferred, but breakthrough can be estimated by extrapolating incomplete breakthrough curves until they intersect with an averaged influent concentration. If no fouling of the adsorbent is considered, this provides a reasonable estimate for when 50 percent breakthrough would occur; however, the shape of the overall breakthrough curve may not be correct if the other transport parameters are not correct. Computer routines or manual methods can be used to optimize the parameter estimates by minimizing the unweighted sum-of-squared differences between simulated output concentrations and the measured effluent concentrations of the system.

For this chapter, the estimates of the parameter values of K and 1/n were optimized assuming a fouling treatment was applied throughout the operating period of the column (see Jarvie et al. 2005 for more information on fouling) and other adsorbent-chemical parameters were estimated from correlations. Data from a system described in Newman and Berry (2019) was used to estimate adsorbent-chemical parameters. Table 15.1 summarizes the best fit parameters for PFAS measured from

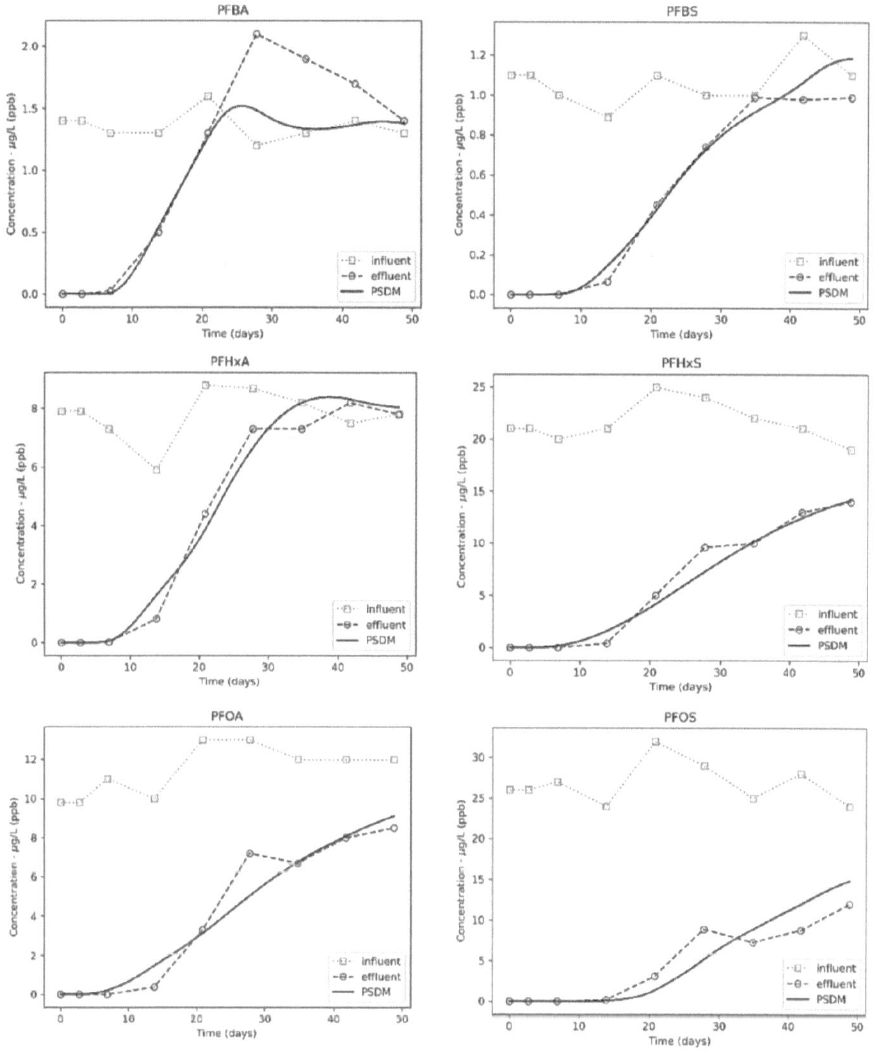

FIGURE 15.1 Examples of Estimated Best Fit PSDM Models for Results Presented in Table 15.1.

this system, which used Calgon Filtrasorb 400 (F400) GAC. Figure 15.1 shows examples of the resulting models for the pilot studies with influent and effluent concentrations included for each. These were fit individually using fouling characteristics of Karlsruhe groundwater (Jarvie et al., 2005), and the values are reported for unfouled or organic-free water. Because they were fit individually, true isotherm values may be different as these implicitly capture some of the multicomponent competitive adsorption behavior and would be expected to have a higher capacity.

TABLE 15.1
Summary Results of Estimated Best Fit Parameters for Column Study (Newman and Berry, 2019).

Compound	CAS Number	Influent Conc. Range (μg/L)	K (μg/g)(L/μg)$^{1/n}$	1/n (-)
Perfluorobutanoic acid (PFBA)	375-22-4	1.2–1.6	17.33	0.375
Perfluoropentanoic acid (PFPeA)	2706-90-3	3.9–5.1	43.68	0.35
Perfluorohexanoic acid (PFHxA)	307-24-4	5.9–8.8	62.99	0.425
Perfluoroheptanoic acid (PFHpA)	375-85-9	1.8–2.2	35.25	0.55
Perfluorooctanoic acid (PFOA)	335-67-1	9.8–13	62.30	0.75
Perfluorononanoic acid (PFNA)	375-95-1	ND–0.082	4.30	0.30
Perfluorobutanesulfonic acid (PFBS)	375-73-5	0.89–1.3	24.78	0.45
Perfluorohexanesulfonic acid (PFHxS)	355-46-4	19–25	80.38	0.75
Perfluoroheptanesulfonic acid (PFHpS)	375-92-8	1.1–1.4	36.16	0.375
Perfluorooctanesulfonic acid (PFOS)	1763-23-1	24–32	374.8	0.3
6:2 Fluorotelomer sulfonic acid (62FTS)	27619-97-2	15–22	148.9	0.45
8:2 Fluorotelomer sulfonic acid (82FTS)	39108-34-4	0.21–0.29	10.01	0.3

However, they can still provide an estimate for predicting treatment in the original or similarly contaminated waters.

15.2.3 Modeling Full-scale Systems

Once the various adsorbent-chemical parameters are known or have been estimated, PSDM can be used to simulate a variety of additional scenarios. Some examples of model results include: (1) the impact of average influent concentrations; (2) an example of specific throughput calculation for a range of empty bed contact times; (3) determination of bed replacement intervals for a full-scale treatment objective; and (4) the impact of different bed depths on bed replacement intervals. Later sections of this chapter will also discuss applications of this modeling for estimations of costs related to different PFAS removal treatment objectives.

Because the PSDM takes as inputs various characteristics related to the modeled system (e.g., size, adsorbent weight, and flow rate), these can be varied as desired to simulate their impact on effluent concentrations. The following examples are intended to reflect modeling capabilities and do not necessarily reflect real treatment objectives.

15.2.4 Specific Throughput

"Specific throughput" is a simple metric for estimating how much volume can be treated for a given mass of adsorbent. Figure 15.2 shows specific throughput as a function of EBCT for three PFAS (PFBS, PFOA, PFOS) with and without considering fouling using parameters presented in Burkhardt et al. (2021). For these plots, the influent of each chemical was set to 100 ng L^{-1}, and specific throughput was calculated based on a 3 ng L^{-1} replacement threshold. This plot was generated by adjusting flow rate to correspond to the desired EBCT for a given system size and adsorbent. This only considers a single compound for each line; however, it demonstrates the utility of such analyses which could be extended to more complicated treatment objectives if both influent concentrations and objectives are known. One observation that can be made is specific throughput tends to plateau, where beyond a certain EBCT minimal gains in specific throughput are realized. When considering fouling (Figure 15.2b), there are minimal gains beyond approximately 20-minutes of EBCT for both PFOA and PFOS. Due to the rapid breakthrough of PFBS, there are minimal gains in specific throughput after about ten minutes of EBCT and performance even slightly declines when fouling is considered. Alternative methods for assessing or understanding system performance are discussed below.

15.2.5 Case Study: Short Bed Contactors

A two million gallon per day (2-MGD) water utility wanted to understand how well their current plant would treat PFOS and PFOA if they replaced their bacterially activated carbon beds with GAC. The ten beds were originally configured for eighteen inches of carbon, but could also be adapted to accept twenty-four inches of GAC. The influent to their system had approximately 50 ng L^{-1} each of PFOS and PFOA, and the utility wanted information on bed replacement intervals to not exceed a combined 70 ng L^{-1} PFOS+PFOA concentration threshold. For this modeling effort, conservative values for carbon capacity from pilot columns at a surface water facility on an industrialized river (Freundlich parameters from Burkhardt et al. 2021 using Water Treatment Model: PSDM) were used to provide an estimate of "worst-case" performance. Figure 15.3a shows results for initially replacing four beds to reduce the blended effluent below 70 ng L^{-1} and two beds at each subsequent replacement. Where the eighteen-inch bed would be expected to need replacement every 72 days and the twenty-four-inch bed would last 96 days. Figure 15.3b shows the expected performance if all ten beds were replaced at once. In this scenario, the eighteen-inch beds would last 162 days, and the twenty-four-inch bed configuration would last 217 days.

15.2.6 Case Study: Assessing Designs for Future Needs

A 19 MGD water utility performed a pilot study as part of their GAC design and planning activities (Burkhardt et al. 2021; Vandermeyden and Hagerty 2020). The objective was to predict PFAS removal at the current treatment capacity of 19-MGD and their expected future capacity of 44-MGD for multiple influent concentrations. Carbons included in the pilot phase were assessed by their ability to remove six

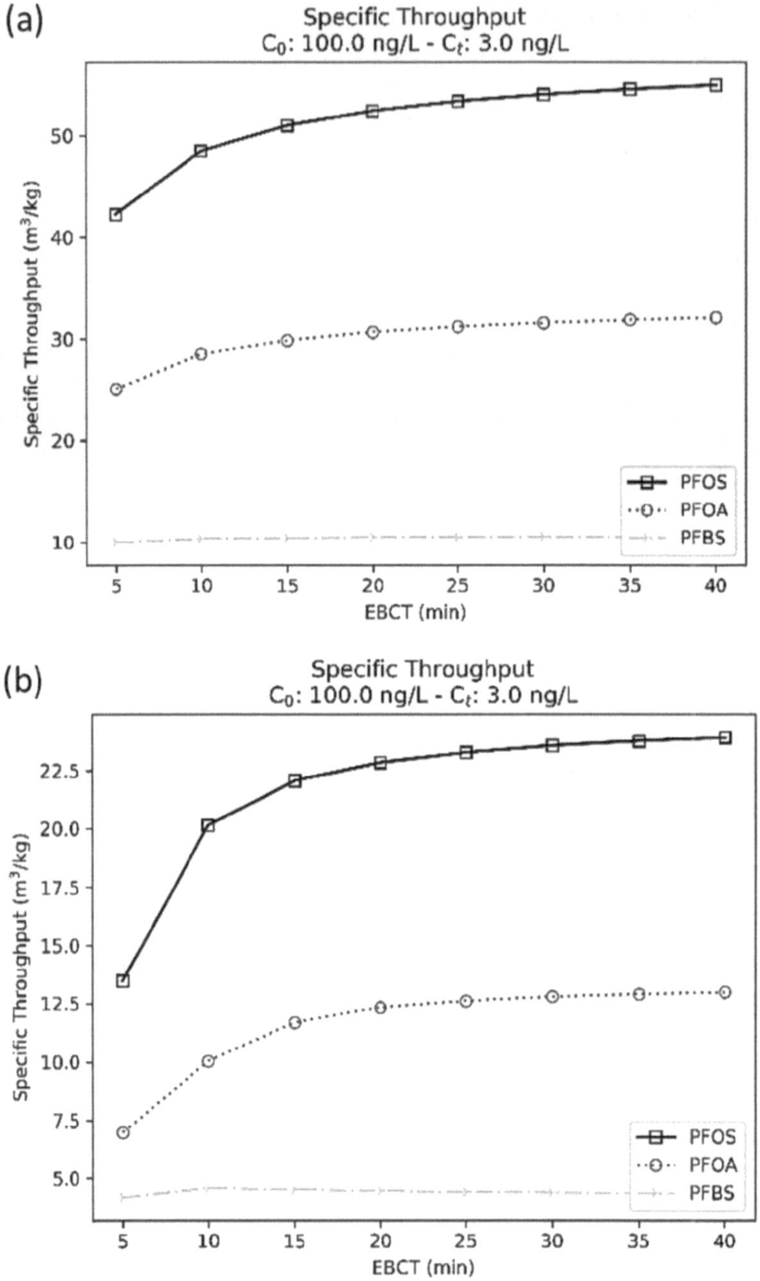

FIGURE 15.2 Specific Throughput Example (a: no fouling, b: Rhine river fouling).

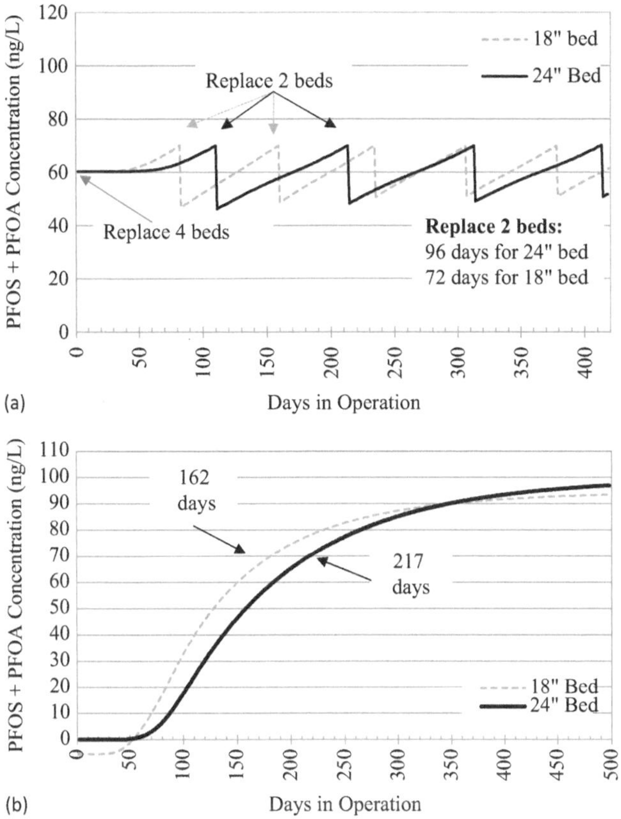

FIGURE 15.3 Examples of bed operations using short bed contactors (a) replacing 2 beds at a time, (b) replacing all beds.

PFAS to a combined concentration of 70 ng L^{-1} (the utility's independently determined assessment objective). Burkhardt et al. (2021) presented a more complete application of these results, but a few examples are presented here for demonstration purposes. Based on pilot column results and available space within the utility, the new contactors were designed to be about 800 ft^2 by 12½ ft deep, which corresponds to a 20 minute EBCT at future capacity. Figure 15.4 shows individual PFAS compound concentrations over time for each of the modeled scenarios. The initial breakthrough order predicted by the model was GenX and PFHpA, then PFOA, then PFNA and PFHxS and finally PFOS. This is driven by compound-carbon affinity, influent concentrations, and background mixture of the water.

Figure 15.5 shows the expected single-bed contactor breakthrough at present and future flow rate maximum capacities, where influent concentrations were modeled as the current average and twice the current average. For this scenario, the 19-MGD configuration experienced approximately 44 minutes of EBCT, and 20 minutes for 44-MGD. Increasing flowrates or concentrations was observed to sharpen the breakthrough profiles.

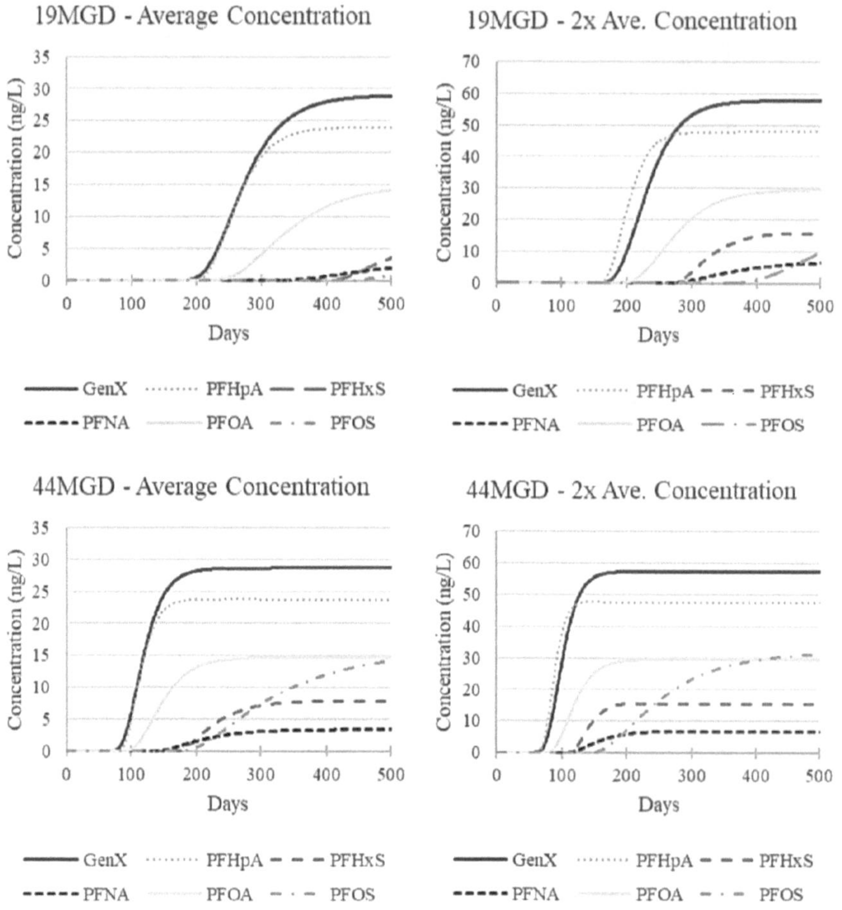

FIGURE 15.4 Examples of Multi-compound Breakthrough Curves for Different Influent Concentrations and Treatment Capacities (using parameters from Burkhardt et al., 2021).

The cumulative PFAS concentrations for each scenario can be used to predict a multi-bed blended effluent by staggering the start times for each bed by some bed replacement interval. Figure 15.6 shows the blended effluent concentration of the six PFAS compounds being considered for current and future production capacities and at the current average and double average concentrations. This analysis shows that at current capacity and current average concentration carbon from two of the eight contactors would need to be replaced every 259 days to satisfy the treatment objective, and 114 days at future capacity. Were the concentration to double, the replacement intervals would need to be more frequent at 93 days and 40 days, respectively. Assessment of the source water highlights that for this system, PFHpA and GenX breakthrough drives much of the replacement interval values. These two chemicals are shown to breakthrough earlier and make up about 55 ng L^{-1} of the utility's assessment objective of 70 ng L^{-1} combined PFAS, for average concentrations or 110 ng L^{-1}

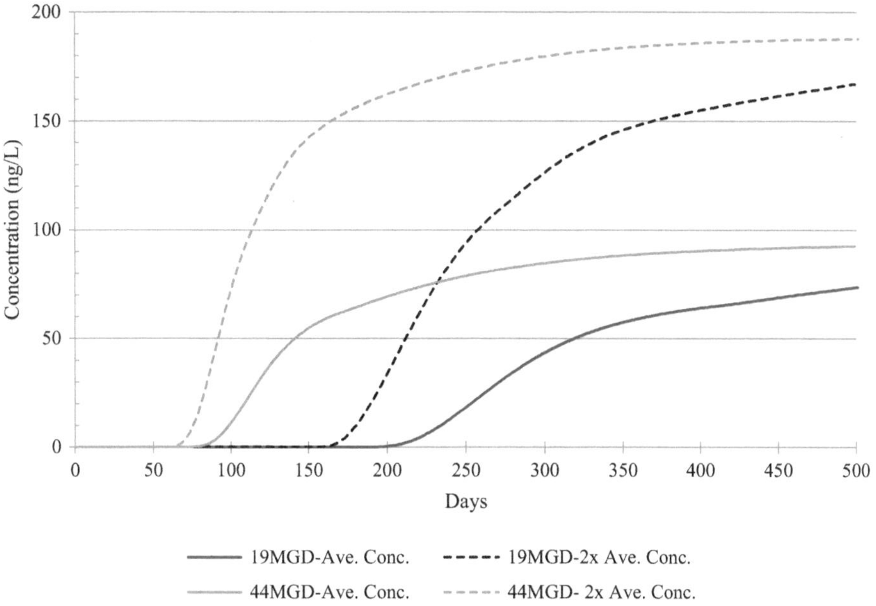

FIGURE 15.5 Example Cumulative Concentrations for Six PFAS Compounds vs. Time (days).

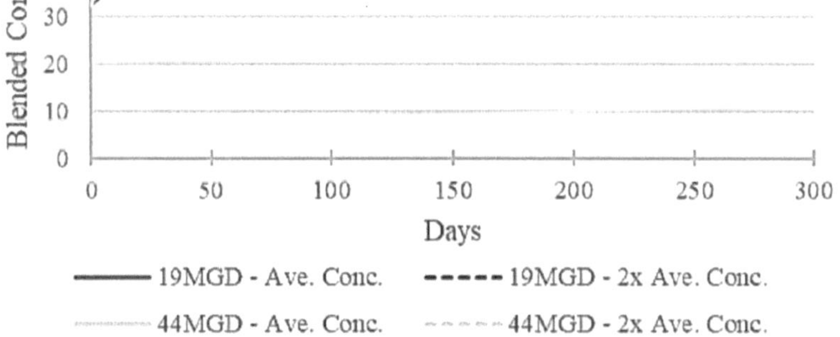

FIGURE 15.6 Cumulative Blended Effluent Concentrations for six PFAS in a System with Eight GAC Contactors with Two Contactors Replaced when Treatment Objective is Exceeded vs. Time (days). Two Total System Throughputs and Two Concentrations are Shown.

when considering doubled average concentrations of these two PFAS. While these chemicals will not drive bed replacement intervals for all utilities—as it will be related to specifics of each utility's source water—this does highlight the value of modeling and piloting to identify which compounds a treatment technology is particularly sensitive to for a given system. (Relatedly, the water utility is also working with their State to identify and reduce upstream contamination sources in order to prolong bed life by reducing loading.)

15.3 COST MODELING

15.3.1 Work Breakdown Structure Model

The publicly available, peer-reviewed WBS cost model for GAC includes pre-defined standard scenarios that reflect treatment for a number of different contaminants, such as atrazine, radon, and various volatile organic compounds. However, the model can be used to estimate the cost of GAC treatment for other contaminants as well, including scenarios involving the removal of different combinations of PFAS compounds. Modeling treatment cost for other contaminants simply involves adjusting default inputs (e.g., bed volumes to breakthrough) to reflect values appropriate for the contaminant and treatment scenario under consideration.

Whether using one of the predefined contaminants or modeling another scenario such as PFAS removal, key inputs to the WBS model for GAC include, but are not limited to, the following:

- Design type (i.e., pressure vessels or open concrete basins)
- Flow rates
- Bed life, measured in bed volumes to breakthrough or months between carbon replacements
- Empty bed contact time
- Number of contactors in series (i.e., parallel or lead-lag operation)
- Backwash residuals and spent carbon management requirements.

The WBS model also allows users to input-specific contactor dimensions and bed depths, and it includes a computer-aided design routine to select optimal (lowest-cost) values for these parameters once all other inputs are complete.

Applying the results of performance modeling in the WBS cost model involves setting the inputs for EBCT and number of contactors in series to match the assumptions used in the performance model, then setting the bed life to match the output of the performance model. To produce a site-specific estimate, a user would select other inputs (i.e., design type, flow rates, residuals management options) to match what is appropriate for the site. To evaluate the cost impacts of various design choices or to compare across sites, a user might vary some or all of these inputs. The cross-site costing example discussed below considers a range of flow rates for comparison purposes. It assumes pressure vessel designs with backwash discharged to a publicly-owned treatment works and off-site carbon reactivation, consistent with typical designs and practices at the sites under consideration.

Based on these inputs, the WBS model generates a cost estimate that includes treatment equipment costs, residuals management costs, building and add-on costs, O&M costs, and indirect costs. It presents these outputs in line-by-line, itemized detail and also as subtotals in the form of direct capital costs (equipment plus buildings), total capital costs (direct capital plus add-on and indirect costs), annual operations and maintenance (O&M) costs, and total annualized cost (total capital cost annualized over the estimated useful life of the process plus annual O&M cost). The costing example discussed below considers the cost model's output for annual O&M cost only, to allow comparison of cost changes for existing systems facing changing removal requirements.

15.3.2 DoD Costing Example

The US Department of Defense (DoD) is currently in the process of identifying all legacy uncontrolled PFAS releases that have contaminated soil, groundwater, sediment, and/or surface water across the portfolio of installations following the CERCLA process (Anderson and Kempisty 2019). During these investigations, potential drinking water exposures are identified and investigated. If exposures above USEPA's Lifetime Health Advisory (LHA) level for drinking water (70 ng L^{-1} for PFOS and PFOA combined) are observed, time-critical removal actions (emergency mitigation authority permitted under CERCLA) are performed to provide an alternative water supply and/or treatment system for the affected population. Numerous examples of DoD-operated or DoD-funded systems exist to date that range from MGD-scale municipal systems to point of entry (POE) and point of use (POU) residential systems (GAO (Government Accountability Office), 2018). While diverse in terms of the various operational and site-specific parameters, these systems have typically been designed and operated to treat PFOS and PFOA to the LHA given the lack of otherwise legally enforceable PFAS drinking water standards. However, due to increasing: (1) regulatory action at the state level; (2) public and legislative awareness; and (3) knowledge of the composition of PFAS mixtures at AFFF-contaminated sites, the DoD (and other entities) are under increasing pressure to more conservatively treat and remove a broader suite of PFAS (Guelfo et al., in preparation). While studies have clearly demonstrated differences among PFAS in terms of their affinity for various sorbents and other treatment technologies due to chain-length and functional moiety (e.g., Gagliano et al. 2020, Zeng et al. 2020), fundamental questions remain regarding long-term O&M costs of these systems under various scenarios, including the treatment of short-chain PFAS. Among other considerations, accurate O&M costs for treatment systems are critical for policy decisions that balance tradeoffs among political and public expectations, cost, and meaningful risk reduction.

Performance model simulations, similar to those described above, were developed using a range of Freundlich isotherm parameters developed from three pilot studies (not shown). This was intended to capture variability in potential outcomes to better highlight relative costs. Models were run with influent concentrations of 50 and 500 ng L^{-1} to establish the impacts of influent concentrations for three PFAS (PFBA, PFOA and PFOS). Bed replacement intervals for lead-lag configurations were determined for three treatment objectives: (1) PFOS+PFOA less than 70 ng L^{-1}; (2) first breakthrough of PFOS or PFOA; and (3) first PFAS breakthrough, estimated as breakthrough of PFBA. Additionally, simulations were performed with and without simulated NOM fouling to highlight its impact. For this exercise, the bed replacement interval was calculated for an exceedance of the threshold leaving the lead vessel, which would ensure lag vessels were below each threshold. Modeling lead-lag configurations was accomplished by calculating the effluent of an initial lead vessel, supplying that to a modeled lag vessel as an influent concentration for some period (pre-swap) and then the system influent for the remainder of the simulation (post-swap). This preliminary "effluent" becomes the lag vessel's influent concentration, but once swapped into the lead position, it experiences only supplied influent concentrations. This simulates the initial preloading of PFAS while in the lag configurations and captures the decreased effective bed life that results (i.e., the single bed replacement interval would be expected to be longer than a lead-lag interval but the carbon is actually in the system for twice the lead-lag replacement time interval when considering both phases of use).

In order to specifically evaluate relative costs of different O&M scenarios, modeled O&M cost outputs from the WBS GAC cost model were normalized to the scenario where breakthrough is defined as PFOS+PFOA ≥ 70 ng L^{-1} (the USEPA LHA level for drinking water and, thus, a common *de facto* treatment target) and are presented herein as cost multipliers relative to the status quo. Density plots of modeled cost multipliers for two different O&M scenarios (with and without consideration of short-chain PFAS treatment) are presented in Figure 15.7 across all model permutations. Modeled cost multipliers for the scenario where breakthrough is defined as any detectable PFOS/PFOA (using 2 ng L^{-1} as a representative detection

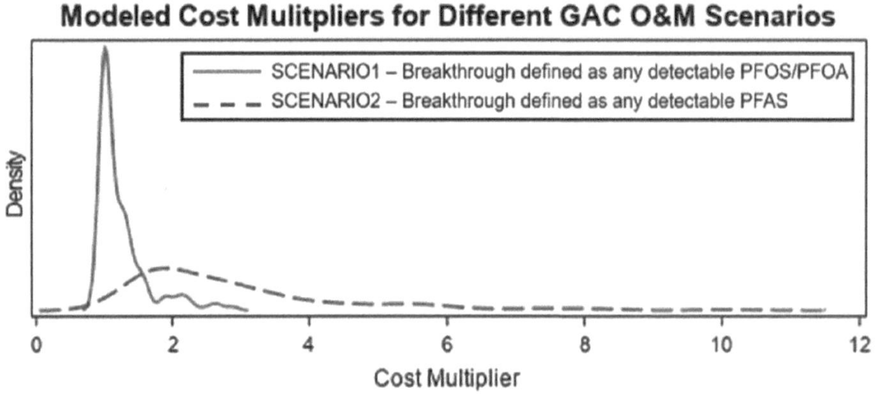

FIGURE 15.7 Density plot of modeled cost multipliers for two different O&M scenarios specific to GAC treatment of PFAS-contaminated water. Costs are normalized to the scenario where breakthrough is defined as PFOS+PFOA ≥ 70 ng L^{-1}. Scenario 1 represents the case where breakthrough is defined as PFOS+PFOA concentrations ≥ 2 ng L^{-1} (a typical laboratory reporting limit), whereas, in scenario 2 breakthrough is defined as PFBA (a conservative surrogate) concentrations ≥ 2 ng L^{-1}. All simulations are based on two vessels in series (lead/lag) with changeout following lead vessel breakthrough. Simulations included 6 water chemistries, 2 EBCTs, 3 flow rates, and 2 influent concentrations.

limit, scenario (1) resulted in a range of cost multipliers of one to generally less than a factor of two relative to the scenario based on the LHA. This is in dramatic contrast to the scenario where breakthrough is defined as any detectable PFAS (modeled based on PFBA as a conservative surrogate and again using 2 ng L^{-1} as a representative detection limit) resulted in much greater costs, as expected, and were more variable indicating that the additional costs required to treat short-chain PFAS by GAC depend more heavily on specific site conditions.

The output of the 72 simulations results were compiled into a data matrix to facilitate analysis of significant effects by factorial analysis of variance (ANOVA) using exclusively categorical main effect variables. Flow rates were discretized into three (low [<1 MGD], medium, and high [>10 MGD]) categories, and the three influent water chemistries and two fouling factors were combined into a single water chemistry variable (resulting in six unique water chemistries). The resulting statistical model included four fixed categorical main effects (influent concentration, EBCT, flow rate, and O&M scenario) and included water chemistry as a random quantitative variable, which partitions error variance accordingly but does not result in a fixed effect for significance testing. Viewing water chemistry as a random variable is reasonable based on an infinite combination of relevant variables in environmentally relevant ranges (e.g., hardness, pH, alkalinity, dissolved organic carbon (DOC), etc.) for which specific combinations (notably in DOC) are known to influence GAC performance.

Table 15.2 illustrates results for all main effect interactions from ANOVA. Only two-way interactions that included the scenario variable were significant. Significant two-way interaction plots are presented in Figure 15.8. Flow rate, EBCT, and influent concentration significantly affected modeled cost multipliers, but the degree to which the effects were observed depend on the scenario. For both the flow rate and EBCT

TABLE 15.2
ANOVA results for all main effect interactions.

Effect[1,2]	F Value	P Value[3]
A*B	0.61	0.4361
A*C	0.05	0.9537
A*D	**7.04**	**0.0082**
B*C	2.65	0.0714
B*D	**39.33**	**<.0001**
C*D	**8.78**	**0.0002**
A*B*C	0.04	0.9636
A*B*D	0.08	0.7736
A*C*D	0.08	0.9659
B*C*D	1.88	0.1538
A*B*C*D	0.00	0.9987

[1] A = Influent concentration; B = EBCT; C = Flow rate; D = Scenarios (as described in Figure 15.8)

[2] Because significant interactions included all variables evaluated, main effect results are not shown.

[3] Bolded entries are significant at $\alpha = 0.05$.

effects, differences between the two scenarios were concordant and differed only in the rate of change; however, the combined flow rate/EBCT ANOVA result of 0.0714 was only slightly outside the significance criteria of <0.05, and EBCT is known to impact treatment performance (as shown above) and can play a role in cost for some systems despite failing to meet significance criteria for this analysis. As flow rate increased, modeled cost multipliers increased nonlinearly with appreciable increases observed above 10 MGD for the scenario that includes short-chain PFAS treatment. Conversely, as EBCT increased modeled cost multipliers decreased with a greater decrease observed for the scenario that includes short-chain PFAS treatment. On the other hand, different effects were observed for influent concentration among the two scenarios. For the scenario where breakthrough is defined as any detectable PFOS/PFOA, higher influent concentrations reduced the modeled cost multiplier, whereas, for the scenario that includes short-chain PFAS treatment modeled cost multipliers increased with increasing influent concentration (Figure 15.8).

The relative costs can be explained by considering the breakthrough behaviors for each compound. When the influent concentrations are simulated at 500 ng L^{-1}, the breakthrough for both PFOS+PFOA ≥ 70 ng L^{-1} (LHA) or PFOS+PFOA ≥ 2 ng L^{-1} cases is controlled by PFOA alone, and even the higher LHA level is only about 10% total breakthrough. These two carbon replacement triggers will occur within a short period of time from each other but are linked. However, due to the lower capacity for PFBA its initial breakthrough (>2 ng L^{-1}) will occur earlier than PFOA and because it is not linked to PFOA, can have differences in observed costs or performance. Ultimately, this cost modeling exercise demonstrated that short-chain PFAS

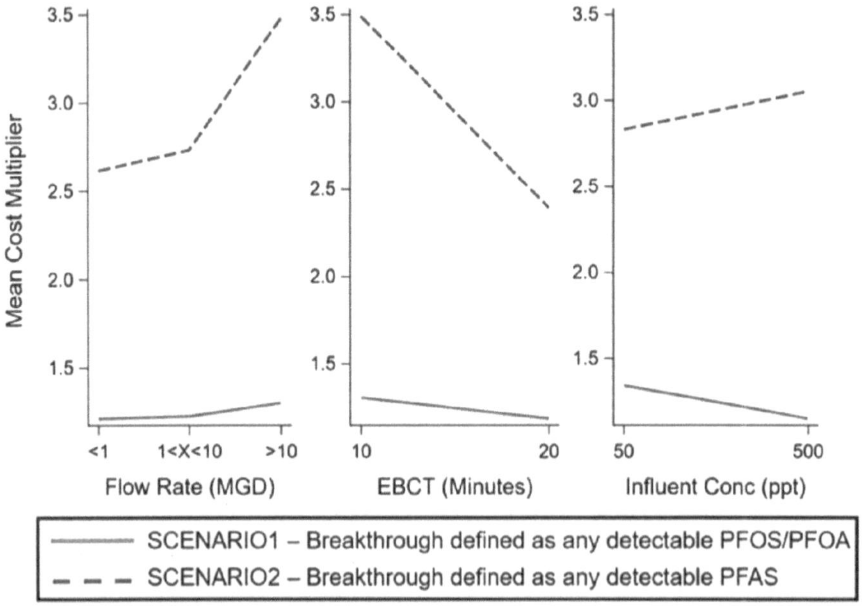

FIGURE 15.8 Two-way interaction plots illustrating statistically significant effects in Table 15.2.

treatment by GAC will likely result in a significant increase (approximately twofold on average, with some conditions exceeding threefold increases) in O&M costs over the system life cycle relative to the cost where only PFOS and PFOA are considered due to more frequent media changeouts.

15.4 CONCLUSIONS

The use of PSDM to interpret and extend breakthrough data from pilot columns or other sources has significant value. Given the numerous species of PFAS and potential treatment objectives, simple estimates of treatment systems may not capture the intricacies necessary to fully predict design conditions. PSDM can be used to predict simple metrics like specific throughput or can be applied to system-specific data to better capture complex treatment behaviors. While pilot or other test data for a specific source water is preferred for predicting performance, pilot studies are costly, and mitigation may be required quickly not allowing for a pilot study. In addition, the use of data from multiple source waters can help capture the uncertainty in some variables. The use of parameters from multiple sources can help provide expected bounds to performance—or associated costs—where utilities can work to understand their "best case" or "worst case" outcomes to inform design or operating decisions.

Applying simulation results to design scenarios can help to identify major driving forces related to achieving treatment objectives. Further, using these results to inform WBS-based cost models can help highlight potential incurred costs for installing, operating, and maintaining a treatment system for PFAS removal. The DoD exercise highlighted that carbon replacement and associated O&M costs for treating drinking water with objectives related to shorter-chain PFAS will be very sensitive to system conditions and water types. The O&M costs were predicted to be higher due to the more frequent carbon replacement cycle, but also more variable between scenarios. The approach discussed herein can be used to highlight factors that may lead to higher costs or identify system characteristics that could lead to a more resilient treatment system on a case-by-case basis.

The modeling tools described and used here are continuing to be developed to inform PFAS treatment in drinking water treatment and remediation applications. The EPA is also adding PFAS treatment data to the Treatability Database as it becomes available.

15.5 SOFTWARE INFORMATION

ETDOT: Environmental Technology Design Option Tool. https://www.epa.gov/water-research/environmental-technologies-design-option-tool-etdot

ETDOT GitHub: https://github.com/USEPA/Environmental-Technologies-Design-Option-Tool

New USEPA developed performance models: Ion Exchange and PSDM models: https://github.com/USEPA/Water_Treatment_Models

EPA WBS Cost Models: https://www.epa.gov/sdwa/drinking-water-treatment-technology-unit-cost-models

EPA Treatability Database: https://tdb.epa.gov/tdb/home

ACKNOWLEDGMENTS

The U.S. Environmental Protection Agency, through its Office of Research and Development, funded and managed, or partially funded and collaborated in, the research described. The cost modeling work was supported under contract EP-BPA-16-C-0001. It has been subjected to the Agency's administrative review and has been approved for external publication. Any opinions expressed in this chapter are those of the author(s) and do not necessarily reflect the views of the Agency; therefore, no official endorsement should be inferred. Any mention of trade names or commercial products does not constitute endorsement or recommendation for use.

The authors would like to thank Matthew Magnuson for providing QSPR parameters used by the PSDM and John Berry for providing pilot data from (Newman and Berry 2019).

REFERENCES

Anderson, R.H., & Kempisty, D.. (2019). Challenges of Managing Emerging Contaminants: Historical Per- and Polyfluorinated Alkyl Substance Use in the U.S. Air Force. In Kempisty, D., Xing, Y., Racz, L. (Ed.). *Perfluoroalkyl Substances in the Environment.* Boca Raton: CRC Press. pp. 113–119.

Burkhardt, J.B., Burns, N., Mobley, D., Pressman, J.G., & Speth, T.F. (2021). Modeling PFAS Removal Using Granular Activated Carbon for Full-Scale System Design, *submitted to Journal of Environmental Engineering.*

Chularueangaksorn, P., Tanaka, S., Fujii, S., & Kunacheva, C. (2014). Batch and column adsorption of perfluorooctane sulfonate on anion exchange resins and granular activated carbon. *Journal of Applied Polymer Science*, 131(3), 39782.

Crone, B.C. Speth, T.F., Wahman, D.G., Smith, S.J., Abulikemu, G., Kleiner, E.J., & Pressman, J.G. (2019) Occurrence of per- and polyfluoroalkyl substances (PFAS) in source water and their treatment in drinking water, *Critical Reviews in Environmental Science and Technology*, 49, 2359–2396.

Dasu, K., Nakayama, S.F., Yoshikane, M., Mills, M.A., Wright, J.M., & Ehrlich, S. (2017). An ultra-sensitive method for the analysis of perfluorinated alkyl acids in drinking water using a column switching high-performance liquid chromatography tandem mass spectrometry. *Journal of Chromatography A* 1494, 46–54.

Dickenson, E. & Higgins, C., 2016. Treatment Mitigation Strategies for Poly-and Perfluoroalkyl Substances. Water Research Foundation Web Report, 4322.

Franke, V., McCleaf, P., Lindegren, K., & Ahrens, L. (2019). Efficient removal of per-and polyfluoroalkyl substances (PFASs) in drinking water treatment: nanofiltration combined with active carbon or anion exchange. *Environmental Science: Water Research & Technology*, 5(11), 1836–1843.

Gagliano, E., Sgroi, M., Falciglia, P.P., Vagliasindi, F.G., & Roccaro, P. (2020). Removal of poly-and perfluoroalkyl substances (PFAS) from water by adsorption: Role of PFAS chain length, effect of organic matter and challenges in adsorbent regeneration. *Water Research*, 171, 115381.

GAO (Government Accountability Office) (2018). *Status to Address DoD Efforts to Address Drinking Water Contaminants Used in Firefighting Foams.* https://www.gao.gov/assets/700/694759.pdf, Accessed July 10, 2020.

Guelfo, J., et al. (in preparation). Environmental sources, chemistry, fate and transport of per- and polyfluoroalkyl substances: state of the science, key knowledge gaps, and recommendations presented at the August 2019 SETAC focus topic meeting, *Environmental Toxicology and Chemistry.*

Hand, D.W., Crittenden, J.C., Hokanson, D.R., & Bulloch, J.L. (1997) Predicting the performance of fixed-bed granular activated carbon adsorber, *Water Science and Technology*, 35(7), 235–241.

Jarvie, M.E., Hand, D.W., Bhuvendralingam, S., Crittenden, J.C., & Hokanson, D.R. (2005) Simulating the performance of fixed-bed granular activated adsorbers: Removal of synthetic organic compounds in the presence of background organic matter. *Water Research*, 39, 2407–2421.

Knappe, D.R.U., Snoeyink, V.L., Roche, P., Prados, M.J., & Bourbigot, M.-M. (1999). "Atrazine removal by preloaded GAC", *Journal of American Water Works Association*, 91(10), 97–109.

Khera, R., Ransom, P., & Speth, T.F. (2013). Using Work Breakdown Structure Models to Develop Unit Treatment Costs. *Journal of American Water Works Association*, 105(11): E628.

Kucharzyk, K.H., Darlington, R., Benotti, M., Deeb, R., & Hawley, E. (2017) Novel treatment technologies for PFAS compounds: A critical review. *Journal of Environmental Management*, 204: 757–764. doi: 10.1016/j.jenvman.2017.08.016.

McCleaf, P., Englund, S., Östlund, A., Lindegren, K., Wiberg, K., & Ahrens, L. (2017). Removal efficiency of multiple poly-and perfluoroalkyl substances (PFASs) in drinking water using granular activated carbon (GAC) and anion exchange (AE) column tests. *Water Research*, 120, 77–87.

Newman, B., & Berry, J. (2019) Case Study: Pilot Testing Synthetic Media and Granular Activated Carbon for Treatment of Poly- and Perfluorinated Alkyl Substances in Groundwater. In Kempisty, D., Xing, Y., Racz, L. (Ed.). *Perfluoroalkyl Substances in the Environment*. Boca Raton: CRC Press.

Ross, I., McDonough, J., Miles, J., Storch, P., Thelakkat Kochunarayanan, P., Kalve, E., Hurst, J., Dasgupta, S.S. & Burdick, J., 2018. A Review of Emerging Technologies for Remediation of Pfass. *Remediation Journal*, 28(2):101.

Sanan, T., & Magnuson, M. (2020) Analysis of per- and polyfluorinated alkyl substances in sub-sampled water matrices with online solid phase extraction/isotope dilution tandem mass spectrometry, *Journal of Chromatography A*, 1626, 461324.

Schaefer, C. E., Nguyen, D., Ho, P., Im, J., & LeBlanc, A. (2019). Assessing rapid small-scale column tests for treatment of perfluoroalkyl acids by anion exchange resin. *Industrial & Engineering Chemistry Research*, 58(22), 9701–9706.

Sontheimer, H., Crittenden, J.C. & Summers, R.S. (1988). Activated Carbon for Water Treatment. *American Water Works Association*.

Strynar, M., Dagnino, S., McMahen, R., Liang, S., Lindstrom, A., Andersen, E., McMillan, L., Thurman, M., Ferrer, I., & Ball, C., Identification of Novel Perfluoroalkyl Ether Carboxylic Acids (PFECAs) and Sulfonic Acids (PFESAs) in Natural Waters Using Accurate Mass Time-of-Flight Mass Spectrometry (TOFMS), *Environ. Sci. Technol.* 2015, 49(19), 11622–11630.

USEPA, 2016a. Drinking Water Health Advisory for Perfluorooctane Sulfonate (Pfos). Office of Water Health and Ecological Criteria Division.

USEPA, 2016b. Drinking Water Health Advisory for Perfluorooctanoic Acid (Pfoa). Office of Water Health and Ecological Criteria Division.

USEPA, Shoemaker, J., & Tettenhorst, D. 2019a. *Method 537.1 Determination of Selected Per- and Polyflourinated Alkyl Substances in Drinking Water by Solid Phase Extraction and Liquid Chromatography/Tandem Mass Spectrometry (LC/MS/MS).* U.S. Environmental Protection Agency, Washington, DC.

USEPA, Wendelken, S., & Rosenblum, L. 2019b. *Method 533: Determination of Per- and Polyfluoroalkyl Substances in Drinking Water by Isotope Dilution Anion Exchange Solid Phase Extraction and Liquid Chromatography/Tandem Mass Spectrometry*. https://www.epa.gov/dwanalyticalmethods/method-533-determination-and-polyfluoroalkyl-substances-drinking-water-isotope. Accessed on March 1, 2021.

USEPA, 2019. *Validated Test Method 8327: Per-and Polyfluoroalkyl Substances (PFAS) Using External Standard Calibration and Multiple Reaction Monitoring (MRM) Liquid Chromatography/Tandem Mass Spectrometry (LC/MS/MS).* https://www.epa.gov/hw-sw846/validated-test-method-8327-and-polyfluoroalkyl-substances-pfas-using-external-standard. Accessed on March 1, 2021.

USEPA, 2020a, *USEPA's Drinking Water Treatability Database*, https://tdb.epa.gov/tdb/home. Accessed on March 1, 2021.

USEPA, 2020b, *GenX and PFBS Draft Toxicity Assessments*, https://www.epa.gov/pfas/genx-and-pfbs-draft-toxicity-assessments. Accessed on March 1, 2021.

USEPA, 2020c, *Environmental Technologies Design Option Tool (ETDOT)*, https://www.epa.gov/water-research/environmental-technologies-design-option-tool-etdot. Accessed on March 1, 2021.

USEPA, 2020d, *Drinking Water Treatment Technology Unit Cost Models*, https://www.epa.gov/sdwa/drinking-water-treatment-technology-unit-cost-models. Accessed on March 1, 2021.

USEPA, 2020e, *Analytical Methods Developed by EPA for Analysis of Unregulated Contaminants*, https://www.epa.gov/dwanalyticalmethods/analytical-methods-developed-epa-analysis-unregulated-contaminants. Accessed on March 1, 2021.

Vandermeyden, Carel, & Hagerty, Vaughn (2020). "PFAS, PFOA, and GenX: A Southeastern North Carolina Utility Story", OpFlow, (Accepted for publication June 2020).

Woodard, S., Berry, J., & Newman, B. (2017). Ion exchange resin for PFAS removal and pilot test comparison to GAC. *Remediation Journal*, 27(3), 19–27.

Zeng, C., Atkinson, A., Sharma, N., Ashani, H., Hjelmstad, A., Venkatesh, K., & Westerhoff, P. (2020). Removing per-and polyfluoroalkyl substances from groundwaters using activated carbon and ion exchange resin packed columns. *AWWA Water Science*, 2(1), e1172.

Index

A

accelerated column test (ACT), 46
acoustochemical, 12
ADONA, 53
aircraft rescue and firefighting (ARFF), 26, 28
America's Water Infrastructure Act, 87, 93
amine, 191, 239
ammonia, ammonium, 130, 133, 145, 293, 298
anionic exchange resin (AER), 318
aqueous film-forming foam (AFFF), 8, 11, 15, 24–46, 100, 5–124, 151, 162, 172–177, 197–201, 256, 276, 307, 321, 331
arrowhead regulatory approach, 70–75
arsenic, 122, 243–245
Association of State Drinking Water Administrators (ASDWA), 14, 95–96
ASTM Method, ASTM Standard, 13, 101–103, 193, 213, 219, 238, 243, 299
atmospheric deposition, 126

B

backwash, 219, 228, 239, 304, 307, 310–311, 330
batch adsorption, 293, 295–296, 303, 305, 308, 310, 312, 319
batch uptake, 298–300
benchmark dose, 166
bench scale test, 196, 201, 240–245, 252, 257, 275, 300, 320
Best Available Technology, 11
bioaccumulation, 9, 24, 30–31, 40, 58, 64, 69, 163, 168–170, 173–179, 212
biochar, 277
biological oxygen demand, 40, 131
biomagnification, 175–176
biomarker, 67
biomonitoring, 4, 13, 67, 69, 145–154
biosolids, 11, 65, 113–114, 120–121, 126, 163, 168, 170, 248–249
biotransformation, 11, 29, 40, 131, 133, 163
blood, xi–xii, 13, 16, 64, 67, 69–70, 100, 146–147, 152, 272
boron-doped diamond (BDD), 202, 275, 284
bottled water, 66, 153
brine, 237, 256, 283, 302
Brunauer-Emmett-Teller (BET), 238

C

cancer slope factor (CSF), 166–167
centralized treatment, 307, 310–311
C8 Health Project, xii, 150
chemical class regulatory approach, 52, 70, 73–74
chemical oxygen demand, 40, 131
chlorinated polyfluorinated ether sulfonate (Cl-PFESA), 4
clay, 121, 135, 252, 276
Clean Water Act (CWA), 93
Clean Air Act (CAA), 115
climate change, 176
coagulation, 12, 131, 193, 272, 294, 304, 306, 309–310, 312
Comprehensive Environmental Response, Compensation, and Liability Act (CERCLA), 15, 321, 331
compressed air foam (CAF), 38
concentrate, 25, 27–31, 35, 39, 42–43, 63–64, 113, 125, 127, 134, 195–198, 200–201, 218, 223, 225, 237, 256, 272, 275, 283–284, 295, 307–308, 310, 312, 318–319
Conceptual Site Model (CSM), 168
concrete, 25, 44–45, 123, 330
Contaminant Candidate List (CCL), 87, 89–90
cyclodextrin, 11, 194, 252, 292–312

D

defluorination, 11, 198–202, 273–285
degradation pathway, 117, 123, 198, 200, 274
dermal, 67–68, 164–165, 171–172
desalination, 195–196, 284
desulfonation, 279, 285
dioxins, 71, 110, 179
disinfection by-product (DBP), 92
dissolved oxygen (DO), 276, 280–281, 285
dithionite, 277
Drinking Water Treatability Database (TDB), 316

E

ecotoxicity, 15
Electric Energy per Order (EE/O), 267
electrochemical fluorination (ECF), 99–100
electrochemical oxidation, 121, 131, 133, 202, 256, 267, 274, 284–285
electron beam, 278
electrostatic, 132, 191, 193–194, 196, 292, 298, 303–304, 312
empty bed contact time (EBCT), 221, 227–228, 242, 244–247, 249, 301–303, 305–307, 310, 320, 324–325, 327, 330, 332–334
Environmental Laboratory Accreditation Program (ELAP), 105

EPA Lifetime Health Advisory Level (LHA/HAL), 91, 94, 101–102, 146, 164–168, 172, 213, 227–228, 230–233, 240, 246–247, 257, 261–263, 267, 331–334
EPA Method, 13, 35, 92, 100–101, 107–108, 154, 166, 260, 299
EPA PFAS Action Plan, 15, 91, 117
epidemiology, 146, 150, 162
EtFOSA, 63, 258–259, 262
ethical practices, 180
ethyl perfluorooctane sulfonamido ethanol (EtFOSE), 127
European Chemicals Agency (ECHA), 32, 71, 73, 74
European Food Safety Authority, 9
exposure pathways, 7–8, 52, 59, 67–69, 164–165, 167–168, 172–176, 179

F

Fenton reaction, 12, 272
Fire Fighting Foam Coalition (FFFC), 30
fish, 9, 41, 64, 163, 165, 168–170, 173–174, 176
Fish Consumption Advisories, 169
fish tissue criteria (FTC), 169–170
fixed bed, 192–194, 219
Flexeon RO system, 216–218, 220, 223, 225, 232
flocculation, 272, 294, 304, 306, 309–310, 312
fluorinated polymer, 5–6, 29–30, 32, 54, 57–58, 65, 72
fluoroelastomer, 55
fluoropolymer, 6, 53, 55–57, 126
fluoroprotein, 24
fluorotelomer alcohol (FTOH), 9, 60, 63, 65, 67, 69, 72, 120, 124–127, 130, 133, 149, 173
fluorotelomer carboxylic acid (FTCA), 123, 125, 127–129
fluorotelomer chain, 6, 29–30, 32, 57, 198
fluorotelomer ethoxylate (FTEO), 54
fluorotelomer iodide (FTI), 124
fluorotelomer phosphate, 124, 127
fluorotelomer polymer (FTP), 123
fluorotelomer sulfonic acid (FTS) and fluorotelomer sulfonate, 31, 41, 63, 107, 125, 128–130, 133, 197, 257, 259, 262, 324
fluorotelomer unsaturated carboxylic acid (FTUCA), 123, 125, 129–130
Food and Drug Administration (FDA), 31, 57, 65, 72
food packaging and food production, 3, 8, 31, 35, 52, 54, 56–57, 60, 65–66, 69, 72, 74, 120, 124–125, 132, 135, 171, 252
food web and food consumption, 9, 31, 58, 64, 66–68, 70, 75, 146–147, 165, 168, 170–171, 173–177, 179, 212
formate, 279

G

gamma ray, 277–279
GenX, 53, 58, 66, 71, 107, 145, 152–153, 296–297, 312, 317, 327–328
granular activated carbon (GAC), 11, 46, 96, 134, 190–193, 199, 201, 213–215, 218–233, 236–252, 256, 283, 301, 309–310, 318–325, 329–335
greenhouse gases (GHG), 121–122
GreenScreen, 25, 40–41

H

habitat, 64, 175, 177
hair, 13, 67
Harmonized Offshore Notification Format (HOCNF), 40
hazardous waste, 199–200, 228–229, 252
Health Risk Reduction and Cost Analysis (HRRCA), 89
heavy metals, 122, 257, 266
hexafluoropropylene (HFP), 55
high pressure membrane system, 121, 272, 318
hydrogen peroxide (H_2O_2), 272, 283, 285
HydroLogic RO system, 216–218, 220, 223–225, 232

I

immune system, 3, 163
indoor air, 7, 59–60, 63, 68–69, 75, 173
indoor dust, 7, 59, 61–64, 68–69, 75, 167
International Civil Aviation Organization (ICAO), 27–29, 36
Interstate Technology and Regulatory Council (ITRC), 4, 52, 94, 162
invertebrates, 31, 41, 131, 163, 170, 173–179
iodine number, 238
ion exchange resin (IEX/IX), 11, 46, 193–194, 236–244, 247–252, 256, 309–310
ionic strength, 194, 299, 303–304
isotope, 278
isotope dilution, 100–101, 103, 105–106, 108, 110, 113
iSpring RO system, 216–219, 223, 225, 232

K

Kennebunk, 246
kidney, 163

L

LASTFIRE, 25, 28, 36–39
leachate, 8, 113, 120–122, 125–135, 163, 201, 252, 292, 295, 298, 306–310, 312
liver, 64, 67, 71, 163
lowest observed adverse effect level (LOAEL), 176

M

maximum contaminant level (MCL), 14–15, 89, 92, 94–95, 164, 309
methane, 122, 130
methanol, 110, 258, 260
microwave, 57, 69, 120, 126, 272
mixture regulatory approach, 70–73

N

nail, 13, 67
nanofiltration (NF), 121, 131, 133–134, 195–196, 236, 256, 274
National Defense Authorization Act (NDAA), 15, 93, 115
National Emerging Contaminant Research Initiative, 93
National Environmental Laboratory Accreditation Council (NELAC), 105–106
National Environmental Management Plan (NEMP), 34
National Fire Protection Association (NFPA), 25, 37–38
National Health and Nutrition Examination Survey (NHANES), 4, 67, 69, 147–154
National Primary Drinking Water Regulation (NPDWR), 88–89, 317
nitrate, 11, 122, 191, 194, 240, 281, 300, 302, 318
nitrilotriacetic acid (NTA), 280, 285
no observed adverse effect level (NOAEL), 176–177, 179
no observed effect concentration (NOEC), 178–179
nomenclature, 4–5, 52, 94, 103, 105

O

Organization for Economic Cooperation and Development (OECD), 41, 52
organofluorine, 199, 260
oxidation/reduction (redox), 12, 266, 272, 282 285

P

packed-bed filtration (PBF), 294, 298, 300–301, 303–307, 309–312, 322
particle-induced gamma-ray emission (PIGE), 115–116
perfluoroalkoxy alkane (PFA), 6–7, 55, 124, 127, 130
perfluoroalkyl carboxylic acid (PFCA), 5–6, 53, 55, 60–64, 69, 123–124, 126–128, 130, 163, 168, 170, 173, 177, 196, 261–262, 273–274, 279–280, 284–285
perfluoroalkyl sulfonic acid (PFSA), 6–7, 53, 63–64, 69, 124, 126, 128, 130, 163, 167–168, 170, 173, 196, 261–262, 280, 284
perfluorobutane sulfonate (PFBS), 63, 71, 91, 107, 110, 113, 120, 123–124, 126–128, 145, 148, 152, 170, 177, 193, 196, 212, 214–215, 226–227, 230, 240, 246, 260–262, 265, 296, 302, 312, 317, 324–325
perfluorobutyric acid (PFBA), 60, 66, 107, 113, 120, 123, 127–128, 130, 152–153, 240, 257, 260–262, 264–265, 304, 312, 324, 331–334
perfluorodecanoic acid (PFDA), 64, 67, 69–71, 95, 107, 128–129, 148, 168, 257, 260–262
perfluoroether carboxylic acid (PFECA), 53, 58
perfluoroether monomer, 55
perfluoroether sulfonic acid (PFESA), 4, 53
perfluoroheptanoic acid (PFHpA), 60, 64, 66, 91, 95, 107, 123, 127–128, 148, 152, 212, 214, 230, 246, 249–250, 260–262, 265, 296, 324, 327–328
perfluorohexane sulfonate (PFHxS), 24, 29, 31, 35, 42, 60–61, 63–64, 69–71, 73, 91, 95, 107, 128, 145–146, 148–152, 167, 212, 214, 224, 230, 246, 257, 260–262, 296, 302, 312, 324, 327
perfluorohexanoic acid (PFHxA), 9, 29–32, 41, 60, 63–64, 66–67, 107, 128, 130, 133, 145, 152–153, 177, 243, 257, 261–262, 265, 274–275, 296, 324
perfluorononanionic acid (PFNA), 10, 63–64, 69–71, 91, 95, 107, 128, 148–152, 170, 212, 214, 224, 226–227, 230, 246, 257, 260–262, 296, 324, 327
perfluorooctanesulfonamide (FOSA), 60, 63, 69, 107, 124–126, 128, 130, 148, 257–259, 262, 264
perfluorooctanesulfonamideethanol (FOSE), 60, 63, 125–127, 130, 262
perfluoropolyether (PFPE), 53, 55–58, 72
perfluoroundecanoic acid (PFUnDA), 64, 67, 148
permanganate, 272, 275–276, 285
persulfate, 201, 256–257, 259, 267, 272–274, 285
PFAA precursors, 15, 57, 61, 71, 73, 100, 116, 146, 257, 259, 261–262
PFAS Leadership Summit, 91
pharmacokinetics, 67, 147, 153
photocatalysis, 121, 134
photolysis, 121, 134
phytoremediation, 11
pica, 171
pilot-scale test, 191, 241
plant uptake, 9, 114, 165, 170–171, 175–176
plasma (blood), 69, 152
plasma treatment, 12–13, 200–201, 257–267, 273, 282–283, 295, 308
point of entry (POE), 11, 213, 217, 219, 221, 223, 225, 228, 231–233, 331
point of use (POU), 11, 213, 223, 231–233, 294, 331
polychlorinated biphenyls (PCBs), 52, 110

polyfluoroalkyl phosphonic acid ester (PAP), 126–127, 130
polytetrafluoroethylene (PTFE), 55, 69
polyvinylfluoride (PVF), 55
polyvinylidene fluoride (PVDF), 55
pore surface diffusion model (PSDM), 229, 320–321, 323–325, 335–336
precipitation, 7, 9, 126, 131, 194
pretreatment, 14, 135, 201, 299, 304, 306, 309, 312
pump and treat (P&T), 121, 134, 256
pyrolysis, 198, 275

Q

Quality Systems Manual (QSM), 101–103, 105–106

R

rapid small-scale column test (RSSCT), 192, 213–214, 219, 221–223, 226–228, 230, 232, 243–244, 252, 300–302, 320, 322
reference dose (*RfD*), 166–167, 169–170
Relative Source Contribution (RSC), 9, 166–167, 169
replacement chemicals, 100, 105, 107
reproductive, 3, 163, 179
Resource Conservation and Recovery Act (RCRA), 228
reverse osmosis (RO), 12, 121, 133–134, 195–196, 213–220, 223–225, 228–229, 231–232, 236, 256, 318

S

Safe Drinking Water Act (SDWA), 87–90, 93–94
screening level, 94, 108, 164–167, 171–172, 174–175, 177–178
sedimentation, 134, 272
selective ion monitoring, 110
self expanding foam, 38
serum, 13, 67, 69–70, 148–154
short bed contactor, 325, 327
Significant New Use Rule (SNUR), 15, 71
siloxanes, 40–41
single chemical regulatory approach, 70–73, 75
solid-phase extraction, 13, 101, 108–109, 294
sonochemical, 131, 201, 256, 267, 276, 284
sonolysis, 46, 134–135, 200–201, 257, 267
Sontheimer correlation, 230
stack emissions, 115, 199
standard operating procedures, 105, 215
Stockholm Convention for Persistent Organic Pollutants, 4, 9, 31–32, 58, 71, 73, 146
Stratmoor Hills, 249–250
subclass regulatory approach, 53, 60, 62–63, 72
Superfund, 15, 212, 215
supernatant, 258

T

taxonomic, 176
telomerisation, 6, 100
tetrafluoroethylene (TFE), 55
thermal decomposition and destruction, 121, 132, 198–200, 282
thermodynamics, 195, 295
titanium dioxide (TiO2), 12, 272
Total Daily Intake (TDI), 176–177
total oxidizable precursors (TOP), 13, 15, 30, 259
total trihalomethanes (TTHMs), 92
ToxConsult 167toxicity reference value (TRV), 176–177
toxicokinetics, 9, 149, 154
Toxic Release Inventory (TRI), 15, 93
Toxic Substance Control Act (TSCA), 15, 53, 71, 93, 96
transportation costs, 256
transport in environment, 7, 13, 32, 61, 67, 114, 127, 129–130, 162–165, 167, 171–172, 256
treatment goal, 239–240, 246–247, 316–318
Treatment Technique (TT), 89, 92, 95

U

Unregulated Contaminant Monitoring Rule (UCMR) and UCMR1, 87–91, 93
UCMR3, 9–10, 66, 91–92, 101–102, 246, 251, 261
UCMR5, 92–93
ultrafiltration (UF), 195, 293, 308–309, 312
ultrasonication, 200–201, 273, 275–276, 285
ultrasound, 46, 272
ultraviolet (UV), 12, 257, 272–273, 275, 277–278, 280–283, 285, 311
urine, 8, 13, 67, 147, 152–153

V

V171, 41, 43
vegetable, 9, 64–65, 168
volatile organic chemical (VOC), 92, 257, 260, 265–266

W

waste stream, 96, 190–191, 194, 197, 199, 272, 283–284, 307, 318
wastewater sludge, 8–11, 101, 114, 123, 127, 131, 162, 215
Widefield aquifer, 212–214, 229–230, 233, 249
work breakdown structure model (WBS), 321, 330–332, 335

Z

zero valent iron (ZVI), 121, 276–277, 285